THE
EXCEPTIONS

THE EXCEPTIONS

Nancy Hopkins,
MIT, and the Fight for
Women in Science

KATE ZERNIKE

**SIMON &
SCHUSTER**

London · New York · Sydney · Toronto · New Delhi

First published in the United States by Scribner,
an imprint of Simon & Schuster, Inc., 2023

First published in Great Britain by Simon & Schuster UK Ltd, 2023

1 3 5 7 9 10 8 6 4 2

Simon & Schuster UK Ltd
1st Floor
222 Gray's Inn Road
London WC1X 8HB

www.simonandschuster.co.uk
www.simonandschuster.com.au
www.simonandschuster.co.in

Simon & Schuster Australia, Sydney
Simon & Schuster India, New Delhi

A CIP catalogue record for this book
is available from the British Library

Hardback ISBN: 978-1-3985-2000-4
Trade Paperback ISBN: 978-1-3985-2001-1
eBook ISBN: 978-1-3985-2002-8

Printed and Bound in the UK using 100% Renewable
Electricity at CPI Group (UK) Ltd

MIX
Paper | Supporting
responsible forestry
FSC® C171272

For FZ and BBZ

Always in memory, especially here

The thought could not be avoided that the best home for a feminist was in another person's lab.

James D. Watson,
The Double Helix, 1968

As for women, God help them.

Barbara McClintock,
letter to Nancy Hopkins,
September 21, 1976

Contents

Part Three

A Note on Names and Language

As is the custom in the settings this book describes, I use first names to refer to most of the major characters. Where two people with the same first name appear in proximity, I have used last names to avoid confusion. In all but one case these are men: while there are two Ruths, there are several Bobs—two of them Bob Bs—as well as two David Bs, and two Larrys, both presidents of Harvard.

The narrative stretches over five decades, and some of the language taken from accounts of the time may strike readers as dated (and grammarians as incorrect). In particular, many institutions referred broadly to "minorities," later refined as "underrepresented minorities," which sometimes included what were described as "Hispanics" or "Puerto Ricans and Mexican-Americans," but not people of Asian descent. Female students were often called "girls," especially before the 1970s, when they became "women students." Women who would later be called administrative assistants referred to themselves as secretaries. I have used the language of the time and tried to be specific about the definitions and reflect how they changed. I include honorifics only where they were commonly attached to someone's name, even if imprecisely: Mrs. Bunting was, by right of degree, Dr. Bunting.

Prologue

In March 1999, a story above the fold on the front page of the *Boston Sunday Globe* reported that the Massachusetts Institute of Technology had acknowledged long-standing discrimination against women on its science faculty. It was "an extraordinary admission," as an article on the front page of the *New York Times* called it two days later, by which point the news had traveled around the world by radio, television, and a fever pitch of emails between female scientists who had long known they were not valued as highly as men but talked about it only among themselves, if at all. Here was one of the most prestigious institutions in the world, synonymous with scientific excellence. The discrimination had happened not in some dark age but in the 1990s, the dawn of a new millennium, decades after legislation and the women's movement had pushed open the doors of opportunity. Most women starting their careers at the time did not think bias would block them. Women who complained of discrimination typically ended up in the deadlock of he-said, she-said. Now the president of MIT was saying it was true.

That admission came about not because of a lawsuit or formal complaint, but because of the work of sixteen women who had started as strangers, working in secret, and gathered their case so methodically— like the scientists they were—that MIT could not ignore them. They upset the usual assumptions about why there were so few women in science and math and unleashed a reckoning across the United States as other universities, philanthropies, and government agencies rushed to address the bias and the disparities that had disadvantaged women for decades. "A climate change in the whole of academia," as an astronomer at the California Institute of Technology called it.

I was the reporter who wrote the story in the *Globe*. I had recognized that it might resonate—though I could not predict how much—because of my father, a physicist who had arrived in the United States in 1956 to work for a small engineering firm in Cambridge populated by MIT graduates and consultants. My parents had moved before I was born, but my father visited me often in Boston on his way to see his collaborators at Lincoln Laboratory, an MIT research center, and he had suggested that I look into the work that a physicist named Millie Dresselhaus at MIT—known as the "Queen of Carbon"—was doing to encourage more women to enter the profession.

I had ignored him, until I heard about the women at MIT. They made me think of my mother, who was around the same age as the oldest of them. My mother had wanted to go to law school when she graduated from college in 1954, but her father surveyed his lawyer friends in Toronto and told her that no one would hire her. So she went to business school instead, up the street from MIT, enrolling in the Harvard-Radcliffe Program in Business Administration, which was the only way women could attend the Harvard Business School. That year the *Wall Street Journal* reported on the program in the middle column of its front page, reserved for offbeat or "light" features. It quoted business leaders marveling that the Radcliffe girls were "just as smart as the boys," but lamenting that "too many marry too soon." ("They're too good-looking, they're just the right age, and there are too many men at the bank.") My mother herself worked in a bank after she finished, quit to get married, and raised three children, but always regretted that she had not gone to law school. Her decision to go when I was seven—I was the youngest of her three—became the defining event of my childhood. She inquired at Yale, where a man told her, "I wouldn't let my wife go to law school." She ended up instead at Pace University.

A year or two after she graduated she was in the law library there and decided to look herself up in the *Harvard Alumni Directory*. There she found her name followed by a series of acronyms: BA, MBA, JD, W/M. Not recognizing the last one, she went to the key and discovered "wife and mother."

My mother was then commuting three hours a day to her job at a law firm in lower Manhattan and still made dinner most nights. I was about twelve and did not fully understand her fury as she came rushing out of the law library, where I was sitting on the steps. She drove home ranting, "W slash M! W slash M?" In time it became a family joke. But I can't say I had fathomed it even by the time I started my own career in Boston. Across the river, Cambridge was no longer the city where my parents had their first apartment; now it was tony restaurants and out-of-reach real estate prices. Twenty-five years after coeducation, I presumed my mother's experience was deep in the past.

The MIT women made me see it was not, at least not in science. They had identified the new shape of sex discrimination, more subtle but still pervasive. I was struck by their ingenuity, and how they had enlightened the men who ran the university. Their experience became a metric for how I thought about my own life and the questions and debates around women that I would write about over the next two decades. In time, what the MIT women had described began to look less faraway, more relevant. So much had changed, and yet.

Then as now, I saw the story as one of remarkable persistence and risk on the part of sixteen women who did not consider themselves activists. Led by a reluctant feminist, they were more pragmatic than revolutionary. They were not interested in publicity; they just wanted to get on with their work. As I explored their story—and the story of women in science before and after them—the word that kept coming up, in different conjugations, was *exception*. Women who succeeded in science were called *exceptional*, as if it were unusual for them to be so bright. They were exceptional not because they could succeed at science but because of all they accomplished despite the hurdles. Many had pushed past discrimination for years by excusing individual situations or incidents as exceptional, explained not by bias but by circumstance. Only when they came together did the MIT women see the pattern. That recognition alone made them exceptional, too.

I had known Nancy Hopkins, the molecular biologist who came to lead them, for twenty years before I realized that she had started her life as Nancy Doe. Like John Doe or Jane Doe, the generic every-woman whose example tells the larger story. The exception who proved the rule.

Part 1

Chapter 1

An Epiphany on Divinity Avenue

It was hard to deny the promise.

It was the second Tuesday in April 1963. Midmorning sunshine splashed the campus of Harvard University, where the trees were budding, and students were just back from a weeklong midsemester break. A Harvard man was in the White House, the youngest man ever elected president of the United States, heralding the dawn of a New Frontier. And here in Cambridge the next generation of ambitious young minds set out in crisp air along the tree-lined paths of the nation's oldest university, any of them on the way to do—*it could be me*—the next big thing.

At eleven o'clock, just north of the wrought-iron gates of Harvard Yard, and just east of the grounds where George Washington had once assumed control of the Continental Army, some 225 undergraduates, many in jackets and ties, filed down the gentle slope of a lecture hall to hear from a professor leading a revolution for the twentieth century. Five months earlier, James Dewey Watson had been in Stockholm with Francis Crick to collect a Nobel Prize for decoding, at age twenty-four, the structure of DNA, a discovery he called, immodestly but not incorrectly, "the secret of life." Watson and Crick's double helix had immediately placed them in the pantheon with Darwin and Mendel for explaining the development of life on earth and sounded a starting gun in the high-stakes race of modern genetics.

Now tenured at Harvard, Watson was about to begin his series of lectures in the introductory biology course for undergraduates. A tutor had written him that the students had done "rather well" on an hour-long test just before break: "They will return to Cambridge full of seasonal and customary liveliness and anticipating meeting you."

Nancy Doe, a junior, was already seated in the second row of the center section, almost directly in front of the lectern. She had arrived early, against her norm, and chosen her seat carefully. Not in the front row, because she didn't want her classmates to think she was a celebrity hound, and not in the back, where she usually hid, because she had read in the student evaluations that Watson dropped his voice at the end of his sentences, and she wanted to hear every word. Tall and slender, she had a sprite-like smile and wide blue eyes that took in everything but gave little clue to what she made of it. Her expression could shift from excited and girlish to wary and jaded in an adolescent minute. She nearly itched with intensity, considered her thick dark hair impossible to manage, her legs in their black wool tights absurdly long. Her mind tended to race, restless until it could alight on the biggest problem she could find, which at that moment was what she was going to do with her life. At nineteen, her future lay wide-open in front of her, but sitting in the wooden fold-down desk, she felt nothing as much as time closing in. Her father had died the previous year, and neither her wide circle of friends, the prerogatives of an Ivy League education, nor her tall, handsome boyfriend had insulated her from mounting dread.

Watson appeared suddenly, as if by a tailwind in a cartoon. Nancy sat up to look. This could not be him, she thought. He looked no more than thirty—in fact, he had turned thirty-five on Saturday, still young for a professor, much less a Nobel laureate. He was well over six feet tall but still gangly like a teenager, with enormous ears, a long bony nose, and round, protruding eyes. His hair was receding and barely tamed, a disobedient squiggle airborne over his forehead, like Tintin. He radiated impatient energy, his eyes everywhere at once—on his notes, on the students filling the room—looking through more than at. Watching, Nancy thought of him as a winged messenger, a wizard in a J. Press suit stopped in the middle of some monumental discovery to deliver the word to these lucky Harvard undergraduates.

Watson was cultivating a reputation as a showman, and he began grandly: What is life? Life, he told the students, came down to one molecule, DNA, which was in every cell of the body. It was made up of four bases, always in two complementary pairs that fit together in sequence, like the teeth of a zipper, to create genes. In those bases was all the information needed to create a living organism. Tear apart the zipper and DNA gave you a template to create an exact copy, the next generation. It was life, and the ability to start a new one.

Nancy had come to class understanding little about the double helix or DNA, little more than that a Nobel Prize was a big deal. From what she had read, she expected that Watson was going to deliver a master plan to explain human biology.

But as he spoke, she realized that Watson knew the answers to the questions that had been preoccupying her over the last year, or at least where to find them. If DNA was in every cell, everything there is to understand about humans must be written in there somehow: not just the color of their eyes, but cancer, and even how they behaved. Watson and the new cadre of molecular biologists were going to be able to figure it all out: a dumb gene, a smart gene, a fat gene, a thin gene, a nice gene, a nasty gene.

Watson was conversational, funny, prided himself on being the liveliest of the four lecturers in Bio 2. His voice was indeed quiet, but his tone was imperative, and she began to feel as though he were speaking directly to her.

It would all become more complicated—the science, Watson— much, much more complicated. But in that moment, the idea that life could be reduced to this one set of rules comforted and thrilled her. The promise of it drowned out everything else: the pressure of the hard wooden seat against her tailbone, the grief over her father, the worries about her widowed mother and her own future.

At the end of the hour, Nancy floated out of the classroom building to join the noontime stream of students emptying out of lecture halls onto Divinity Avenue. They passed hulking Memorial Hall and crossed Kirkland Street and Broadway, oblivious to the four lanes of traffic waiting for them to pass, then funneled through the ornate gates to disperse onto the diagonal paths of Harvard Yard. Normally, on a sunny day like this, rather than walk back to the Radcliffe dining hall, she would meet her boyfriend, Brooke, on the other side of the Yard. They'd grab the special at Elsie's Sandwich Shop—roast beef with Russian dressing on a roll—and join the other young couples on the grassy banks of the Charles. But on this day, she wanted to be alone with her thoughts, to give her brain time to absorb what it had just heard. She took her time, avoided eye contact. She did not want anything to break the spell.

All year Nancy had been casting about for what to do with her life. She was adamant that it be something serious and meaningful, but she had little idea what that would be, beyond a diffuse desire to reduce human suffering. She imagined she would get married and have children—few young women her age would do otherwise—and she knew that she had to do so before she turned thirty, after which childbirth was thought to be dangerous. That gave her ten years to accomplish the professional goals she had not yet determined, and now, a year to figure out what those goals might be. Otherwise, she feared, she would too easily slide from graduation to marriage, a dog, children, the suburbs. A fate she thought of as a kind of death by privilege.

She had grown up in a rent-controlled apartment on 120th Street and Morningside Drive in New York City, in a building owned by Columbia University. Her mother had gone to Teachers College there and taught art in the city's public schools, and Nancy's father was a librarian at the New York Public Library. She had a sister, Ann, who was eighteen months older. Their maternal grandmother, who had emigrated from England, lived in the same building; from her Nancy acquired a slight accent that would for her whole life flummox people trying to put a finger on her background.

Since kindergarten she and Ann had been scholarship students at Spence, the elite girls' school on Manhattan's Upper East Side, both of them in the thick of the small class of girls in their years. Spence instilled in its girls an understanding that they were privileged, and that with privilege came responsibility. They were cultivated to do important things, to go on to Seven Sisters colleges and be the best students there, to be leaders, though leaders of a certain kind: in the Junior League, or charity work. And they should be well mannered in their pursuits: the school taught its girls never to chew gum on the bus or speak loudly in public, to defer to elders and to not boast. Arriving at school each morning, they curtsied to a uniformed doorman, which their teachers told them was good training should they ever be presented at court to the queen.

Both Nancy and Ann were known as exceptionally bright, especially Nancy. She could hear a song on the radio and immediately play it on the piano; when the woman who played at morning assembly at Spence quit, Nancy took over. She had little interest in reading, but loved math, saw a beautiful language in its order. She thought of it like eating candy: a sweet burst of pleasure in solving each problem. The telephones in their building at 106 Morningside Drive ran through a plug-in switchboard, and the operator who worked it once complained to Ann that she'd taken thirty messages from Spence girls looking for Nancy's help with the night's math homework.

Nancy's experience at Spence also taught her not to put too much value on money. Her school friends with their governesses and duplexes on Fifth Avenue wanted their playdates to be at Nancy's house, where her mother would be on the floor making papier-mâché dolls from spent light bulbs. It seemed to Nancy that her classmates' parents were out every night—she read about them in the society pages of the *New York Times*—and always getting divorced. She and Ann called their parents by their first names, which evolved into made-up terms of endearment, and even their classmates called Nancy's mother Budgie and her father Diegles. Proper Manhattan never strayed north of Ninety-Sixth Street, but to Nancy, her neighborhood was like a small town in the city; she and Ann trick-or-treated between apartments in their building, played on the deep sidewalk that faced Morningside Park. At Easter they went to watch the parade of hats and finery along Fifth Avenue, inventing a game to see who could pat more mink stoles.

On weekends, Budgie led the family on adventures around the city: to the medieval Cloisters, or the Museum of Modern Art, where Nancy was entranced by the Kandinsky. She recalled her childhood as unusually happy, rich, not in money but in education and family.

Still, she fixated early on particular anxieties. The radio played often in the apartment, and from a young age Nancy had heard news reports about the aftermath of World War II, the emaciated and orphaned children returning home from concentration camps, a little boy who'd had his eyes poked out by Stalin's guards for the sin of putting flowers on his father's grave. How could humans act so cruelly? The Cold War air raid drills at school, sending Nancy and her classmates diving under their desks for cover, made her think these terrors could come right to her in New York City. She worried about losing her tight-knit family, couldn't imagine life if any of them died. Her father had had rheumatic fever as a child, which left him with a weak heart. When Nancy was ten, her mother had skin cancer—doctors cured it, but Nancy knew from the way her mother whispered the word that it was reason to be fearful.

Her father was quiet, New England in his ways. His grandfather had been a distinguished chief justice of the New Hampshire Supreme Court and his legacy dominated the family lore. Justice Charles Cogswell Doe had been an eccentric—he kept the courtroom windows open in the winter and insisted that his four children wear only navy blue. But he had a reputation for granite integrity. From his example, which Budgie summoned regularly, Nancy understood that the worst thing you could do was tell a lie. When the calls seeking help with math homework became too frequent, she decided it would be faster if she just did the homework problems for her classmates on the chalkboard at school. When the teacher got wind of this and confronted her, Nancy replied that she had done no such thing. But she couldn't sleep that night and returned to school the next morning to confess.

Budgie was the more outgoing parent, the driver of her girls' efforts and success. She had impressed upon Nancy and Ann the need for women to make an independent living, though she also told them that having children was the most satisfying thing they could do. She herself had grown up hearing the story of her grandmother who had been left widowed with eight children and impoverished after her husband, a doctor in rural England, fell from his horse returning from a house call. Budgie recalled the despair of the Great Depression, seeing people jump from windows after losing their savings and jobs. She had given up her ambitions to be a painter when she realized that she could not make a living as an artist. She had a first-generation American's faith in the transformative power of education and had early on decided that the girls should go to Harvard and that the best way to get to Harvard was through Spence.

Ann had gone off to Cambridge first. Nancy skipped tenth grade— Budgie worried she'd be languishing at home—and followed the next year, in September 1960.

Of course, the Doe girls could not be Harvard students; they were admitted to Radcliffe, which had been founded in 1879 after Harvard rebuffed women's repeated attempts to apply. "The world knows next to nothing about the natural mental capacities of the female sex," Harvard's president Charles Eliot declared in his inaugural address in 1869. "Only after generations of civil freedom and social equality will it be possible to obtain the data necessary for an adequate discussion of women's natural tendencies, tastes, and capabilities."

Radcliffe started as an experiment, dreamed up largely by daughters and wives of Harvard professors, to educate young women "with the taste and ability for higher lines of study." For decades, Radcliffe students relied on Harvard professors who were willing or interested enough in the extra paycheck to cross Massachusetts Avenue to teach women in separate classrooms around Radcliffe Yard. Harvard allowed 'Cliffies into its lecture halls only during World War II, when the number of young men going off to war put the university at risk of losing tuition dollars. This arrangement of "joint instruction" continued throughout the twentieth century. And during Nancy's years in the early 1960s, everyone, from the presidents of both institutions to the Radcliffe alumnae to the Harvard Undergraduate Council, still agreed that full integration would be a step too far. By quota, Radcliffe was allowed to admit one girl, as they were still called, for every four men of Harvard. It was an exclusive set; while Radcliffe graduated more Black women than the other colleges of the Seven Sisters, there were still only two or three in a class of roughly three hundred each year. There were none in Nancy's class.

The president of Harvard, Nathan M. Pusey, was the first since the founding of Radcliffe to have a daughter but showed relatively little interest in the education of women; he declined to attend the dedication of a new graduate center at Radcliffe in 1956, noting in his papers that it conflicted with the Harvard-Penn football game. In speeches and reports Harvard officials referred to the university as "she" with the reverence one would a goddess or an ocean liner. Actual women had to enter the Faculty Club through the back door and eat in a separate dining room. They were not allowed in the main library, or Harvard's undergraduate dining halls except as someone's date (the Harvard man had to pay for her meal). For the most part, Radcliffe girls aligned with the expectation that they be the ornamental sex: "We know that beauty is only skin deep, but you don't have to look as though you lived only for things of the mind," a Radcliffe student handbook from the 1950s tsked, explaining the rules against pants downstairs in the dormitories.

Those rules persisted through Nancy's time at Radcliffe. But the institution had begun to rethink women's education, starting with the arrival of Mary Ingraham Bunting, a microbiologist who in 1960 became Radcliffe's fifth president and its first with a PhD.

Bunting, known since childhood as Polly, quickly saw that Radcliffe women could not help but feel like second-class citizens. Life in their dormitories was "detached and thin" compared to Harvard's house system, with its live-in tutors, guest speakers, and bounty of student activities. Bunting began publicly decrying what she called the "climate of unexpectation" for American girls, steered away from education and into early marriage by "hidden dissuaders," "the inherited influences, the cultural standards which produce, for example, the belief that a scientific career is somehow 'unladylike' or that marriage should be enough of a career for any woman." Among the high school students scoring in the top 10 percent on ability tests, 97 percent of those who didn't go on to college were girls. Those who did go on, she argued, were squandered by a society that did not embrace their accomplishment or potential. The women of America—emancipated, educated, and enfranchised—were a "prodigious national extravagance." While it was no longer unusual for women to desire and obtain college degrees, "we have never really expected women to use their talents and education to make significant intellectual or social advances," Bunting wrote in the *New York Times Magazine* in 1961. "We were willing to open the doors but we did not think it important that they enter the promised land."

Bunting had been thinking about these ideas well ahead of Betty Friedan, whose bestselling book, *The Feminine Mystique*, had been published in February 1963, two months before Nancy took her place in Watson's lecture hall. Friedan's book raged against a culture that had locked intelligent and well-educated women into lives of "quiet desperation"—kept from full participation and their full potential—by convincing them there was fulfillment in polite children, passive sex, and a perfectly waxed kitchen floor. Friedan had asked Bunting to collaborate on the book, and Bunting had met with her several times but soon concluded that Friedan was too angry, too intent on blaming men for women's problems.

Bunting didn't blame men. She thought the limits on women hurt everyone, as she wrote in the *Times Magazine*: "A dissatisfied woman is seldom either a good wife or a good mother."

And she didn't think women necessarily had to have careers. In fact, she suggested they work on the fringes, "where there is always room," rather than compete directly against men. She urged her undergraduates to marry and have children and also to find "something awfully interesting that you want to work on awfully hard." Having children would be a pause, not the termination, of intellectual pursuits outside the home. She envisioned the successful path of work and family ambitions like the new interstate highway system connecting postwar America, with women finding on- and off-ramps along the way.

She faulted previous generations of educated women for encouraging a negative stereotype of smart women: "For the most part, they became crusaders and reformers, passionate, fearless, articulate, but at times, loud," she wrote in her annual president's report in 1961. "Today, several generations later, the bitter battles for women's rights are history. The cause has been won. The stereotype has disappeared and with it, the hard prejudice. But not altogether. For there is still prevalent a form of anti-intellectualism which insists that whatever her aspirations, a woman must eventually choose between career and marriage, and that if she attempts to combine the two, both will suffer and the marriage probably the more keenly."

Bunting was fifty when she became president of Radcliffe, and saw her presidency as a perch from which to promote a happier image of a life that combined family and professional pursuits. She herself had raised four children and four goats, served on school and library boards, and grown all her own vegetables in between part-time positions at Bennington and at Yale, where her husband had been on the medical faculty. She had carried on despite adversity, taking on the job as dean of Douglass College, the women's branch of Rutgers, after she was widowed at forty-four. And she saw Radcliffe as "a promising instrument for the attack that is called for." As some trustees urged a merger with Harvard, she argued that it was a "negation of fact" to assume that women and men should be educated the same way. Women needed the same rigorous coursework as Harvard men, but their futures were in many ways more complicated, and they needed help planning "wisely and largely" so they would not lose their way. Bunting started what she called a "campaign versus apathy" that included a lecture series around Radcliffe Yard, thesis presentations with professors, but also "living room talks" of how to be a mother and wife.

Bunting noted approvingly the year Nancy arrived that the number of Radcliffe sophomores declaring English as their major was the lowest in a decade, and there had been a significant decline in history as well. Though those were still by far the most popular majors, Bunting wrote in her annual report, "The trend is welcomed as an indication that Radcliffe students are becoming a little more adventurous and imaginative and perhaps serious in their choice of major fields." The number of Radcliffe students who married before graduation had fallen steadily—it had been 25 percent in 1955; it was about half that now. Perhaps because they were competing for fewer spots, Radcliffe women in Nancy's class were increasingly more accomplished than Harvard men—"fearfully bright," as *Time* magazine pronounced in its cover profile of the woman the magazine and her students called "Mrs. Bunting." The women arrived with higher SAT scores and were far more likely to graduate with honors.

There were rumblings of the revolution that would bring coeducation and increased racial diversity at the end of the decade. Spring of 1963, the year Ann graduated, was the first time Radcliffe women would be given Harvard diplomas. (Graduation was still separate, and President Pusey continued to send a faculty member in his stead.) Radcliffe relaxed the social rules that had required girls to sign out of their houses and secure permission to be out past one o'clock in the morning. (While Radcliffe's young women debated how this would affect dating and sex, Bunting thought the rules had disadvantaged young women who wanted to work all night in science labs.)

Still, Radcliffe girls who wanted a professional life had few role models on campus beyond Polly Bunting. The undergraduate faculty at Harvard had 295 tenured men, and 2 women—one, Cecilia Payne-Gaposchkin, had spent thirty years as a lecturer before being granted tenure. (Her salary had been listed on the budget under "equipment.") "What would happen if there was a genuine effort to see that one woman's name was present on every slate considered, as is now customary in government with respect to Negroes, would be interesting to observe," Bunting wrote in her annual report. Radcliffe administrators, she confessed, could themselves barely come up with candidates on the rare occasions they were asked. "Certainly it would be encouraging to Radcliffe students and probably revealing to Harvard men to have a greater number of able women scholars on the Faculty," she wrote, "provided that they are qualified."

It never occurred to Nancy to count the women in Bio 2. There were 47, and 178 men, about the same as the proportion of Radcliffe students to men at Harvard, and the number had been growing; the percentage of women declaring biology as their major had more than doubled over the previous decade; it was now the fifth most popular concentration.

Nancy had switched her major to biology only a few months before. She had arrived at Radcliffe intending to major in math, but her freshman adviser told her the first week that she was too far behind to possibly catch up; she would need two years of calculus, and Spence had not even offered it. She chose architecture as her major because she liked art and math and thought it might be a way to combine them, but the math classes didn't move her the way she thought they would. Math didn't seem to relate to anything she cared about. Her latest idea was to become a doctor. But the Bio 2 lectures so far had left her thinking she wasn't cut out for medicine, either. Physiology fascinated her; how the heart pumps blood, the muscles contract, the kidney maintains the salt balance, was all so intricate and beautiful. But she didn't think she could spend her time with sick people or tell parents their child was dying. She wanted to figure out what caused the disease in the first place, and how to fix it.

Watson and Crick's discovery of the double helix had been described as a flash flood, arriving so quickly that few saw it coming, and forever reordering the scientific landscape. Scientists had barely agreed that DNA was the stuff of heredity, and they did not understand how it passed along traits. The double helix explained these mechanics— it was in the sequence of the always matching base pairs, and in the ability of DNA to make an exact copy of itself. Having understood this, the infant science of molecular biology was on its way to identifying the code that gave form and function to all of nature.

Watson and Crick, atheists both, believed that genes could replace religion as the organizing principle of the universe. Nancy was ripe for conversion. For the month of April, her life became all about Bio 2, Tuesdays, Thursdays, and Saturdays at eleven. Watson fed the students the story of DNA like a mystery: How did it relay its message so that a gene knew to code for a protein, and a bacterium knew how to digest a certain sugar? How did genes know how and when to turn on and off? He started with the question of how cells grow and ended with what happened when they did not get the signal to stop growing and became cancer.

Nancy found herself on the edge of her seat, almost falling off if not for the wooden arm on the desk. She quickly came to guess what the next question was going to be before Watson asked it—an affinity she would later understand was called "good taste in science."

She was hooked. So at the end of Watson's lectures, she went to his office to ask if she could work in his lab.

In contrast to the puritan buildings of Harvard Yard, the Biological Laboratories building was whimsical and grand, not unlike Watson himself. Hidden on a path off Divinity Avenue, it was a massive U-shaped brick building surrounding a large quadrangle. Its facade had been drilled with friezes of animals representing the world's four zoological regions, a sable antelope and Asiatic wapiti among them. Guarding the entrance on either side were two massive bronze rhinoceroses, sized to match the largest known of the specimen and named for England's Queens Victoria and Elizabeth, Vicky and Bessie for short.

From his third-floor office overlooking the rhinos, Watson was busy overthrowing the old world order. He chafed against the traditions of Harvard, its refusal to give him a $1,000 raise that year despite the Nobel, its unwillingness to bestow biology with as much stature as physics or chemistry. He found the biology department fusty and lumbering, too focused on fields like ecology and zoology, which he considered extinct, hobbies at best. Like many scientists of his generation, he had been captivated reading Erwin Schrödinger's *What Is Life?*, which posited that biology could be understood like physics and chemistry, as a set of universal laws. Molecular biology—and particularly the understanding of DNA, the most significant development of the new field—offered the possibility of understanding the chemistry behind the cellular processes that make up the living world. Why would you waste your time on taxonomies or the competition of the species when you could be figuring out how things worked in every last cell? He trained distinct disdain on Edward O. Wilson, the evolutionary biologist who rivaled Watson as a wunderkind but had been tenured a few months before him. Wilson, a genteel Southerner, in turn saw Watson as a rapacious megalomaniac with no time for collegiality or polite conversation, even a hello in the hallway. Wilson called him "the Caligula of biology," brilliant but "the most unpleasant human being I have ever met." After several tense years and frosty faculty meetings, Watson had succeeded in splitting the department in two, shipping the old-line biologists off to Harvard's Museum of Comparative Zoology across the street and recruiting physicists and chemists to join his own labs to further decode the workings of DNA.

While other students seemed afraid of Watson, the lectures had made Nancy bolder, and she knocked confidently on his door. She entered to a large airy office, one wall all books, the other, all windows. Diagonally in front of her was a piece of art Watson had purchased with his prize money from the Nobel, a life-size wooden likeness of a Papua New Guinean man, naked and roughly anatomically correct from the waist down. Watson knew she was a neophyte, but agreed to work with her without discussion or argument: "Sure," he said, then jumped up from his sparsely covered desk and breezed around her to push through a door into a small room with a lab bench running along its length. He told her she would share it with two other Radcliffe students.

They shared the lab room with twelve Harvard undergraduates, but still it was an unusual concentration of women. Watson created room for women partly because he liked having them around. He was looking for a wife and thought he might find one at Radcliffe. He also thought they made life more interesting. At the Nobel ceremony in December 1962 he had convinced Princess Christina of Sweden to apply to Radcliffe and, once at home, arranged with Presidents Bunting and Pusey for her to attend. (Pusey had little patience with some of the antics of his young Nobelist, but wrote back, "She sounds like a most attractive young lady and I appreciate the interest you have taken in this matter.") And Watson had an eye for talent: as his first female graduate student, he had taken on Joan Steitz, who was two years older than Nancy and would go on to be a Yale professor and one of the most highly regarded biologists of her generation. Steitz's first choice had rejected her because she was a woman: "You'll get married and you'll have kids—then what good would a PhD have done you?"

Watson wanted to fill the lab with fun people who knew when to laugh and remained upbeat even when experiments went nowhere. He liked unconventional thinkers and saw Nancy that way. He was not interested in her romantically, considered her more handsome than pretty—his taste ran more traditional and blonder. He was fascinated that she had gone to Spence, which he considered one of the nation's best schools. He was proudly Irish, from Chicago's South Side and a family richer in intellect and culture than land or cash—Orson Welles was a distant cousin, and Watson himself had been a radio "Quiz Kid" and entered the University of Chicago at sixteen. His nose was still firmly pressed against the glass. He was intrigued that Nancy's friends were debutantes from old-line families, Winthrops and Pratts. That, and her accent, which he mistook for Long Island lockjaw, made him think she was from New York society, not a building with an unfashionable address and no doorman.

The students in the lab were supposed to be doing their own experiments, but only those who had worked previous summers or semesters in labs were doing so. Nancy was still watching more than she was doing, but she was learning about science and how scientists worked. She found them open to her questions—and she had lots of them, her brain popping with ideas for experiments even if she didn't yet know how to carry them out. The pastimes that had once consumed her—Saturday excursions to Crane Beach on the North Shore, parties in Eliot House with Brooke and his roommates—no longer held her interest. Only the lab did.

Watson could be demanding and at times dismissive, turning on his heel to leave if you bored him, which Nancy tried hard not to do. He soon began ducking through his door into her small lab space more regularly with "What's new?" or "Lunch?" and calling her "kiddo." Often he would deliver a bit of information, or a joke, transmitted in his staccato: dot dot dash dot dash. He'd be gone in a flash, the door still swinging on its hinges. She'd had other instructors who related well to students—Erich Segal, who later wrote *Love Story*, had been among them—but Watson was unusual. She knew his extraordinary accomplishment, yet still he spoke to students on their level, griped about professors the way they did. He soon became a friend and her idol, one of the most important people in her life.

Jim, as Nancy now called him, lived in a railroad apartment in the only house on Appian Way, near Radcliffe Yard, a narrow white clapboard building where he threw crowded parties that mixed professors and students and the occasional celebrity—one party was for the princess, another for Melina Mercouri, the Greek star of the hit film *Never on Sunday*. If it was unusual for faculty and students to socialize like this, science was unusual, and the lab like a family, its members climbing every year onto the rhinos in front of the building for goofy group portraits. Jim's widowed father lived in an apartment downstairs. Jim doted on him—they shared a lifelong passion for bird-watching and the Democratic Party—and took him out to dinner each week. Jim began inviting Nancy and sometimes Ann, who was living in Cambridge after graduation, or their mother, who was living in New York but visited often after the death of her husband.

The Watsons loved to argue and talk politics, which Nancy's family had never done. It was her first stirring of a political consciousness. Jim and his father considered themselves New Deal socialists, had left the Catholic Church but kept its concern for the poor; Nancy knew only that her mother had liked FDR and that her father had been a New Hampshire Republican.

She was the happiest she had been since childhood. The lab had opened up a world she had not known existed, much less thought she might inhabit. The field was so new, the universe of molecular biologists so small, and Watson so central to it that it was all unfolding in front of her. He and his friends were still identifying the genetic code, the three-letter labels for the twenty amino acids that translate DNA to protein. They had formed what they called the RNA tie club to share notes that might advance the understanding; members were initiated with a woolen necktie embroidered with a helix and a pin with a three-letter code. Those members included some of the biggest names in science: Sydney Brenner, Edward Teller, Richard Feynman.

Every afternoon at three, students and scientists from the lab would crowd into a small square room for tea and chocolate chip cookies fetched from Savenor's by a receptionist on her old Raleigh bike. Jim had adopted the tradition from his time at the Cavendish Lab in Cambridge, England. Nancy, her fellow students, and some technicians filed in first, followed by the postdocs and then the faculty members. Jim typically entered last, silencing the room as the others waited to hear what gossip or discovery he had learned that day. The information he shared could be monumental: one day, it was that the genetic code was universal—the transfer of information from genes was the same whether the creation was a virus or a fly's wing, the leaf of a plant or the human brain.

Nancy saw how small facts from individual experiments added up to larger insights that together could answer the big questions. She began to see the elegant logic of biology; it was profound, mysterious, and yet totally sensible.

She had found her place in the world, her purpose. Bio was her "awfully interesting" thing, though she quickly concluded that Mrs. Bunting must be withholding some secret to explain how anyone with a family could also have this consuming a career. The students and postdocs worked around the clock, coming in at least six days a week and often a seventh if they needed to check a tube or skim a gel from a machine when trying to isolate a protein or a compound. Watson set the example. One Saturday night, Nancy noticed a light on under the door to his office. She pushed through to see what he was working on. He was at his desk writing what would soon become the first and classic textbook of the new field, *Molecular Biology of the Gene*. He looked up at Nancy, and his face told her that he did not want to be interrupted.

Nancy's roommate, Deming Pratt, saw few charms in Jim. Deming thought his gestures awkward, his clothing too baggy. He didn't even look you in the eye. Nancy was amused by his aspirations to the world of Long Island's Gold Coast and membership at the Piping Rock country club, picked up during his summer stints at Cold Spring Harbor Laboratory. Deming, who was of that world, considered it unseemly, almost Gatsbyesque, though she did grant Jim's request to attend her twenty-first birthday at her family's estate. (There she noted that he couldn't waltz.)

Nancy's boyfriend, Brooke Hopkins, sometimes joined her outings with Jim, but professed little interest in her new world. Brooke and Nancy had met at a freshman mixer her first week, and he was the kind of boyfriend she had dreamed of in girls' school, where her classmates had convinced her that good grades would scare boys away. Six foot five, a rower with dark reddish hair and brown eyes, he came from a conservative Social Register family in Baltimore, had gone to Hotchkiss and gotten into Harvard on mediocre grades and, he presumed, family legacy. He was doing his best to rebel, listening to jazz and reading Freud and James Baldwin. He had been recruited by the Porcellian Club, the most elite of Harvard's all-male social clubs and the proving ground for the future leaders of American social, political, and business institutions, who, it was understood, were almost always white and rarely Jewish. He'd joined Fly Club instead, thinking it more progressive.

Ann divided the men of Harvard into those who wore dark socks and those who wore white: the dark socks were worn by the boys who had gone to boarding or elite private schools, the white socks were worn by the boys from public schools, Bronx Science or Stuyvesant if they were lucky. These white-sock boys were becoming Nancy's friends in the lab, and she felt more at ease with them than she did the boys with dark socks. Science was becoming her social world, too.

The lab brought many famous visitors, none more eagerly anticipated than Francis Crick. Crick was twelve years older than Jim, and Jim revered him. Jim quoted him constantly, aped his confidence and his English intonations. ("Both young men are somewhat mad hatters who bubble over about their new structure in characteristic Cambridge style," a visitor from the Rockefeller Foundation to the lab wrote shortly after their discovery of the double helix. "It is hard to realize one of them is an American.") It was close to graduation in the spring of 1964, and Watson had planned a big party in his apartment on Appian Way in Crick's honor, inviting students along with some of Harvard's most prominent faculty. Nancy made a mental note not to drink too much; Watson served a potent cognac-and-wine punch, and she wanted to stay sober enough to have an intelligent conversation with Crick. She went to do some work in the lab early that afternoon in the hope she'd get the chance to meet him there first.

Francis Harry Compton Crick had grown up the son of a shoemaker in the British Midlands, but carried himself with the satisfaction of an aristocrat, a man used to being acclaimed as a genius, as almost everyone who knew him now did. He was tall, handsome in an imperious way, with a prominent nose and chin—a profile worthy of a coin—and an arch smile. Nancy would have recognized him from the photograph in Jim's office of the two men gazing at their six-foot model of the double helix, had she time. But Crick arrived so suddenly behind her in the little side lab that she realized he was there only because his hands were on her breasts.

"What are you working on?" he asked, his voice as loud as its reputation.

Nancy froze. Crick was married, and she had heard he was a womanizer; a former grad student from Watson's lab had told her that he knew a postdoc who had slept with him in England, that Crick's beautiful and cultured wife knew about his infidelities but looked the other way. But Nancy hadn't dwelled on the gossip, and she hadn't considered this would come to her. She'd heard similar rumors about Jim, and he had never made a move on her. She'd never even seen Jim with a date.

She wriggled on her lab stool to break free of Crick's clutch and stammered, trying to find a way to move the conversation onto science. She thought first that she didn't want to embarrass him, second that she didn't want to embarrass herself. She had to be able to face him that night.

She told him she was studying bacterial viruses but that she really wanted to study the repressor function of genes, though she realized that was too challenging a problem right now. She kept talking, holding out hope he would see her as a serious person.

Then just as suddenly, Jim burst through the door, clapped his hand on Crick's back to greet him, and steered him back into his office. Nancy was alone again. If Watson had seen anything, he didn't say so. Nancy went to the party that night and held her liquor. It was crowded, with dancing, as usual. Jim urged her to stay to the end, which she did, hanging on the small circle talking science around the famous duo. Then, because no Radcliffe girl would walk through Cambridge alone at one o'clock in the morning, Watson and Crick, purveyors of the secret of life, the great men who generations of biology students would know by their conjoined surnames, escorted her along the brick sidewalks of Garden Street and back to her dorm.

It had been, Nancy thought as the heavy door closed behind her, a spectacular day.

She could have done without Crick groping her, but after the evening at Watson's she had almost forgotten about it. What filled her head instead was newfound assurance, confidence that she had found the meaning she had been looking for. She liked her private-school friends well enough, liked to go to their parties. But these other people were on a higher plane—the ideas they discussed, the questions about life they were poised to answer. She wouldn't quite say she felt at home in their world, at least not yet, but even to be at the edge of their conversations—it was the most exciting thing that had ever happened to her.

Chapter 2

The Choice

Nancy never felt relaxed in Jim's presence, but she felt a kind of order with him around, it made her feel unusually at peace. She saw him as a smarter version of herself, the first person beyond her mother who understood how her mind worked.

One day he breezed through the door into the small room where she worked: "You should be a scientist. You're like me. You have a one-track mind."

Nancy was thrilled, but still doubtful. She asked one of the postdocs why he thought Watson had taken an interest in her. "We're all curious to know if a girl can make it to the top in science," he said. "We think you might be the one."

Now Jim had decided Nancy should go to graduate school to get a PhD. While Nancy was inclined to do whatever Jim thought she should do, on this she was not persuaded. Not for lack of interest, but lack of time.

A year of working in Jim's lab had convinced her that she wanted to do molecular biology—that she had to do molecular biology. She was confident she could do great science, even science worthy of a Nobel Prize. She thought other people might be smarter than she was, but she had an eye for the next big question, the next big experiment to be done. All she needed was more practice in the lab.

She also had what she called an understood engagement with Brooke, who was planning to go to graduate school in English at Harvard and become a professor. They had now been together nearly four years, and she loved him, saw him as one of the only people she felt truly understood her. They were both searching, intense—according to their mothers, too intense for each other. Brooke did not share Nancy's interest in science; he had scored poorly on his math SATs and concluded he was no good at it. He was putting faith in Freud and psychoanalysis just as she became increasingly convinced that human behavior would ultimately be explained by chemical reactions. But Nancy believed they complemented each other. Brooke introduced her to jazz and to movies and books she'd never heard of—she still didn't think herself much of a reader. She thought he was gorgeous, the way his skin glowed in the sun, the reddish-blond hair on his muscular arms and long legs.

The choice between Brooke and science seemed to Nancy an impossible one. Even the word *choice* implied far more freedom than she felt. She was expected to have children and to accomplish important things outside the home. She didn't see how she could do both, not at the same time, not in science. The scientists Nancy had watched at Harvard were almost all male and worked all hours, with their wives and children occasionally dropping by in the evenings or on weekends to say hello. The women in the lab were mostly technicians, and she didn't need to go to graduate school to do that job. The only woman she saw with a PhD was Ruth Hubbard, who was not a professor and worked in the lab of her husband, George Wald, who was seventeen years her senior. Nancy assumed there were no women because the women had chosen children over science.

So Nancy had set herself a new deadline: she would do a Nobel Prize–winning experiment by the time she turned thirty, when she'd have to turn to raising children. Jim, after all, had made his big discovery at twenty-four. (She didn't need to win the actual prize, but the idea of a discovery that pushed the field forward enough to deserve one—that excited her.) Maybe she would stay in science as a technician after that, since technicians could work regular hours. But she thought graduate school was a waste of time if she wasn't going to go the traditional path of PhD to postdoc to assistant professor, associate professor, full professor—all that time spent building up a portfolio of research and publishing papers to create enough of a reputation that senior people in the field would write letters declaring that you should win tenure. Nancy just wanted to do experiments. Studying and taking courses in graduate school would steal from the time she could be in the lab. Getting a PhD took between three and five years, then it was another two or three for a postdoc, and five to six years to get tenure. The math didn't work for her.

She didn't think about this as a case of limited options. She saw it as reality. She had read Betty Friedan's *Feminine Mystique* as well as Simone de Beauvoir's *Second Sex*, published a decade earlier. But at twenty, Nancy did not see her own life in the historical and cultural oppression they described, the portrayals of marriages filled with rote sex and hours of housework done solely to justify buying the latest laborsaving appliance. She assumed that by diagnosing the problem the books had solved it—they had made it possible for her and the other young women of 1964 to have a life outside the home. That Nancy could not have both children and a full-time career was, she thought, a problem of choosing to do science. She admired Polly Bunting—Nancy had been reading about her for years, and everyone, including Jim, had said how impressive Mrs. Bunting was. Yet something didn't add up; Nancy could not understand how Mrs. Bunting had been able to be a microbiologist and a mother both. (She did not know that Bunting had worked part-time in the lab at Yale and started her "day" around six o'clock in the evening, when her husband took over watching their children.) Nancy had gone to the living room talks at Radcliffe and heard nothing there to explain it, so she assumed Mrs. Bunting had some special talent or circumstance that had allowed her to have it all.

The women in Nancy's Radcliffe class were struggling with their own versions of the same dilemma. Early marriage could sometimes seem the only acceptable option: premarital sex was taboo; state law banned birth control and abortion. In the fall of 1963 Harvard administrators cracked down on "misuse" of parietals, the limited hours that women could be in the Harvard houses (provided they kept the door open and one foot on the floor). The dean of students warned that it would be "scandalous" if the public discovered that students were having sex in the Harvard houses: "It's our positive duty to deal with fornication just as we do with thievery, lying and cheating—by taking severe disciplinary measures against the offenders."

While some Radcliffe women had come to college prospecting for a husband—President Pusey liked to say that 60 percent of those who married, married Harvard men—many also wanted to work. The largely unspoken wisdom, though, was that they would have to choose between the two, and in doing so they would inevitably let someone down: their parents, their intendeds, Mrs. Bunting, themselves.

"The young women of today are a race of culturally induced schizophrenics," the Radcliffe yearbook for 1964 declared. "They are reared and trained to be the equals of men, and have heard innumerable stories of women who carved out places for themselves in science or politics. Yet these women are also fed the Great American myth of house and home, of children and of a husband with pipe in mouth, paper in hand, and wife on his lap."

The problem, the yearbook editors proclaimed, "is unique to our generation." Every generation thought it was unique, but this one could feel the ground moving beneath its feet. The Harvard man in the White House had been assassinated that previous November; students had glimpsed President Kennedy in Cambridge just thirty-four days before he was shot, in the stands at the Harvard-Columbia football game wearing sunglasses and puffing on a small cigar. The arrival of the Beatles in America in February had lightened the mood on campus, but now the country was fracturing again over whether to pass the Civil Rights Act. The world was changing just as these young women prepared to enter it, and no one was sure what shape it would take.

The yearbook editors dismissed *The Feminine Mystique* as a "temper tantrum" against the circumstances and circumscriptions of their mothers' generation, "a world of men who were convinced that any form of occupation was unfeminine, and that women should never leave the home." It was no longer news that a woman was "a functioning human as well as a female and should be treated accordingly." Feminism was a phenomenon of the past—as outdated as suffragists chaining themselves to lampposts to demand the vote. Friedan and de Beauvoir had pushed for the right for women to defer marriage and to have a career, and they had won.

With new options, the Radcliffe class of 1964 faced a paradox of choice. "Mrs. Bunting advances the hope that the graduates of today will enter the vanguard of female leadership tomorrow, but just how to join this vanguard is far from clear. There seems to be a necessary choice between two roles: doting wife or dynamic executive. Unwilling or unable to choose, most girls do their best to avoid the decision. They drift in one direction without ever fully examining the alternatives." Some became science majors, pledging "eternal dedication to the Bunsen burner flame of knowledge," others dabbled in fine arts, avoiding any plans for the future. But the science major "still flirts with her lab partner"; the English major "enjoys arguing about Barry Goldwater."

"She is too *aware* to think that her life is a failure if she graduates without an engagement ring, but she knows that any career she undertakes is likely (hopefully?) to be interrupted in a few years. She graduates, then, untrained, unwed, and uncommitted."

While the yearbook editors scoffed at Friedan's generation, they diagnosed many of the problems she had in their own peers, noting that the share of women going on to graduate work and professional training had receded to what it was in 1910, that the average woman continued to marry soon after graduation if not before. (The average age of marriage for women had slid to twenty, and the age of childbirth to twenty-five; by 1964 half of all American women had their last child by age thirty.) "While she has acquired the same background for jobs as many men, the great American myth has also seen to it that she is not emotionally prepared to compete for a career."

What was their solution? The Radcliffe graduates of 1964 could not be "sent back to the parlor and taught sewing." They could not become, as in Friedan's bleak picture, "just a housewife." Yet they limited their aspirations. Theirs would be the double role: bringing up children and augmenting their husbands' incomes. They were fighting to emerge after having babies to become an executive assistant to "some publisher," or to "pursue a career on a part-time basis." They rallied their classmates to convince the world of their full capabilities, to show that they could be intelligent, articulate, and feminine all at once. The next hurdle for their generation, they argued, was to enlist their classmates across Massachusetts Avenue in the challenge: "The sad truth is that purely feminine effort will not win the right to do significant part-time work. The Harvard graduate of 1964 must also be made to realize that he can not expect a wife in his grandmother's image, that the business world has ceased to be a masculine retreat, and that women should be welcomed for the particular contribution they can make. A large part of education to the future lives of women is, in fact, the education of men."

Nancy considered Brooke educated in this respect, as men went. Still, she, too, understood that childcare would fall primarily to her. (Census surveys showed that 60 percent of women left the workplace on the birth of their first child, while 16 percent were given paid maternity leave and 14 percent unpaid leave. Five percent were fired.)

A change in Brooke's plan prompted her to reconsider Jim's idea that she go to graduate school. Brooke had applied for a fellowship to Oxford University and won based on the strength of his senior thesis. She had been thinking of getting a job in Boston, but no one—including Jim—had offered her one. With Brooke going away, she figured she, too, might as well go to graduate school, if only to defer a more permanent decision.

By custom, scientists did not do their graduate work at the same institution where they had earned their undergraduate degree. Jim's first choice for Nancy was the Rockefeller Institute for Medical Research in New York, perhaps the most selective program in the country, and the place where Oswald Avery had first identified DNA as the molecules that carried hereditary traits from one cell to another. Rockefeller accepted students not by the usual application, but solely on the recommendation of someone with stature or connections—or, as with Jim, both.

He wrote a brief note to Rockefeller's president—"A bright and extremely pleasant Radcliffe girl now working in my lab might be a suitable graduate student at the institute"—and sent Nancy to her interview instructing her to dress well and carry the *New York Times*. Nancy borrowed a brown skirt suit from her sister and arrived in New York feeling smart in its fitted jacket and midlength flare. The Rockefeller campus sat on a bluff above the East River, with fountains and shimmering gardens secluded behind tall iron gates. Its formality, so unlike the Biolabs at Harvard, daunted Nancy. Oil portraits—all of men— hung on the walls of the dining hall. Maintenance men in crisp uniforms swept errant pebbles from the paths of the manicured grounds. Rockefeller's president, Detlev Bronk, was also the former head of the National Academy of Sciences and descended from the family that lent its name to the Bronx. Nancy had been told that he made up his mind whether to admit you in the time it took to walk the vast length from the doorway to the desk in his office. Sitting across from her, Bronk asked whether she thought she could succeed there and said that if her answer was yes, she could attend. She told him she could. But she never heard from him and assumed he had told Watson he could not take a chance on such a novice. She couldn't really blame him.

She applied to Yale, a more conventional application. Still, she knew the most important criterion was experience in a lab and the referral from the lab head. In his letter, Jim recommended Nancy as "bright and cheerful," but noted that she had only recently decided to concentrate in biology, "so her factual knowledge is still meagre."

"If sufficiently motivated, she might do very well in graduate school for she learns quickly," he wrote. "Up till now however, she alternates between serious phases and the carefree mood" of a young woman "immersed in a pleasant social atmosphere. It is my hope, nonetheless, that she will realize the advantages of using her brain and acquire a graduate degree. If this transformation happens, she would be a first-rate student."

Yale accepted her, and Jim introduced her to the chairman of its Department of Microbiology, Edward Adelberg, who was visiting Harvard to give a talk. "He'll take care of you," Jim told her. Nancy agreed that Adelberg seemed like someone who would.

She hated the prospect of being apart from Brooke. But she felt elated as she celebrated graduation at a lunch with her mother, her sister, and Jim—three of the four most important people in her life. Brooke moved to England shortly after, and Nancy to New Haven.

Jim was right that Nancy would be a first-rate student in graduate school, and also about her tendency to swing between extremes. Ambivalence about marriage and science gnawed at her.

Yale was not the charming place she remembered from her weekend visits in college, cheering at Brooke's crew races or the annual game in the football rivalry with Harvard. She took most of her classes not on the main campus but in the medical school area, in a part of the city considered so unsafe that she had to rush to walk home before dark. The house where she rented a room was populated by women in their seventies, single or widowed and living on Social Security. For the most part these were the only women she saw during the week. She counted few if any in her classes. The gray Gothic quadrangles of Yale's old campus may as well have been a monastery, devoid of women except on weekends, when buses arrived from Vassar and Smith to deliver exquisitely dressed dates, appearing like extras on the set of a movie suddenly in color. Nancy worried jealously about Brooke; she had heard that one of her Radcliffe classmates, a summa cum laude in English who looked like Grace Kelly, was at Oxford pursuing him.

Nancy recognized her impatience and threw herself into her studies. She earned all As and mastered biochemistry, the subject she had feared would sink her. She moved to an apartment in a semidetached house on Science Hill that she inherited from a friend of Brooke's from prep school. It was closer to the new Kline Biology Tower, where molecular biologists were setting up their labs. Knowing she would have to prove she could read and write two languages to get her PhD, she set about learning German—she already knew French—and with just two weeks of tutoring from a new neighbor who was an instructor at Yale, picked up enough to pass a timed test translating a scientific article about DNA. The fall she arrived, Dorothy Hodgkin, a British scientist and mother of three, won the Nobel Prize for using X-ray crystallography to identify the molecular structure of penicillin and vitamin B_{12}. She was only the third woman to win the Nobel in chemistry, after Marie Curie and her daughter Irène.

In her second year, Nancy had to find a lab where she would do the research for her PhD. She understood from working in the Harvard Biolabs that the only questions worth exploring were the biggest ones—Wally Gilbert, a physics professor Watson had recruited to biology, had said that experiments to prove unimportant facts required just as much work as those to prove important ones, so why not aim for the home run? And what Nancy had told Frances Crick in that awkward encounter on her lab bench was true: she wanted to understand the repressor.

The earliest molecular biologists had studied genes in phages, the viruses that invade bacteria. Viruses were small bundles of genetic material wrapped up in protein that multiplied once inside the cells they attacked, and phages were simple viruses, which made them accessible for understanding the structure of genes. Ultimately the scientists wanted to understand the role of genes in more complicated systems—how a fertilized egg developed into an organism. But that challenge remained too enormous, and far off. The big question standing in between was what controls the way genes function, so that a particular gene knew when and how to express itself, or not? If every cell had the same DNA, how did one know to become a blood cell, and another a liver cell?

In 1961 two French scientists, François Jacob and Jacques Monod, had proposed that the answer was in genes that produced molecules whose job was to prevent other genes from expressing. The molecules were referred to collectively as repressors. But no one had proven that hypothesis—how repressors worked or even what they were made of, though there was speculation it was protein. Because the amount of repressor in any cell was so small, finding it was a challenge. In four years, no one, not even Monod's lab, had been able to isolate a repressor. It remained the holy grail.

The question of the repressor had captivated Nancy ever since Watson raised it in his Bio 2 lecture her junior year at Radcliffe. She understood that someday—in one hundred years, maybe—scientists would be able to look at human genes and see that cancer was caused by failures of gene expression. That question—why some cells became cancerous—captivated her most of all.

She asked around Yale to find someone who might be interested in studying the repressor. She was referred to Alan Garen, a biochemist who worked on gene expression. He agreed that the repressor was a critical challenge, but he was not sure he wanted to tackle it. Maybe if he was younger, he told her. Garen was thirty-nine. Nancy, at twenty-one, bemoaned that she'd ended up in the geriatric ward of molecular biology. "You spoiled me," she wrote Jim.

A conversation with Jim was usually all it took to renew her interest in science. He told her Garen was first-rate—Jim's highest praise—and that she should try working in his lab at least a few months, especially since Garen had agreed to take her. She could learn from him. So Nancy signed on.

She immediately judged Garen too cautious. "He is a very nice man, very sweet and all that," she wrote Jim after her first day of work. Garen had rejected one of her ideas for an experiment. "So I'm doing something which is no doubt related to the repressor in the distant future but so distant as to be discouraging. Doing experiments is extraordinarily time-consuming and on the whole—tedious. It seems to me that only if they are directly related to the most interesting questions does all that work become tolerable. I'm interested in science because I honestly want to know the answers to questions like control, cancer, and how does a fertilized egg produce an ear etc. etc. At this rate I'll be dead before any of those questions are answered."

A few months later she wrote again, sounding more optimistic and more resolved. The Radcliffe classmate who had been pursuing Brooke had married his roommate from Oxford instead. Garen had recruited a new biochemistry professor who had done his postdoc in the lab with Jacques Monod, and Nancy thought she might work with him instead of Garen, though she still wondered whether graduate school was worth it. "When overcome by such doubts I recall the famous advice from great man to aspiring young scientist: 'When you get tired of studying—study harder' and I return to my Schrodinger equations with renewed vigor."

The lab was quiet, she missed the intellectual hum of the Harvard labs, the technicians and scientists willing to answer her questions, the cross-fire critiques of the tearoom seminars.

Garen was reticent and worked mostly with his wife, sharing lunch with her in his office. He exchanged only polite conversation with Nancy. Occasionally while working, she effused about how experiments were run in the Harvard Biolabs. She thought maybe Garen was looking at her askance, and one afternoon he burst out, "I know how Jim Watson does science!" He told her that a postdoc in Watson's lab had scooped him on a major discovery, stealing the answer to a question that Garen had set up an elaborate experiment to answer. Nancy chose not to push for more information. She knew that science was fiercely competitive. She wasn't sure she understood enough to judge both sides of the story, but the postdocs at Harvard were her friends. She was inclined to be on their side. She began to wonder whether it wasn't science she loved but the way science was done in the Biolabs.

She had heard that Mark Ptashne, who had been a PhD student in the Biolabs and an instructor in a course her senior year, was working on the repressor. Watson had recently helped secure him a junior fellowship in Harvard's Society of Fellows, a prize given to a handful of exceptionally promising young men presumed to be on their way to a great discovery or advance. Ptashne was competing against Wally Gilbert, and Gilbert had the benefit of a postdoc to help him, but Ptashne was working alone. He had produced promising results but was having trouble replicating them. Nancy wrote Mark to ask why it was so hard to isolate repressors, whether Jacob and Monod could have been wrong about it being the key to gene expression. They exchanged letters about possible experiments.

Meanwhile New Haven seemed to be conspiring against her. She was walking to meet another graduate student for dinner one rainy spring night when a Yalie on a bicycle rounded a corner and struck her full speed and head-on, landing Nancy flat on her back, head cracked open on the sidewalk. She flagged a ride to the infirmary and arrived with blood dripping down her new tangerine corduroy coat. An intern stitched her up, then came back to check on her several times, finally climbing into bed on top of her. She told him truthfully that she had a splitting headache, and he left. Returning to her apartment, she fell into the old worries about marriage and science. She slept for two and a half days and woke up determined that she had to get out of New Haven, at least temporarily.

It was May, the end of the term, and she decided to travel with a friend from Radcliffe who was going to Europe for two months. On the beach in Mykonos she concluded that she was interested only in the science being done at Harvard. She understood she could not go to graduate school there—if Jim had wanted her there, surely he could have arranged it. But there was a way she could still work there. Ptashne's fellowship was going to fund his lab for three years, no strings attached. It could pay for a technician, and that's what Nancy would become. She didn't care about the PhD. She'd rather be low status working on science she was passionate about than getting a fancy degree doing work that bored her.

Returning to New Haven in August, she wrote Jim a five-page letter seeking advice, or permission. She acknowledged that she had learned a lot in her first-year classes but recounted the dreariness of her time in Garen's lab: "At the end of that year I had not acquired a single new fact that was of use, nor had I begun to tackle the next big problem that still lies ahead, namely learning how to get along productively on my own in a laboratory." The next two to three years, she told him, looked miserable.

"I realize that graduate school must be like this for everyone since its only goal is to get the degree which announces to everyone that you are now indisputably qualified to do what you want to do," she wrote, in a flash of cynicism. But was it worth going through it? She saw two cons:

(1) Unfortunately—and try as I may—I have always found it difficult to suffer for long on the promise that you'll get something good at

the end of it. This is a very serious fault and will probably "get me" in the end—but it still has to be accepted and faced. My enthusiasm for science has already suffered seriously from the boredom and frustration of last year. If you lose the enthusiasm and are left with only the suffering—you quit—and that's what I'm afraid will happen.

(2) I'm a girl. By the time I finish here it will be just about time to get married if I'm ever going to and not long after, time to have a family. Despite the words of people like Dr. Trinkaus ("I'm married and look what I've accomplished") and also the people who hold up the exceptional Dorothy Hodgkin as an example—I think if you're going to be any kind of a reasonable wife that marriage would probably put an end to really serious science. Thus it would be nice to be very interested in what I'll be doing in the next five year period.

She saw only one pro: "You have to have a degree to be taken seriously or ever allowed to <u>do</u> anything. And this is a very major point since I would still <u>very</u> much like to make science a career or at least a lifelong 'deep interest.'"

Mark Ptashne was working on the subject that really interested her—"and that I have had on my mind (my one-track mind) for the last 3 1/2 years," she wrote. She promised she would be more than "a passive technician"—"very much interested not only in doing what told to do but also in pushing on.

"IF doing this would not be a total dead end to my ever doing science, IF you think that I could really learn while doing the job, and IF you would approve the idea, then I think it would be better than staying here."

She added two postscripts to her letter:

"I am totally serious in all I have said—not just toying with the idea."

"I have not at any time, am not now, and never will be in love with Mark Ptashne—my motives are <u>only</u> those stated above."

On the last point, Watson was unconvinced; in an unusually formal meeting in his office, he made her promise not to distract Mark from the important task in front of him. With that assurance, he told her reluctantly that he would support her decision to drop out of graduate school.

Nancy worried she had embarrassed him at Yale, that she had proven herself the dilettante or aspiring debutante he had taken her to be. Still, Brooke would be back at Harvard and so was the science. She was ecstatic to be returning.

Chapter 3

An Immodest Proposal

To most people working at the highest levels of science, Nancy did risk looking like a dilettante by dropping out. Anyone who wanted to be taken seriously would have stayed on the academic track. Nancy didn't see it that way; she was dead set on using every minute she had to understand the repressor. Unlike the young men in her graduate school classes, she could feel the clock ticking, not only to do the science she wanted to do, but to start the family she was expected to have. A future on the usual path of academic science was unimaginable.

There were few alternative paths in elite science. It was a rigid hierarchy, with the professor as lab head on top, controlling the grant money, and so the jobs. It was built on relationships, which helped keep it a man's world. Faculty were hired on referrals from a graduate professor at one university to a department chair at another (both almost certainly male). Members of the National Academy of Sciences determined which results were important enough to be published in the academy's *Proceedings*, one of the most significant scientific journals. Awards and recognition from professional societies also relied on recommendations, often from past honorees. The American Academy of Arts and Sciences, founded in 1780 by John Adams, John Hancock, and other "scholar patriots" to celebrate and advance knowledge in their newly founded nation, elected Maria Mitchell as its first female member in 1848, a year after she became the first American scientist to discover a comet. She died in 1889 and the academy didn't elect another woman until 1942, and then only under pressure from some members. It wasn't that the other elected scholars thought women weren't accomplished enough, as one member explained to another: "It isn't really that they object to them while a paper is read on the main floor. They object to them in the room above sitting around the ginger ale, the beer and the pretzels and cheese."

When the dean of Harvard Medical School asked for permission to hire Dr. Alice Hamilton as the first woman on the university's faculty in 1919, Harvard president Abbott Lowell practically rolled his eyes. "If she really is the best person for it in the country," he wrote the dean. Hamilton really was the best; the medical school had been looking to hire a professor of industrial medicine, and she had all but founded the field after she diagnosed lead and mercury poisoning in the factory workers she treated at Jane Addams's Hull-House in Chicago. Still, Harvard hired her on three conditions: she was not allowed in the Faculty Club, she could not march in academic processions, and she was not eligible for faculty tickets to football games. She would never get tenure. Hamilton came from privilege—she earned her medical degree at the University of Michigan by way of Bryn Mawr and Miss Porter's School. For disadvantaged women and especially women of color, the requisite education remained out of reach: while the first white woman to get a doctorate in math in the United States was in 1886, the first Black woman was not until 1943.

During World War II, universities hired women to fill the places of male scientists who had gone to work on the war effort. But after the war, as the G.I. Bill flooded campuses with an unexpectedly high number of returning servicemen, universities tried to improve their prestige, which meant replacing the women with men. The number of female professors declined considerably even at women's colleges. Many universities also restored anti-nepotism rules preventing the hiring of spouses, which almost always meant wives. Male scientists worked around this by using outside grants to hire their wives as research associates, and for women who wanted to work in science, those jobs were often the best they could expect. It was a dead end for promotion, with rare exceptions: Gerty Cori worked as a research associate at Washington University in St. Louis for sixteen years before she was made a full professor in 1947, the year she and her husband won the Nobel Prize for their discoveries about how sugars are metabolized. She had earned one-tenth of what her husband did, even though they had earned the same degree from the same medical school and worked side by side.

Women who dared complain were told that they had to work harder and better than the men. A female chemist at Mount Holyoke who wrote a book of career advice scolded two colleagues that what was needed was "a rather higher quality of work than might be expected of a man whose choice of profession is assured." Others argued that women lacked seriousness of purpose. Edwin G. Boring, a professor of psychology at Harvard, asserted in an article in the *American Psychologist* in 1951 that women failed to advance because they were unwilling to work eighty-hour weeks and did not share the "professional fanaticism" of men who were more successful.

The biggest hurdle to hiring was that women were expected to have families. As an article about women in science in the *Atlantic* explained in 1957, "Why take a woman when, first thing you know, she is going to get married, have a family and quit? Here is the center of today's professional resistance." It was one thing, the article noted, when women had worked a few years out of college and families had two children. The postwar boom was different: "Marriage is likely to come during or immediately after college, family plans include three to five children, help is scarce and costly."

In 1957 the Soviet Union launched Sputnik, beating the United States in the race to put a satellite in space and renewing American calls for "scientific womanpower." But those calls did little to stop the decline in the percentage of scientific jobs that went to women. There was still a sense that science just wasn't for them. Betty Lou Raskin, a chemist who ran the radiation lab at Johns Hopkins, noted in an article in the *New York Times Magazine* in 1959 that the Russians graduated more female engineers in one year than the United States had in its entire history. "In our society it is somehow 'unfeminine' for a girl to try to find out how, why or what makes this world tick, but it's very ladylike indeed for her to fly around it serving cocktails," she wrote. She recalled a recent talk to seventh graders where a "ponytailed red-head" had asked her if she was a "real scientist" or an "actress scientist." The girl wanted to be a biologist, and her mother had told her that lady scientists didn't care how they looked, so the girl was dubious that a real scientist would wear the pretty hat and suit Raskin did.

A headline in the same magazine later that year worried, "For Bright Girls: What Place in Society?"

"Is the woman with considerable technical training necessarily less 'feminine,' less warm, understanding and attuned to the roles of wife and mother?" the article asked. "Does the technically trained woman today face difficulties in her work any more disturbing than those confronting her counterpart in a traditionally feminine field?" To ask the question was to answer it, but the magazine reported that "opinions abound": a father whose daughter was a college sophomore aiming for a career in scientific research spoke proudly of her summer job as a mother's helper to a family with three children. "She's great with those kids," he said. "And a good cook, too." A "youthful mother" with a doctorate in biology offered that a girl with high intelligence couldn't be expected to "simply put her brains on ice." Still, she couldn't say much to recommend her career choice: "Advanced technical education at this point in history does, she feels, limit the number of a woman's choices of potential husbands, appropriate jobs, enjoyable friends."

Women and girls simply could not escape the message that their highest and best use was in service to their husbands and children. When Maria Goeppert Mayer, a research physicist at the University of California, won the Nobel Prize in 1963—the fourth woman in science, after Gerty Cori sixteen years earlier—the headline in the *San Diego Tribune* led with her domestic role: "S.D. Mother Wins Nobel Physics Prize."

By 1960, women made up 33 percent of American workers, but accounted for just 7 percent of those in a national register of scientists. Women earned about 40 percent less than men with the same educational credentials. At the twenty top research universities, women accounted for just 4 percent of the full professors in science.

In 1961, a report from the National Science Foundation, established by Congress a decade earlier, issued yet another call to encourage more women into science: "As the need for specialized knowledge and skills becomes more pronounced with every technological advance, the concern that we may be wasting valuable human resources becomes more serious."

In view of "occasional doubts still raised as to women's ability to do scientific work," the agency felt it necessary to point out "that women's intellectual capacities are now generally recognized to be inherently no less than those of men."

In 1963, the year Nancy fell under the spell of DNA in Jim Watson's classroom, there was unusual effort toward allowing women to live fuller lives outside the home, and in science. Friedan's book became a bestseller in February. In June, President Kennedy signed the Equal Pay Act, prohibiting lower wages for women doing the same work as men—they had been earning sixty cents for every dollar a man did. The bill had been based on a recommendation from the President's Commission on the Status of Women, which Kennedy had appointed two years earlier. The commission had been the result of political maneuvering; it allowed Kennedy to signal support for women without alienating labor leaders who opposed the Equal Rights Amendment because they feared "equal treatment" would mean eliminating long-standing protections such as maximum hours that applied to women only.

Kennedy appointed Eleanor Roosevelt as honorary chairwoman, but the commission's real power was in its vice-chairwoman, Esther Peterson, who had been the first female labor lobbyist in Washington. The recommendations in its final report, issued in October 1963, were revolutionary. They included paid maternity leave, subsidized childcare, federal laws preventing discrimination in hiring and promotion, a guaranteed basic income, extending the required minimum wage to women who worked in hotels and laundries, and eradicating state

laws that gave husbands control over their wives' earnings and forbade married women from owning businesses. The report, written by
Peterson, envisioned a world in which "each woman must arrive at her
contemporary expression of purpose, whether as a center of home and
family, a participant in the community, a contributor to the economy,
a creative artist or thinker or scientist, a citizen engaged in politics and
public service." It, too, became a bestseller.

Polly Bunting had served on the president's commission, and in
October 1963 she joined Esther Peterson and others in planning a symposium at the American Academy of Arts and Sciences to "raise new
questions" about the subject they described grandly as "The Woman in
America." The conference was held at an Italianate mansion overlooking the sweeping gardens of Faulkner Farm in the Boston suburb of
Brookline. It featured some of the nation's most prominent academics
and writers, including the developmental psychologist Erik Erikson,
who gave the keynote, and David Riesman, the Harvard sociologist
whose 1950 bestseller, *The Lonely Crowd*, had diagnosed the anxiety of
newly affluent postwar Americans, unmoored from traditional institutions and unduly influenced by mass media and their peers.

Among those invited to discuss the issue was Lotte Bailyn, age
thirty-three, who had earned her doctorate in social relations from
Harvard six years earlier and had since been freelancing on a series of
research projects and raising her two sons, the older four, the younger
ten months. Lotte didn't understand the idea of a conference about
"The Woman in America"; she was young enough to think it peculiar
that anyone would regard women as different from all other Americans, though experience had begun to show her that it happened
all the time. She was a little surprised to find herself included—she
guessed that it might have been because of her husband, a rising star
in the Harvard history department, or maybe because there were so
few female academics to choose from in the Boston area. But she was
pleased. She was in awe of Riesman and Erikson—and of another
sociologist scheduled to speak, Alice Rossi.

Lotte proposed as the title of her discussion topic "The Role of
Choice in the Psychology of Professional Women." For her the choice

had been obvious: growing up in her family there was little doubt that she would choose a professional life—even or especially as a woman. Her paternal grandmother, Sofie Lazarsfeld, had become a psychoanalyst under Alfred Adler in Vienna in the 1920s despite having no formal schooling beyond the age of fourteen (the city had no high school for girls). She took a job with the Viennese Marriage Advice Bureau, which led to a newspaper column and books advising women on how to advance their independence while preserving their marriages and a robust sex life. *Woman's Experience of the Male*, later translated into English and still being updated in the 1960s, was considered one of the first explicit feminist tracts. ("Only men who are sure of their sexual potency are attracted by the independent woman," Lazarsfeld wrote in one of its less racy passages. "Indeed, it is possible to gauge the degree of a man's potency from his attitude on the question of gainful employment for women.")

Lotte's father, Paul, had a doctorate in mathematics when he met her mother, Marie Jahoda, in an association he started for young Social Democrats. Marie, known as Mitzi, got her doctorate in social psychology shortly after they married; she claimed that Lotte as a toddler had eaten ten pages of a draft of her dissertation. Together the young couple joined the small team that conducted what would become a groundbreaking study of people's relationships to work, *The Unemployed of Marienthal*, moving into a once-thriving village south of Vienna where a textile mill that had employed most of the residents had closed just before the Great Depression. The study pioneered a new style of research and writing, producing a narrative portrait that combined immersive reporting—conversations with patrons in bars and priests—with data on how fast people walked, how many books they took out of the library, what they bought and sold in local shops. The conclusion had been counterintuitive. While political theory held that unemployment would fuel rebellion, instead it bred apathy, as residents lost their sense of time and purpose ("Nothing is urgent anymore; they have forgotten how to hurry") and turned on friends and neighbors.

Mitzi and Paul divorced soon after, and Lotte remained in Vienna until Mitzi was arrested for being a member of the Social Democrats, shortly before the Nazis came to power. Lotte went to live with her father and new stepmother in the United States, where Paul Lazarsfeld was becoming one of the most influential figures in modern sociology. His work analyzing the real-time reactions of radio listeners became a forerunner of focus groups. His research on voter behavior shaped modern polling. (One of his best-known studies, conducted in Elmira, New York, in 1948, found that women tended to follow their husbands in the way they voted.) He founded the Bureau of Applied Social Research at Columbia University, which created the field studying mass communication; there he identified the role of opinion leaders in influencing what people bought and thought, and diagnosed the narcotizing effect of mass media on political activity. Mitzi later joined her daughter in the United States and became a leading social psychologist, studying race relations in public housing, the effect of McCarthyism on civil servants, and nonconformity in group behavior. Most recently she had developed a widely cited theory of Ideal Mental Health, identifying the five necessary elements as time structure, social contact, collective purpose, social identity, and regular activity.

Lotte's parents were giants in her eyes—her boisterous father was "like some wonderfully benign hurricane," as her husband described him; her mother liked to perform headstands in public. Lotte was shy by comparison; she listened carefully and measured her words, which were still accented by her early childhood spent in Vienna. But she had been confident as she started college at Swarthmore, where she was a math major, and where most of her female friends assumed they would go on to graduate schools—and did. It was only when she arrived in Cambridge that she realized that women might not have the same opportunities as men. At Harvard she was told that she was one of her department's top recruits, but because women were technically Radcliffe students, she was eligible for only a relatively small Harvard scholarship of $100. She was irritated to be barred from Lamont Library, and there were no dormitories for female graduate students. She was evicted from her first apartment when the landlady complained to Radcliffe's dean that Lotte was hosting men in the evenings.

She met her husband, Bernard Bailyn, known as Bud, and they married after her first year at Radcliffe. When Bud bubbled to his dissertation adviser about their engagement and Lotte's professional plans, the adviser paused, then asked, "Does she type?" Her own adviser regretfully told Lotte that Harvard would like to hire her as an instructor, but that involved being a tutor in one of the Harvard houses, where women were not allowed. So instead she took a job on a research project that one of her male classmates had turned down.

Bud and Lotte collaborated in 1959 on one of the first historical studies to analyze data with a computer, a study of the shipping economy of early Boston. Bud became a full professor at Harvard two years later. Lotte continued through a series of short-term research gigs, raised her boys and ran the house north of Harvard Square, retired with the women into a separate room at dinner parties, and acquired a silver tea set to host the History department spouses' teas. While she'd thought about women's experience, only now for the American Academy symposium did she write about it.

Her paper argued that within a certain class, women started out with more choice than men, because while men were expected to work and earn a living, women could still choose whether to do so. But once women decided to work, their choices narrowed, and they faced constraints men did not. Employers, colleges, and training programs continued to limit or deny spots to women. A husband's job limited a woman geographically, if only because it brought in more money than any work she might do. Employers knew this and exploited it, forcing women to accept conditions male workers would not: lower salaries and jobs beneath their qualifications and experience.

Then there were the emotional choices. The working mother could not help but hear the whispers from neighbors that she was neglecting her children. They undermined her confidence and aggravated her guilt. She assuaged this by hiring childcare to cover only the hours she absolutely had to be at her job and doing more than she needed to around the house and the school and her community, just to prove she was not a lesser mother for working. As Lotte wrote, "It is often more difficult for her, for example, to say 'no' to the PTA chairman when asked to participate in a project than it is for the woman who fulfills only the family role." A child who woke up with a cold on a weekday was enough to send a working mother swirling into crisis about whether she had made the right "choice."

Lotte identified the pattern that would decades later be pathologized as "opting out." "When faced with discouragement in her work," she wrote, "it is not difficult for her to discover that her children really do need her all of the time, or that dilettantism is quite as acceptable as the serious intellectual effort that has bogged down. And, as a matter of fact, a decision to desist would surely have strong social support."

The men at the conference saw things differently. Riesman, the Harvard sociologist, seemed untroubled by the culture of unexpectation. He acknowledged the stigma against young women who wanted careers, "who, as it were, go through the sound barrier and throw traditional canons of femininity out; to do this they must have more brass than the average boy." Inevitably they created "a stereotype of the aggressive woman." But contrasting his mother's generation with his daughters', he argued that modern young women had happily "moved away from feminist (that is, masculinist) ambitions." They no longer felt the same drive to enter engineering and architecture, law and business, economics and archaeology. What they had gained "is the right to enjoy their own children." Educated, emboldened, and relaxed by the reassurances of Dr. Spock, they would be "frightened at any thought that they did not love their infants enough or were tired of them and wanted a surrogate to help care for them." They were freed to get married and develop their cooking—"not simply in the steak-in-the-backyard pattern of the middle strata, but in the exploration of casseroles, wild rice, and gourmet soups." For those who chose to work for a time after college, a "throttling down" was to be expected. Though Riesman acknowledged that this, "in the usually vicious circle, allows men so minded to deprecate women as incapable of the highest achievements," he did not count himself among those men. Rather, he praised this generation of young women for displaying "less resignation and inhibition" than previous ones. "There is an effort to lead a full, multidimensional life without storming the barricades at home or abroad."

Erikson called for redefinition of old sex roles in the face of "intensified industrial competition." But he grounded his arguments in what he called "the biological rock bottom of sexual differentiation." Simply put, women were different and would always be different. Built for the pain of childbirth, they "took pains" to understand and alleviate suffering. They observed and thought more deeply than men, just as girls in lab studies had narrated scenes differently than boys did when presented with the same set of toys; the girls told stories about "inner" spaces and low walls, the boys "outer" spaces and high towers. Anatomy was destiny: men had "an *external* organ, *erectible* and *intrusive* in character, serving the channelization of *mobile* sperm cells"; women had "*internal* organs," with "vestibular access, leading to *statically expectant* ova." Erikson supposed an exciting opportunity lay in this: women would ultimately adapt traditionally male jobs to "the potentialities and needs of the feminine psyche." Such a "revolutionary reappraisal may even lead to the insight that jobs now called masculine force men, too, to inhuman adjustments." While women could match men's performance and competence, they would fully succeed when they shaped the workplace to fit their nature: "True equality can only mean the right to be uniquely creative."

To Lotte, this seemed like essentialism, and absurd. In the ladies' room at the American Academy symposium she ran into the sociologist Alice Rossi, who agreed. They found it especially bizarre that Erikson had argued women would practice law differently. And the very notion of the gathering was odd: Would anyone hold a symposium titled "The Man in America"?

Rossi was a decade older than Lotte and had studied under Lotte's father at Columbia. Now also a mother of three, she had been content holding positions as a research assistant at three different universities, most recently at the University of Chicago, where her husband was a professor. Then she was told that as a research assistant she could not submit a grant proposal to the National Science Foundation. She asked a male faculty member if he would submit it under his name. He did—then took the money, took over her project, and fired her. While her earlier work focused on housing and civil rights, the incident prompted Rossi to begin writing about women and feminism.

The paper she presented at the American Academy symposium that day in October 1963 became a sensation. Titled "Equality Between the Sexes: An Immodest Proposal," it argued for a kind of social androgyny, where work could become more a part of women's lives and children more a part of men's, where a girl's "intellectual aggressiveness" as well as "her brother's tender sentiments" would be "welcomed and accepted as *human* characteristics."

Motherhood, Rossi argued, had become a full-time job for the first time in history, as fewer women worked farms, craft shops, or printing presses alongside their families, and some could afford not to work outside the home. But child-rearing did not last forever, and women were living out the rest of their increasingly long lives "in a perpetual state of intellectual and social impoverishment." Home pursuits had been "needlessly elaborated" even as technological advances had eliminated many of the time-sucking chores of the past. This pattern stunted marriages as well as women, made children excessively dependent and directionless. (They needed, Rossi argued, "a healthy dose of inattention.")

Rossi's arguments echoed Bunting's and Friedan's. But Rossi pushed beyond, for a "wholesale re-examination" of American society. She argued for a new network of childcare centers, especially as "Negro and Puerto Rican women," who took care of many white women's children, would move up to "better paying and more prestigeful jobs." The suburbs deadened women's aspirations: "A woman can spend in the suburb, but she can neither learn nor earn." Housing should be developed closer to cities so that mothers and fathers could be closer to where the jobs were. And "sex-typing" of jobs had to end, to broaden ambitions beyond "the narrow confines of the feminine models girls have." In Rossi's utopia, a woman would see pregnancy and early motherhood as "one among many equally important highlights in her life, experienced intensely and with joy but not as the exclusive basis for a sense of self-fulfillment and purpose in life."

The essays from the conference at Faulkner Farm were published the following year in *Daedalus*, the academy's journal, and Rossi's was greeted as a scandal. She was called a monster and an unfit mother; someone sent her husband an anonymous card mockingly expressing condolences on the "death" of his wife.

Rossi was not one to retreat. She published a version of the essay in *Redbook* magazine under the title "The Case Against Full-Time Motherhood." And in October 1964, she honed her arguments to the question of women in science.

She had been invited to speak at a two-day symposium on women in science and engineering that had been organized by female students at MIT. The audience was a group of nearly three hundred students from campuses across the country and six hundred high school and college administrators. The women at MIT wanted to introduce the girls to the difficulties they might face in science and to show them that those difficulties were not insurmountable—and that the rewards were high. They gathered in Kresge Auditorium, a newly built, glassy futuristic dome designed by Eero Saarinen, a testament to postwar imagination and ambition on campus. Bruno Bettelheim, the prominent Freudian psychologist, opened the proceedings, and Erik Erikson again gave the keynote. Also on the dais were several women: Polly Bunting, on leave from the Radcliffe presidency that year to serve on the Atomic Energy Commission; Katherine Goble Johnson, the trailblazing black physicist and mathematician who had calculated the trajectories that sent the first men into space (she would be celebrated decades later as one of the "Hidden Figures" of early NASA exploration); and Lillian Gilbreth, an industrial engineer sometimes called "the first lady of engineering" but most popularly known as the mother to the regimented brood of *Cheaper by the Dozen*.

Rossi argued that marriage and children were the central reasons more women didn't become doctors and engineers and scientists. Women were bowing to the expectations of society and, in particular, of their male peers. She based her talk on the results of an ongoing survey of men and women who had graduated from American colleges three years earlier, in 1961. Most of the women were married and planned either to be "housewives" or to hold jobs in fields dominated by women—as teachers, social workers, nurses. Just 7 percent aimed for jobs in what Rossi called "heavily masculine" fields such as science, medicine, economics, and law. (They were also the least likely to be married.) Men in the class of 1961 were far more likely than women to say that a woman should not pursue a job that would be difficult to combine with child-rearing, that she should not take a part-time job while her child was a preschooler, and that she should not take a full-time job until her child was "all grown up." Men were also far more likely to say there was "no need at all" for the recommendations from the Kennedy Commission on the Status of Women.

The survey proposed a hypothetical: A woman had graduated from college with honors in biology and taken a teaching job in an elementary school to help put her husband through law school. Now that the husband was established in a good job, she wanted to get a graduate degree in biology before having children. Should she give up or postpone her career ambitions to have children? Postpone childbearing to pursue the career? Do both simultaneously? Three-quarters of the women surveyed thought the woman should not give up her ambition. But that same percentage of women said "most husbands" would want her to.

"Regardless of whether women themselves wish to secure advanced degrees and pursue a research career, their sympathies are strongly on the side of the women who do wish to," Rossi told the audience. "I do not think women will seek higher degrees in any great number in fields like the sciences and engineering if, by doing so, they are apt to be punished socially and psychologically instead of being rewarded, as men are, for their efforts and achievements."

To cultivate in girls the analytic abilities that science required, parents and schools had to encourage them to be independent and self-reliant: "A childhood model of the quiet good sweet girl will not produce many women scientists." This started at home, where mothers were always available, with a desk in the kitchen if they had one at all, and fathers were sequestered in a study or den. "'Be quiet, Father is thinking' is far more apt to be a family injunction than 'Be quiet, Mother is thinking.' "

"Will Science Change Marriage?" asked the *Saturday Review* in a lengthy article about Rossi and the MIT event. The writer, John Lear, cautioned at the outset that Rossi's ideas might provoke "violent reaction from my own sex." But he tiptoed out enough to suggest that a change in attitudes might start with fathers, who were more inclined to encourage their daughters toward careers in law, science, medicine, or engineering than they were to encourage their wives. (Why? Lear quoted Rossi: "The father would not have to live with the consequences.") "If fathers who are now willing to support their daughters' later ambitions could accept questioning young girls underfoot as an intellectual challenge, more than half of that stage of the conflict would be won," Lear wrote. "Dr. Rossi's study shows that mothers already are cheering scientific and scholarly awards to women even louder than they applaud literary or artistic awards to women."

But few could envision such a social change. The *Boston Globe* account of the symposium did not mention Rossi, focusing only on the "pitfalls" of women in science, "the main one being babies." Headlined "Over the Din of Babes at MIT / Women Told of Science Role," the article quoted Bunting—"the distinctive features of a woman's career are family responsibilities"—and Bruno Bettelheim: "We must start with the realization that as much as women want to be good scientists and engineers, women want first and foremost to be womanly companions of men, and to be mothers."

More scathing was the editorial board of the MIT student newspaper, the *Tech*, which sensed scheming in the gathering: "MIT men have never ceased to wonder—and they are not alone—how a woman can cling to her equality and her femininity at the same time. How many times has a lab technician tripped over his test tubes trying to open the door of the refrigeration room for some female colleague, deliberately overladen in hopes of just such service? How many millions of bacteria have been murdered by contamination with nail polish?"

The status quo was working well for men in science, as the head of the biophysics department at Penn State indicated in a 1964 article in the journal *Science* about how to remain in the lab while also leading a department. "You must have a laboratory assistant, preferably female," he wrote. "A female is better because she will not operate quite so readily on her own, and this is exactly what you want."

Chapter 4

At the Feet of Harvard's Great Men

The Cambridge that Nancy returned to in fall 1966 had changed; the growing movement against the Vietnam War roiled the city and the Harvard campus. In November, eight hundred students confronted Secretary of Defense Robert McNamara on his way to speak at the School of Government that Harvard had recently renamed in memory of President Kennedy, assassinated three years earlier. The demonstrators swarmed his car behind Quincy House, demanding that he debate the war. McNamara jumped on the hood defiantly and offered them five minutes. The confrontation ended only when campus security guards and his host—a future congressman named Barney Frank— spirited him away through steam tunnels running under campus (to his appointment with Professor Henry Kissinger). The event landed on the front page of the *New York Times* when twenty-seven hundred students signed a letter of apology to McNamara. The next October, two hundred students barricaded themselves inside the Mallinckrodt chemistry building to protest the visit from a recruiter from Dow Chemical, the maker of the napalm that American forces were using to destroy forests and villages in Vietnam. Holding placards that read DOW SHALT NOT KILL, they kept him hostage in a conference room for nine hours.

In his annual report that year, Harvard president Nathan Pusey departed from his usual recitation of promotions, retirements, and budget challenges to decry the "belligerent nonsense" of the student activists. "I find it painful to accept in Harvard men either such behavior or the reasons now being given by some of their contemporaries in justification of it," he wrote. "Safe within the sanctuary of an ordered society, dreaming of glory—Walter Mittys of the left (or are they left?)—they play at being revolutionaries and fancy themselves rising to positions of command atop the debris as the structures of society come crushing down."

Mark Ptashne, now twenty-six and Nancy's new boss, was a sympathizer of the revolutionaries—"a fellow traveler," Jim Watson called him. He had grown up in Chicago and then Minneapolis, where his father worked in a family business that manufactured snowsuits. His parents were active in leftist politics; Mark recalled listening under the family piano as their friend Paul Robeson, the actor and Black activist, sang spirituals. His parents' skepticism about authority predisposed Mark to the same, and he had learned to question everything in science as well as politics. He had done his undergraduate degree at Reed College in Oregon—left of Harvard in culture as well as coast—and now at the Biolabs arrived each morning in a black leather jacket with his motorcycle helmet in one arm and his violin in the other. He traveled to the Cultural Congress in Havana to meet Fidel Castro and was confident enough that as an untenured junior fellow he led the Harvard faculty in a resolution urging the United States to withdraw from Vietnam. E. O. Wilson, the evolutionary biologist, classified him as one of Watson's "younger shock troopers." Wilson had the office next door to Mark's at the Biolabs but was decamping to a new lab in the Museum of Comparative Zoology across the street. He was sitting at his desk one day shortly before the move when Mark walked in unannounced with a construction supervisor to measure the space to take over.

Nancy thought Mark was brilliant, the funniest person she had ever met. He was short and wiry, moved quickly, and spoke in insistent, almost conspiratorial tones. He made self-deprecating jokes about his height, but even with six inches on him, she thought of him as large—Mark had a way of filling the room. With his cynical quips and deep sideburns, he was the opposite of Nancy and her girls'-school graces—

he read her accent as Philadelphia Main Line—but they shared a single-minded fascination with the question of the repressor.

Biologists understood that genes were turned on and off in response to different conditions. What they didn't yet understand was how this worked. Even simple organisms had hundreds of genes; why were only a few of them expressed at a given time? How did a gene know how much of a certain protein was needed, and when?

Working at the Institut Pasteur in Paris, the French scientists Jacob and Monod had hypothesized the presence of special regulatory genes that made molecules whose sole function was to block the expression of other genes—repressors. But no one had isolated a repressor molecule to test the idea or understood how the repressor worked. The amount of repressor in any cell was extremely low, and it had no enzymatic activity, so the molecules could not be detected using a standard biochemical assay. Several scientists had attempted the challenge and given up.

Ptashne first heard of Jacob and Monod's work as a college student. It dazzled him, this idea of a master switch; he declared that the only reason to go into science was to figure out the repressor. He saw it as the key to the still-unfathomable question of how a fertilized egg became a more complex organism—a fruit fly, or a human—governed by tens of thousands of genes.

Mark had been racing Wally Gilbert for nearly a year and a half and nearly around the clock to be the first to isolate the repressor. Wally was eight years older than Mark, and another of Jim Watson's hand-picked stars; he had been tenured in biophysics when Jim recruited him to work in the lab next to his on the third floor. Wally looked the part of a professor, with his pipe and tweed blazer. He spoke in the considered manner of someone delivering a monologue, and mostly with Mark, who considered Wally a close friend. Nancy was too intimidated by Wally to say anything to him, and he didn't seem to notice her. But he and Mark talked constantly, often long into the night after dinners at the Gilbert home with a crowd that mixed science and politics; Wally's wife, Celia, was the daughter of I. F. Stone, who published the eponymous weekly newsletter that was then crusading against the war in Vietnam.

Crick had once said that your best friend in science is the one who is most critical of your data before you publish it, and Mark and Wally seemed to take this to heart. They rushed to be the first to find the flaws in each other's arguments in the tearoom seminars, sharpening their hypotheses as two dozen professors, postdocs, and students leaned around the big rectangular table as spectators. In the lab, they jousted over who was hogging the scintillation counter.

Jim invited an ABC News crew to produce a documentary about the rivalry between them, narrated by the stentorian voice of George C. Scott, soon to star as General Patton in the Academy Award–winning biopic. "This is the story of a battle," Scott intoned over noir footage of Wally pulling open the heavy bronze doors of the Biolabs after dark. "When the battle is won, it will be more important than the atom bomb. When the battle is won, man will have the power to determine what manner of people shall inherit the earth." The film presented the elusive repressor as a discovery on the path and of the magnitude of the double helix: the key to understanding, as Scott's voice-over explained, "why one cell creates a genius and another gives man cancer." If the race to unlock the repressor succeeded, "man will ultimately control his own heredity."

Nancy watched Mark and Wally elbow each other out to get in front of the camera. She appeared in the film briefly and silently, working over a microscope as Mark delivered instructions to her, whispering to Jim in the background as Mark and Wally sparred in the tearoom. She chafed at the competition. Other scientists around Boston often dropped in to see the progress of the promising young Ptashne, and Nancy thought she noticed relief on their faces when they heard that another of his experiments had failed to replicate the rumored good results. Benno Müller-Hill, the German postdoc working with Gilbert, came up from the lab on the third floor frequently to gloat about how well their work was going. At least Nancy could leave at night to catch a late dinner or a movie with Brooke. She couldn't understand how Mark, who practically lived in the lab, could remain so polite.

Jacob and Monod had first speculated that repressors were RNA molecules, but Mark and Wally believed they were proteins. They were trying to isolate different repressors using different approaches. Wally was trying to isolate the repressor that controlled the ability of *E. coli* bacteria to digest lactose, a milk sugar—the lac repressor. Mark was trying to find the lambda repressor, which turned off the genes of a virus, known as phage lambda, that infected *E. coli*. (*Phage* was the general name for viruses that infect bacteria, and lambda was a specific phage.)

Mark had designed an experiment that was elegant on paper but complicated to carry out. He and Nancy irradiated *E. coli* cells with huge doses of ultraviolet light, then infected them in two separate vials with two viruses, one that could produce repressor molecules and a mutant that could not. They added different radioactive amino acids to each vial, which would then tag any newly made proteins. The radiation damaged the bacteria so much that it was hard to get the cells to produce enough repressor protein to make the experiment possible. Nancy had to pick a colony of bacteria and grow it overnight to do the experiment; Mark came to think she had a unique power to look at the colonies under the microscope and know which bacteria were healthy enough to produce the protein. But Nancy knew it wasn't some secret or intuition: it was hard work and a learned skill. Initially Mark stood behind her, walking her through each step—and, she thought, criticizing her every move. She learned to fear his withering look if she made even a small mistake, but came to appreciate that science required this level of precision. Soon she could work alongside him without instruction.

Her job as a technician came with mundane administrative duties: every day she went to the food truck on Divinity Avenue to fetch Mark's lunch—an egg-salad sandwich, with an éclair for dessert on days when his morale was flagging. Still, she was learning to be a bench scientist: to be unrelenting, methodical, and meticulous; how hard it was to replicate experiments. It was crushing when they failed, thrilling when they worked.

Wally was clearly ahead when Nancy returned to Cambridge in

September 1966; she scolded herself for frittering away those weeks on the beach in Mykonos when so much work had been happening here. By late October, when their experiments had started to produce more consistent results, Nancy told Mark they had to hurry up to beat Wally and Benno. Mark told her they couldn't; Wally had had the first hint of success isolating the lac repressor in May and had spent the summer and early fall working to confirm his results. He and Benno were preparing to publish, which would mark them as first in the discovery. Mark had nothing but admiration for Wally; Wally was the only scientist Mark knew, he told Nancy, who worked harder when experiments were going badly than when they were going well.

Watson submitted Wally and Benno's paper in October, and it was published in the *Proceedings of the National Academy* in December. By the last week of that year, Mark and Nancy had isolated the lambda repressor. Watson sent off Mark's paper two days after Christmas, and it was published in the same journal in February.

Mark and Nancy immediately pushed ahead to the bigger question, of how the repressor worked. Did it block the transcription from DNA to RNA by binding directly to the DNA? If so, it would have to recognize a tiny specific sequence of DNA lying in a sea of similar sequences. There was no precedent in molecular biology to believe this was possible. But Mark—and Nancy—believed it was. Again, the experiment he designed was intricate: they mixed the radioactive repressor protein with DNA from the lambda virus and with DNA from a closely related virus that lacked the specific sequence they knew was essential to the activity of the repressor; they then spun the two mixtures in an ultracentrifuge machine at thirty thousand revolutions per minute to see whether the repressor bound to lambda DNA but not the DNA from the related virus.

It took weeks to get the setup right. The experiment was finally ready to go on a Friday when Mark had to attend a seminar, so he left Nancy to run it alone. She spun the mixtures to move the DNA down the tubes, collecting samples from each and dripping the samples onto a filter paper. She washed the filters repeatedly, then placed them in glass vials with special fluid that would detect whether any repressor had moved with the DNA. She carried the glass vials upstairs to the fifth floor to load them in the scintillation counter, then closed the giant cover on the machine and waited for it to read the amount of radioactivity in each vial. The machine spit out the results onto a rolling slip of paper with the furious noise and energy of a teletype machine. The samples had to be in the machine for ten minutes each, and if she opened it too soon, light from the room could contaminate them. She'd never realized ten minutes could be so long; she decided to wait in Mark's lab one floor down and sat on her hands. Up and down the stairs she went every ten minutes to collect the next data point, which she plotted on graph paper.

Suddenly she could see the pattern: the repressor had bound to the lambda DNA but not to the DNA of the related virus. The hypothesis had been correct.

She ran to the seminar room to tell Mark, but finding the door closed, she went back to the lab and began to pace. When Mark finally returned, she thrust the graph paper at him. He saw with a quick look what they had done. He whooped and dashed out of the lab, waving the graph paper, Nancy running behind him. They ran to the end of the hall and, finding only one young professor to tell the news, turned to run in the opposite direction to Watson's lab another flight down.

They found Jim and Wally putting on their coats to go out to dinner. Mark showed them the graph paper, and Jim's face lit up. Wally, though, regarded it passively and looked pained. He could see what had happened: the proof that the repressor bound to DNA would become a foundation in the understanding of gene expression. His younger colleague had beaten him to be the first in the world to prove it.

Nancy and Mark went back to their lab on the fourth floor, their excitement now muted by Wally's reaction. When they returned to the lab Monday morning, Nancy was eager to do more experiments. Mark was subdued, looked sick. He told her Wally had spent the weekend trying to repeat the results of their binding experiment. He had reported promising results and believed he might replicate them in time to beat Mark to publication. Nancy was staggered; how could the person Mark so revered, his friend, try to take away his moment of triumph?

Watson agreed to hold off relaying Mark's results to journals to give Wally time to catch up. Science, Watson argued, should not be winner takes all, especially when two people were so close at the finish line. Wally was the senior colleague, and Watson believed Mark respected him enough that he would understand. Mark took days off for the first time Nancy could recall, claiming illness. Nancy complained to Jim that Mark was being treated unfairly, but Jim gave her a rare scolding, saying it was not a technician's place to involve herself. Jim had grown irritated trying to moderate the competition between his young stars and complained to others that he would never again allow two scientists in the same department to work on the same problem. But the competition had also worked to Jim's advantage: he could now boast that Harvard was taking over from the Institut Pasteur as the center of gene regulation.

After two months, Wally was having difficulty purifying the radioactively labeled protein, so Jim sent Mark's paper off for publication. The journal *Nature* ran it immediately, in April 1967. Mark told Nancy he understood why Wally had done what he had. The competition had pushed them both on. His conversations with Wally had helped Mark design a better experiment, and only Wally, Mark said, was smart and persistent enough to attempt to repeat such a hard experiment so quickly.

Nancy could not so easily forgive. She told him her Spence friends would never try to undermine a friend like this: "They're so nice."

"Your Spence friends don't do anything," Mark said. "It's easy to be nice if you don't do anything."

Still, Mark was stung. His friendship with Wally never fully recovered. The fallout made Nancy think she might not have the stomach for the competition that she now understood was part of science. Mark had a point: she had been raised not to boast. She considered herself shy. Every year the students and faculty in the Watson and Gilbert labs climbed on top of one of the massive bronze rhinos that stood guard at the door of the Biolabs for a group portrait, the women in the group laughing and holding their shift dresses close to their legs as the men hoisted them. Nancy hated the ritual and avoided it—the rare Biolabs tradition she did not celebrate. She could see how important it was to have an advocate in the lab, a powerful person like Watson. Jim would make sure that Mark got tenure, that Wally's career flourished—he later won a Nobel Prize—and that both got credit where credit was due. And the powerful people in the lab were men.

Watson had arranged for Mark to give a talk about his results, as a consolation. Nancy edged into the back of the room to listen. The crowd understood that the experiment had been a triumph, and a complicated effort.

"Who did all this work?" someone asked.

"I did it myself," Mark said.

He knew this was not true, though in the moment he didn't realize it fully. To him the question seemed preposterous, so he responded defensively. Nancy herself thought nothing of it.

But late that evening, Mark called Nancy at home to apologize. At dinner after the seminar—at the Faculty Club; women were not allowed—Watson had chewed him out for not acknowledging her critical role. The call—and the flowers Mark sent later—surprised her. She had known the rules for technicians when she dropped out of graduate school, and she had never expected credit. Mark had designed the experiment, that was the hardest part. She knew he respected her; he had told her she was smarter than any other woman at Harvard and at least half the men.

The success of the binding experiment had changed Nancy's life, and it elevated her status in the Biolabs. Soon after, Watson came up to Mark's lab on the fourth floor and briskly informed her that it was time for her to go back to graduate school: "You've had your fun." She could finish at Harvard. "You've already done enough work for at least half a PhD thesis."

With Jim's support, her admission was all but guaranteed. He returned within a couple of days and told her she'd been accepted.

As a graduate student, Nancy was still working in Mark's lab, with him now as her doctoral adviser. But she no longer fetched his lunch. She was free to come and go as she wished, which meant she worked much longer hours than before. She began designing her own experiments. She presented her results at a seminar in the tearoom—first nervously, then confidently, as she survived the questioning and realized that her preparations had paid off, that she belonged at the front of that room. She went to scientific meetings where people from other universities sought her out to talk science and hear about her work, and she published papers under her own name.

She had done the significant experiments she had aspired to when she left Radcliffe—and she was only twenty-four. She now worried less about losing her own identity if she became Brooke's wife. When she had moved back from New Haven, she had insisted on renting her own apartment on the third floor of a crumbling Victorian on Sacramento Street. But in the fall of 1967, with the repressor experiments behind her, she moved into his apartment—larger and on the more stylish side of Harvard Square. They covered sofas from Goodwill with Indian bedspreads and hung a big paper lantern over the bare light bulb in the bow window. And Nancy Doe became Nancy Hopkins. She and Brooke announced their marriage with a party; the guests included the Grace Kelly look-alike who had pursued Brooke at Oxford, with Mark and Jim trying to flirt with her. When Nancy had arrived at Radcliffe seven years earlier, she thought there were few fates worse than being known as a smart girl. Now she cringed watching these two brilliant men so awkward in the company of a beautiful woman, and this beautiful, brilliant woman wishing that someone would talk to her about literature.

Brooke had a big group of friends, none of whom was a scientist, but Nancy liked them, and her life outside the lab. Cambridge felt like a small town at the center of a world that was suddenly younger and more exciting. She and Brooke heard Bob Dylan play and Allen Ginsberg read poetry in a bookshop in Harvard Square, went to movies at the Brattle Theatre. Sometimes Brooke joined her for dinner or a movie with Jim. But more often Nancy saw Jim, or Jim and his father, by herself.

Walking back from lunch one day, Jim bought her a wedding gift, a life-size leather pig she had been admiring in the window of the London Harness Company on Church Street. ("I don't want to give you something boring like a dish," he'd told her. "I want to give you something unique that will remind you of me whenever you see it.") They laughed on the way home at the double takes from people wondering about the Nobel laureate walking through Harvard Yard carrying a stuffed pig. Jim opened up more to Nancy now as a graduate student. He was increasingly fixated on finding a wife and over dinners would review the merits of the various "candidates," as he called his potential girlfriends. Nancy preferred it when the conversation stayed on science.

On a summer day in 1967, Nancy and Jim drove in his dark green convertible MG to Woods Hole, on the southern tip of Cape Cod, to visit Albert Szent-Györgyi, a Hungarian-born biochemist who had won the Nobel Prize in 1937 for discovering ascorbic acid (vitamin C). Szent-Györgyi wanted Watson to meet his new wife; Watson wanted company and Nancy's fresh eyes on a memoir he had been writing.

As they set off from Cambridge, Jim handed Nancy a stack of loose typed pages, the manuscript for his account of how he and Crick had hit upon the structure of DNA at Cambridge University in 1953. Nancy wondered how much of this she was expected to get through; she was a slow reader. Within a few pages, she could tell the book would be a bestseller.

Watson had imagined writing a nonfiction novel in the style of Truman Capote's *In Cold Blood*, which had come out the year before. He wanted to capture the personalities and the drama behind the race to understand the essence of how life had been replicating for 4 billion years: Watson and Crick, the untenured and excitable rubes, competing against the erudite but stodgy Brits, Maurice Wilkins and Rosalind Franklin, and the famous American, Linus Pauling, working offstage in California.

In Watson's telling, it was a rollicking quest by two improbable heroes—Crick, age thirty-six, did not yet have his PhD—fumbling but determined as they sought to subjugate mysterious phosphates and polynucleotides. The challenge sparkled socially and intellectually; conversations about science spilled over into games of tennis, lunches at the Eagle pub, bottles of Chablis and claret and parlor games on holiday in Scotland. Watson had titled the memoir *Honest Jim*, and he spared no one. Crick talked too loudly and too much, swung between intensity and absentmindedness, except when it came to his "enthusiasms about young women," which were many: "A discourse of only one or two minutes on the emotional problems of foreign girls was always sufficient tonic for even the most staid Cambridge evening." Crick's wife got her news from *Vogue* and had told her husband that gravity ended three miles into the sky. Wilkins was the Eeyore of the crowd, too cautious, and droning on about his frustrations with Franklin, who was secretive and sour, possessed of a good brain but needing to "calm down." Rosy, as Watson called Franklin behind her back, played the ice princess, with her "acid smile" and lack of patience for pleasantries or the men she had to work with.

Watson cast himself as the jester of this court, girl crazy but inexperienced. He arrived in England without "the faintest trace of a respectable idea" about solving the structure of DNA. But he was willing to work hard, spurred on by visions of fame and a Nobel Prize. And he remained clear-eyed against the myopic older scientists—"not only narrow-minded and dull but also just stupid"—who could not see that DNA was the most important problem in biology.

Bouncing in the bucket seat of the MG heading southbound on Route 3, Nancy struggled to keep the pages from flying loose in the wind of the convertible. She gasped and laughed out loud at Jim's impertinent descriptions of some of the world's most important scientists. She had to scold him to keep his eyes on the road instead of leaning over to see what prompted her reactions. The book was irreverent and chatty, but the science was clear and exciting, weaving gossip with schematics of adenine, cytosine, guanine, and thymine. Watson and Crick had coaxed Pauling's son to England in the hopes he would leak his father's progress, then celebrated when they realized that the elder Pauling had erroneously proposed an impossible DNA structure, with three strands instead of two. Pulses raced, tempers frayed, spirits soared and crashed as Watson and Crick twisted and schemed over molecular models made from cardboard and wire.

In one critical chapter, Watson went to London to see Wilkins, who like Franklin was trying to determine the structure of DNA using the more traditional and time-consuming method of X-ray crystallography. Not finding Wilkins in his lab, Watson wandered into Franklin's. She became so irritated by his surprise visit that she strode from behind her lab bench and nearly hit him. Watson slouched off to Wilkins's office to commiserate about the "emotional hell" of working with her. "The thought could not be avoided," Watson wrote, "that the best home for a feminist was in another person's lab."

Wilkins unexpectedly let on that Rosalind had several months earlier taken X-ray photographs indicating DNA to be two strands in a helical shape. He showed Watson one of her photographs—photograph 51. It was an unremarkable and even routine exchange in the moment. But on the train home Watson realized that "the cat was out of the bag." Out of stubbornness or lack of imagination, Rosy had failed to recognize what her own photographs showed: DNA was helical. Back at the lab, Watson told Crick—mildly hungover from a dinner party—that the structure of DNA was within reach. More dinners, more tennis, a struggle to sort out the arrangement of the base pairs around the helix, but soon the two men had built their model and sent off news of their discovery for publication. Wilkins and Franklin published papers in the same journal at the same time identifying the structure in their photographs. In the book's telling, all was well that ended well: Wilkins and Franklin recognized the genius of Watson and Crick's breakthrough, and Watson realized Rosalind was a "first-rate" scientist, not a "misguided feminist."

Watson had been writing the book for several years. Now, in 1967, he told Nancy he was having trouble getting it published. Harvard University Press had recently rescinded its contract with him after Crick and Wilkins had threatened to sue. In a six-page letter, Crick accused

Watson of violating friendship, privacy, and decorum, calling the book "misleading and in bad taste," a slop of gossip, ego, and faulty memory. "Science is not done merely by gossiping with other scientists, let alone by quarreling with them," Crick scolded. He condemned the book as a "vulgar popularization of science" and reported that a psychiatrist he had shown it to had concluded that "the book could only be made by a man who hates women."

Watson now thought he had a new publisher and was making revisions to satisfy lawyers. He was showing the manuscript widely. Richard Feynman, a 1965 Nobel laureate in physics, told him not to change a word, seeing the book as an exploration of the power of science to transform petty people into great, as they "see together a beautiful corner of nature unveiled and forget themselves in the presence of the wonder."

"From the irregular trivia of ordinary life mixed with a bit of scientific doodling and failure, to the intense dramatic concentration as one closes in on the truth and the final elation (plus with gradually decreasing frequency, the sudden sharp pangs of doubt)—that *is* how science is done," Feynman wrote. "I know, for I have had the same beautiful and frightening experience."

The book was published in February 1968 as *The Double Helix* and instantly triumphed, remaining on the *New York Times* bestseller list for seventeen consecutive weeks (in the company of a book from another Cambridge luminary, Julia Child, author of *The French Chef Cookbook*). *The Double Helix* was serialized in the *Atlantic*, translated into more than two dozen languages, and became a model for popular writing about science. "One has only to pick up the N.Y. Times or turn on the TV to hear of *The Double Helix* or see the celebrity himself," Nancy wrote to Jim's father, now wintering in Florida. "Any criticism at this point could only be construed as sour grapes."

For all the complaints about how Watson portrayed his competitors, few objected to his sexist portrayal of Franklin as the scolding schoolmarm who refused to wear lipstick or pretty clothes.

"By choice she did not emphasize her feminine qualities," he wrote. Listening to her "nervous" lecture on crystallography—"not a trace of warmth or frivolity in her words"—his mind had wandered from X-ray diffraction patterns of DNA: "Momentarily I wondered how she would look if she took off her glasses and did something novel with her hair." Franklin's friends would later argue that Watson had stolen photograph 51 and denied her proper credit in the discovery. But at the time, the incident was accepted as a consequence of scientific competition. Franklin could no longer speak for herself; she had died of ovarian cancer in 1958, five years after the discovery and five years before the Nobel Prize went to Watson, Crick, and Wilkins. (Her death spared the Nobel Committee a debate about who most deserved the prize: while it can be won by a maximum of three people, the rules also dictate that prizes cannot be awarded posthumously.)

Watson had added an epilogue saying that his initial impressions of Rosalind Franklin were wrong. He had realized "years too late the struggles that the intelligent woman faces to be accepted by a scientific world which often regards women as mere diversions from scientific thinking." Just before the book came out, he told Nancy that he did not expect it would satisfy the critics. Feminists, he told her, would hate the book.

Nancy thought the feminists must not know what science was all about. She herself had never heard of Rosalind Franklin until she read Jim's book. She had no reason to doubt Jim's portrayal of her. *The Double Helix* affirmed what Nancy had seen of science firsthand: this was a competitive business.

Hers wasn't simply the reaction of a twenty-four-year-old under the influence of a powerful mentor. This was the way the world saw women in science in 1968. They were rare but not exquisite; more abnormal. If women were smart enough to be working at the highest levels or at the most prestigious laboratories, they were probably unfeminine and unfriendly, definitely not much fun.

There was little variety of women in science or professional life to challenge the stereotypes. Alice Rossi and Betty Friedan and others had founded the National Organization for Women in 1966. But women were still a minority in the rooms where things happened. There were twelve women in Congress—2 percent. While 40 percent of students on college campuses were women, elite schools such as Yale and Princeton and Dartmouth remained all-male. Harvard finally allowed women into Lamont Library in 1967 after a trial period the previous year eased worries that they would distract or disrupt the men. But Harvard still clung to other quotas and restrictions, and few were agitating against them. (When Radcliffe students began to push for a merger with Harvard in 1969, an assistant dean at Harvard, Francis "Skiddy" von Stade, wrote the Radcliffe dean of admissions to object: "I do not see highly educated women making startling strides in contributing to our society in the foreseeable future. They are not, in my opinion, going to stop getting married and/or having children. They will fail in their present roles as women if they do.") The faculty of Arts and Sciences at Harvard included only one woman with tenure, an anthropologist named Cora Du Bois, and she held a professorship endowed through Radcliffe that was earmarked for a woman. (She retired in 1969, leaving no tenured women at Harvard.)

In the Biolabs, Dr. Ruth Hubbard worked as a research associate in the lab of her husband, George Wald, studying the molecular basis of vision. In 1967, they shared the prestigious Karrer Prize, but only after Wald wrote to the prize administrators asking if his wife might be included in the recognition: "She has been my closest collaborator in much of the work I shall be talking about, and is responsible for some of its most important aspects." When Wald won the Nobel Prize later that year, Hubbard's name was not on it. (He won with two other men.) Hubbard, a Radcliffe graduate, considered herself "lucky" that Harvard allowed her to work there at all. Only on reflection in her retirement decades later did she consider what more she might have aspired to if Radcliffe had had its own faculty: "By proudly offering its students the privilege to sit at the feet of Harvard's Great Men, it lost the opportunity to awaken in us the expectation that we might someday become Great Women."

Nancy, too, considered herself lucky. She had left college wanting to study the repressor and had participated in the experiments that defined it. She wanted to marry Brooke, and she had done that. She also wanted to study cancer, and now it looked as if there would be time to do that as well. The disease had been one of the unfathomable mysteries that terrified Nancy as a child. She recalled how her parents had whispered about her mother's skin cancer. Doctors had been able to excise it. But even now, twenty years later, surgery remained the only real hope for anyone with cancer; if a surgeon couldn't cut it out, the disease became a death sentence. Even with chemotherapy only some blood cancers could be treated. Cures were unthinkable. Cancer was still a black box. So many scientists had failed to understand how it developed that it was considered almost folly to study it.

In 1911 a virologist named Peyton Rous had discovered that cancerous tumors in chickens were caused by a virus. Yet for a long time scientists didn't have the molecular biology to appreciate or exploit this new knowledge—to understand why viruses caused cancer in animals, or even what viruses were. The exploration of phage and the discovery of the structure of DNA had unlocked what had been great mysteries about the internal workings of genes. The discovery of the double helix had yielded the formulation that Francis Crick called the "central dogma" of molecular biology—that information flows one way from DNA to RNA. But in 1970, two scientists working independently of one another, Howard Temin and David Baltimore, discovered an enzyme called reverse transcriptase, which allowed the opposite: RNA copied into DNA, which then inserted into a host cell's DNA. The DNA then began replicating, forever changing the genome of the host cell. Because they defied the direction of the central dogma, RNA viruses became known as retroviruses. More important, they included many of the viruses that were known to cause cancer in animals, so scientists now had hope that by studying them, they could understand how genes turned normal cells into cancer cells. Animal viruses had relatively few genes, so it was possible that it would not take long to decode their structure and understand how they caused the damage they did. No one was certain these viruses caused cancer in humans, but even if they didn't, it would be a world-altering discovery to understand how any gene turned made cells cancerous.

Nancy, with her experience in bacterial viruses, was ideally trained to start studying animal viruses. Again, Jim would help her. In 1968 he took over as the director of the Cold Spring Harbor Biological Labs in Long Island; he was moving there with his new wife, Liz, but keeping his appointment at Harvard. Jim was going to start a program at Cold Spring Harbor to study the DNA virus known as SV40, which caused cancer in monkeys. Nancy could split her time between Long Island and Cambridge, which was fortunate because Brooke had finished his PhD and taken a job as an assistant professor of English at Harvard. The position would keep him in Cambridge for at least five years. Nancy would be in Cold Spring Harbor much of the time, but they wouldn't have to be completely apart. Brooke agreed she had to do it.

Studying cancer was still considered risky. "Cancer is a graveyard for scientific careers," a well-known phage geneticist told her. "You'll never get grants." Nancy didn't necessarily agree, and she didn't worry about it; she had only a few years left in science before she was going to have children. As at Yale, she decided she might as well work on the problem that most interested her. In her excitement about studying cancer, she hurried to write up her PhD thesis—because she had already published papers, she had only to compile them and write an introduction.

In 1971 she won a prestigious research fellowship, from the Jane Coffin Childs fund, to pay for her work at Cold Spring Harbor. (The letter notifying her of her acceptance started, "Dear Mrs. Brooke.") Jim had recommended her for the award without reservation—"most strongly"—citing her "outstanding research performance" in Mark's lab. She was "extremely bright," he wrote, "knows how to choose an exciting problem, and sticks with the problem until it is solved."

Turning the corner of Massachusetts Avenue one morning on her walk from her apartment to the lab, Nancy caught a glimpse of the wrought-iron gate into Harvard Yard and allowed herself a rare pause for self-congratulation: she was twenty-seven years old, still shy of her deadline, a Spence girl with a doctorate in science from Harvard. She wouldn't have imagined this when she arrived in Cambridge ten years earlier. The world looked wide-open.

Chapter 5

Bungtown Road

Matinecock Indian tribes, Yankee whaling merchants, and the yachtsmen and summer people of the Gilded Age had all discovered the intrinsic riches of Cold Spring Harbor before the biologists did in the late nineteenth century.

The "good little water place," as the Matinecock called it, possessed perfect traits for the exploration and exploitation of nature: brackish marshes and freshwater springs, forest and field, abundant small game running along the slopes to the shoreline, and a well-protected harbor. Marine creatures, simple in structure, were considered models for testing still-new Darwinian theories of evolution, and institutions around the globe were rushing to set up seaside zoological stations. In 1890, after shipwrecks and the discovery of petroleum had doomed the Long Island whaling industry, the Brooklyn Institute of Arts and Sciences took over the empty warehouses along the inner harbor to create an outpost for the study of biology, calling it the Cold Spring Harbor Biological Labs. Forty miles east of New York City, it ran summer sessions to train high school students and an early generation of American science teachers, who spent days with nets and pails gathering specimens and nights singing songs about embryos and fertilization around campfires.

In 1898 the labs hired a new director, Charles Davenport, who had trained and taught at Harvard and wanted to build a year-round laboratory devoted to genetics. He found a benefactor in Andrew

Carnegie and expanded the grounds, turning the shingle shacks built for whaling families and a firehouse into dormitories and building a new Italianate "Animal House," where biologists performed breeding experiments using local cats, chickens, and fish. Within a decade, Davenport had trained his ambitions on accelerating the creation of what he believed to be ideal humans. He insisted that all traits—from intelligence and skin pigmentation to "sea lust," shiftiness, and frigidity—were genetically inherited and "subject to our control." Through careful breeding he could weed the population of undesirable characteristics—entire undesirable populations. He turned the labs into a boomtown for eugenics, financed by Mary Harriman, the widow of the railway magnate E. H. Harriman and, according to the *New York Times*, "the richest woman in the world." Researchers observed residents of public institutions and "wayward girls" and cataloged their traits in a new Eugenics Records Office Davenport built on the Cold Spring Harbor campus. It was questionable science, and insidiously influential. Davenport's attempts to classify "socially unfit traits" in certain immigrant groups became the justification for the "natural origins" quotas in the Immigration Act of 1924, which sharply limited the number of Jews and other Eastern and Southern Europeans allowed to settle in the United States.

Davenport's term as director ended the year the act was passed, and the lab was renamed the Long Island Biological Association, surviving through the generosity and gala fundraisers of its Gold Coast neighbors: Vanderbilts, Morgans, Tiffanys, and Marshall Fields. (The Carnegie Institution maintained a genetics department on the grounds.) The year-round staff was small, and the labs turned most of its efforts to producing an annual summer symposium. As the organizers wrote to advertise the first gathering: "Modern quantitative biology is so young, and biology in general has become so specialized, that it is very desirable that productive men should have adequate opportunity to expose their work and ideas to the appreciative criticism of the relatively few men in the country who really know what a given investigator is doing and why he is doing it."

By 1941, the lab was broke and looked to be on the brink of extinction; the financial support from wealthy locals had dried up as the United States moved closer to entering World War II. But that year proved seminal, for biology and for Cold Spring Harbor. A new lab director, Milislav Demerec, sold the land that included the Eugenics Records Office to finance a new sixteen-thousand-square-foot lab building and decided to focus that summer's two-week symposium for the first time on genetics—a theme that would dominate the summer symposia and biology for the next decades.

The announcement of the symposium in *Science* noted that there would be thirty-six participants—anyone else could "attend and take part in the discussion of papers without further invitation"—and that several participants "as well as a number of scientific men" would make the lab their summer headquarters. Two of those who did, an Italian microbiologist named Salvador Luria and a German physicist named Max Delbrück, spent that summer doing their first experiments together on bacteria. Cold Spring Harbor, so recently the incubator for bogus ideas about traits and heredity, had become a wartime refuge for two men who had fled fascism—Luria was Jewish and had escaped Italy as Mussolini came to power, and Delbrück had fled Nazi Germany, where his family had been active in the resistance. Their experiments that summer would plant the seeds of a revolution in the scientific understanding of genetics.

Until then, genetics had relied on studying mice, maize, or fruit flies; they had provided the basic understanding that traits in living organisms were determined by genes. But no one knew what genes were made of, much less how they carried out the function of heredity. Bacteria could be studied on a molecular basis, but it wasn't clear that bacteria had genes like those in higher organisms. The work on phages would show that they did, and that those genes could mutate. Viruses were packages of genes wrapped up in protein—elegantly simple structures—and those that infected bacteria, the bacteriophages, were the simplest of all. And bacteria and phages alike were easy to grow. All that made them the new model organisms for studying how genes function on a molecular level.

Delbrück and Luria's work spawned the so-called Phage Group

of scientists working at universities across the country—sometimes called the Phage Church—and set the stage for great leaps in molecular biology. Adherents were inspired by a slight book written by the physicist Erwin Schrödinger in 1944 titled *What Is Life?*, which posited that biology might be governed by universal laws like those that explained chemistry and physics.

In 1945 Delbrück—known as the pope of the Phage Church—organized Cold Spring Harbor's first yearly phage course to spread the new gospel. Luria's first graduate student at Indiana University, Jim Watson, attended the course for the first time in 1948, at age twenty. In the summer of 1953 Watson returned in triumph for his first public discussion of the discovery of the structure of DNA.

The labs produced many significant discoveries after World War II—scientists used mice to show that some individuals are more susceptible to cancer, discovered a treatment for Addison's disease, and developed a method to mass produce penicillin. But it was the summer symposia that made Cold Spring Harbor legendary. They became the event of the year for biologists and geneticists who, for a week, sometimes two or three, would sleep in tented cabins or in concrete-sided dorm rooms, crowd into small seminar halls to hear the latest in the field, stay up late working in the labs, and in the evenings hold picnics and square dances on the lawns. The volumes of the proceedings became the bible for recent converts to molecular biology.

Nancy attended for the first time in 1967, the summer after she began working for Mark, and returned along with some of Jim's other PhD students. To her, it was heaven. There was one entrance, an easy-to-miss hard turn off the two-lane state highway stretching along Long Island's North Shore. One main artery led into the hundred-acre property: Bungtown Road, named for the plugs, or bungs, that once stoppered barrels filled with whale oil. Dotted on either side were laboratory buildings, cabins, a mess hall, and a new one-story motel and lecture hall that had been built in the 1950s to accommodate the increasing size of symposium crowds. The labs were perpetually in debt, supported only by grants to the small number of year-round scientists, and the property had the look and musty smell of a mothballed summer colony. Floors sagged and ceilings leaked, vines shrouded old trees, and cornstalks grew up along two sides of the Animal House. To one side of the road was the placid harbor, now plied only by pleasure craft.

But for the weeks of the summer symposia Bungtown Road became an information superhighway for biologists, clogged with people walking and talking deep in scientific discussion. It was like a festival or, in the comparison one of Nancy's friends made, Rick's Café in the movie *Casablanca*, the gin joint where those in the know, revolutionaries caught up in passionate cause, gathered to exchange intelligence. Nancy could listen to the great names in the field discuss their work at a seminar, then grab them to ask a question along the road or across the communal tables where everyone ate. Just like in those first days in the lab at Harvard, her brain exploded with ideas for experiments. She wanted to talk science all day long, and here she could, from breakfast through cookouts and well into the night.

Everyone was on a first-name basis; Jim encouraged this tradition, believing that science flourished when the atmosphere was casual and everyone felt on the same level. He also believed that scientists could be most creative if they could blow off steam: volleyball games at Cold Spring Harbor quickly became competitive; a dispute between two geneticists over an out call lasted several years. Most buildings had no air-conditioning, so participants cooled off by swimming off the half-mile sand-spit beach extending to the outer harbor, clamming, or canoeing into the village across the way for ice cream cones.

The symposia attracted about three hundred scientists, many young like Nancy, except that overwhelmingly they were men. Nancy made close friends among them, and with a couple of women who shared her obsession with genetics. At night, they watched other young women who drank too much and flirted in the bar in Blackford Hall. Nancy and her friends agreed quietly that those women would not last long in science. There was power in being young and female—the bright and rare thing in a sea of men—but you had to stick to science. If your colleagues thought you were sleeping around, they would never take you seriously. Nancy wondered about the wives left taking care of the children at home and pitied them a bit. She felt protected by her own marriage, but at the same time relieved that she didn't have to worry about getting home for dinner or feel guilty if she was late to meet Brooke. Her friends at Cold Spring Harbor became some of her closest friends in science, and during the academic year she spoke to them often weekly, Cambridge to California or New York.

When Watson took over as director in 1968, Cold Spring Harbor Labs had again been near demise; the endowment had dwindled and no longer covered the $3 million it cost to run the plant each year. Harvard allowed Watson to remain a full professor, splitting his time between Cambridge and Long Island, because it recognized the importance of keeping such a vital institution alive; he took no salary as director. Just as the study of phages had reinvigorated the lab in 1941, the promise of understanding how cancer developed now provided it a new mission and new money. Jim was not running experiments in the lab anymore but had become more the scientific impresario, raising money, advising the government, and identifying promising areas and big questions for exploration. Finding the genes that cause cancer in viruses—viral oncogenes—was the biggest challenge of all. Jim feared that retroviruses might be so highly contagious that they could jump from animals to humans. So the program at Cold Spring Harbor would study DNA viruses that cause cancer in monkeys. He started a $5 million fundraising campaign and secured a $1.6 million grant from the American Cancer Society. Within a year he had doubled the lab's income and within two years nearly doubled the staff.

By 1971, the first year of Nancy's postdoc there, construction crews were busy renovating and winterizing the lab spaces and living quarters, their noise interrupting the usual quiet of winter along Bungtown Road. Nancy commuted from Cambridge by train most weeks, five hours each direction, working often around the clock and living in a large dorm with other graduate students and postdocs, men and women. They dined communally every night, sharing a cooking rotation that meant one night Indian, another French, and always one night roast chicken, because that was the only dish Nancy knew how to make. They found frequent occasion for parties, and on Saturday nights there were often dinners at the homes of senior scientists, who were almost all men. Their wives prepared indulgent meals and afterward got up to clear the table, leaving the scientists to talk shop. Nancy was often the only female scientist, and she felt like a scientific tomboy, uncertain how to behave. She didn't want the women to think she was not one of them or, worse, a snob, but she wanted to talk science and knew she'd miss an exciting idea or fact if she left the men. She tried to sit strategically on the edge of one group or the other, talking with the women but listening to the men.

On those weekends Brooke often came to Cold Spring Harbor, but the evenings usually ended badly. At Harvard he was a popular assistant professor, but here he felt insecure. He had never felt confident about math or been interested in science. He tried to make small talk at the parties, but scientists turned away from him when they realized he didn't speak their language. He and Nancy decided it was better for her to spend more weekends in Cambridge. She was torn when her weekends with Brooke ended, but she couldn't wait to get back to the lab.

Studying animal viruses was slower, more difficult, and more intricate than the work she had done with bacterial viruses. For starters, you had to grow animal cells before you could study whether a virus might cause cancer. Bacteria grew almost anywhere there was water, sugar, and some essential salts. They divided quickly—two bacteria could reproduce to make more than a thousand in five hours. Once you infected them, it was a matter of only hours until you could measure the growth of the virus. Animal cells preferred to grow in whole bodies; it had taken scientists decades to figure out how to grow them in dishes in the lab. They required serums and growth factors and ultrapure water to grow, at least twenty-four hours to reproduce, and as much as two weeks to measure the growth of a virus. Not any cells would do; you wanted to select the ones that were as flat as possible before you dropped a virus on top to see if it made them cancerous.

One summer as a PhD student Nancy had taken a course on animal cell culture at Cold Spring Harbor taught by Bob Pollack, who as a postdoc had done what still seemed like a miracle: he had taken animal cells that a virus had made cancerous and turned them back into normal cells. Watson had recruited Pollack to Cold Spring Harbor to direct a new lab studying cancer reversion, and Nancy was now working there. One day when Pollack came into the lab, Nancy was experimenting with a drug that caused the nucleus of the cell, considered its brains, to protrude. The nucleus typically remained attached by only a thin stalk. When Nancy had spun the cells in a centrifuge at low speed, the nuclei spun loose, leaving the original cells like the proverbial chicken with its head cut off. Nancy told Bob she had just replated some of the cells without nuclei onto a new lab dish.

"That's impossible," he said. Cells weren't supposed to attach to a new surface without the nucleus, yet here they were. They had even assumed the same shape they had before they lost their brains. Nancy was pleased with her discovery, even if she hadn't realized it was a discovery at first. Little was known about how cells acquired their shape, or the role of the nucleus when a virus replicated inside the cell. She and Pollack and another colleague began doing other experiments on these enucleated cells and published several papers together. Just a year into her new field Nancy was earning recognition.

She found herself sought out by one of the most famous residents of Cold Spring Harbor Labs, a geneticist named Barbara McClintock. McClintock, now nearing seventy, was the most highly decorated woman in American science and one of the most significant geneticists in the world. In 1944 she had been the third woman elected to the National Academy of Sciences, at the young age of forty-one—the academy was in its eighty-first year and even by 1970 had elected only ten women, among more than a thousand men. In 1970 President Nixon selected her as the first woman to receive the National Medal of Science. As a graduate student at Cornell in the 1920s—the only woman in a group of male corn geneticists—McClintock had developed techniques that allowed her to differentiate the distinct shapes and structures of chromosomes, the long strands that carry the genes. She moved on to identify the linkages between genes that were inherited together, and to map how they changed over generations. She and a graduate student she supervised—another woman, Harriet Creighton—then became the first to understand that chromosomal material is exchanged in the production of sex cells. Other geneticists quickly recognized Barbara as the world's "foremost investigator" in the study of chromosomes.

Yet she had been unable to get a job on the faculty at Cornell, which had no female professors outside the department of Home Economics. (She did her graduate work in the botany department because the genetics program did not take women.) She took a job as an assistant professor at the University of Missouri but left because she soon realized she would not get tenure; her job was paid for by a grant given to a man, and a dean had told her she'd be let go if the man went to another university. Male colleagues who tried to get Barbara a job at Iowa State were told the director would not hire a woman.

The Carnegie Institution had finally made a place for her at its genetics institute at Cold Spring Harbor in 1942, and there McClintock had another breakthrough discovery, though it was not realized at the time. Observing the patterns of maize, she discovered that chromosomes could break and fuse, and that genes could even jump from one region of the chromosome and reinsert themselves in a different one. Until then scientists had believed that genes sat fixed along the chromosome, like a string of pearls, and always replicated in the same pattern. She suggested that these genes deactivated and reactivated the new genes around them. But when she presented her discovery at the 1951 summer symposium at Cold Spring Harbor, it was received with a chill—shrugs, disbelief, and even snickering. It was two years before the double helix blew open the understanding of gene structure, so McClintock's discovery of "jumping genes" arrived too early to be appreciated, like the heredity in Gregor Mendel's garden peas or the cancer-causing viruses in Peyton Rous's chickens. Some colleagues also argued that her writing was dense and hard to follow. Whatever the reason, she refused to give seminars at Cold Spring Harbor ever again and stopped publishing her papers. She rightly felt that a scientist of her stature deserved a greater show of respect. "She is sore at the world because of her conviction that she would have had a much freer scientific opportunity if she were a man," one prominent geneticist wrote in a letter to the Rockefeller Foundation, whose grants financed

McClintock's work in the 1930s when universities would not. A patron at the foundation wrote more sympathetically that she "very actively resents the fact that she is not given scientific opportunities" because of her sex, "since she has brains enough to realize that she is much more able than most of the men with whom she comes in contact."

Early on, Barbara had resisted thinking she would be treated differently as a woman in science. She was surprised to be stigmatized as a career woman and a spinster; "I had gotten myself into this position without recognizing that was where I was going," she told a biographer. She pronounced herself "stunned" to be elected to the National Academy of Sciences: "I am not a feminist but I am always gratified when illogical barriers are broken—for Jews, women, Negroes, etc. It helps all of us."

She thought her problem was more than her gender. It was that she was a maverick and an oddball. That was an image she cultivated.

Quirky and petite—just five feet, two inches and at most one hundred pounds—Barbara wore round rimless glasses and had her hair cropped short; she'd chopped it in college before flappers made the bob fashionable. As a younger woman she had slit her skirts and sewed them into makeshift pants so she could walk her cornfields without getting tangled. Now she dressed in men's white shirts and khaki pants, and over them a lab coat that she wore until threadbare and patched with iron-on tape. She lived alone in an apartment on the grounds, where she refused to have a telephone, telling anyone who wanted to reach her to send a letter instead.

She had been born in Connecticut to a *Mayflower* descendant and a surgeon and was told that her parents had named her Eleanor but changed it when she was just four months old because they believed Barbara was a more masculine name, one that captured the independence she had already shown as an infant. Her father gave her boxing gloves when she was four, and her mother made her bloomers so Barbara could play boys' sports. From her earliest days she nurtured a vision of herself as gender-free—feeling at home neither as a girl nor a boy—and once complained of her body as "a nuisance." She insisted on going to college despite her mother's worries that too much education would make her unmarriageable. She was popular at Cornell; she was elected president of the women's freshman class, rushed by a sorority (which she declined to join, because it was exclusive), and played banjo in a jazz group. But she had chosen never to have a serious romantic relationship, saying she could not accept the loss of freedom that such attachments entailed.

Still, Barbara believed that such eccentricity would have been tolerated and even celebrated in a male scientist. Evaluations of her work read like psychological portraits: "She is admittedly nervous and high-strung," one of the Rockefeller funders wrote. Another noted, "Says she has bought no clothes for years and looks it."

Around Cold Spring Harbor, she had become a sort of spectral figure, wandering the cornfields muttering that she was the only one who understood how to water them. Maize genetics had fallen out of fashion, supplanted by the fixations of this new crop of molecular biologists. Yet Barbara stayed the course of her science. She was most often seen wandering the wooded paths or walking between her lab in the Animal House and the beach on the other end of Bungtown Road, her arms full of leaves and black walnuts and Queen Anne's lace she collected and analyzed for genetic patterns. She was a regular in the front row of the symposium rooms, with incisive questions. While she could be prickly, especially with a succession of lab directors— Watson called her "the Katharine Hepburn of science," headstrong and a bit imperious—she was also known as warm, with a bawdy sense of humor. She kept her eye out for young people, especially women, she saw as gifted.

Nancy was twenty-five when she met Barbara. Nancy thought her odd, and ancient. But David Botstein, one of Nancy's new friends, told her that Barbara had done groundbreaking work in corn genetics, so Nancy eagerly accepted Barbara's invitation to visit her lab.

It was crowded yet intricately well organized. Long lab tables were piled high with data sets, papers, and journal articles that Barbara had underlined using a straightedge. There was a television set, on which she liked to watch political talk shows, and a cot where she sometimes spent the night, though she wasn't much of a sleeper—from what

Nancy could tell Barbara worked fifteen-hour days in the lab. She had perfect recall of scientific literature going back decades; as she talked, she would get up and go over to the tall bank of black filing cabinets along the wall, where she could quickly pull out a paper and find the page and position of the figure she had just mentioned.

She told stories of scientific discoveries and about her solo drives across the country to Caltech in a Model T—she prided herself on her ability to repair any car. She now drove a behemoth Pontiac, its dashboard nearly swallowing her tiny figure. She told Nancy she often got funny looks, but the big car gave her the ability to fight back if someone cut her off in traffic. "I just give them the finger," she said. She allowed how happy she had been to reach an age when men were no longer interested in her "in that way," that it made her life in science so much easier when she no longer had "to deal with all that nonsense." (She referred to sex in purely reproductive terms, calling it "servicing the female.")

While Nancy could understand that Barbara McClintock's discoveries were extraordinary, she sometimes found the science hard to follow—some of it involved organisms she'd never heard of. She found herself itching to return to her young friends in molecular biology, who could speak the same language as her. But the longer Nancy spent at Cold Spring Harbor, the more aggressively Barbara began to seek her out. Barbara seemed to be reliving her life in science and told stories about all the times discrimination had held her back. On more than one occasion, she had been told to hand over her results to men, who went into a room to discuss them and closed the door, leaving Barbara waiting outside. Even as her stature outside academia grew, her department at Missouri never passed along letters from other universities seeking to hire geneticists with her background. She was president of the American Genetics Society, but it was as if she didn't exist.

The stories would continue into cocktails at five o'clock, when Barbara would bring out a tumbler of brandy and a dish of dry-roasted peanuts. One evening she went to the file cabinets and pulled out a letter on yellowed paper. It was from a famous geneticist to a scientist seeking to hire a professor; Barbara was the best person in the field, the geneticist had written, but it was a shame that of course she could not be hired because she was a woman.

Nancy could hear the bitterness in Barbara's voice as she read. She understood Barbara's rage—the letter upset Nancy, too—but she didn't feel it. She didn't want to consider that the stories could be true. Some seemed too far-fetched, as if they couldn't actually have happened. Jim had told her Barbara was difficult. Like Rosalind Franklin. Nancy could see why anyone would think this. She herself wasn't having a problem. There were men who didn't talk to her and she sensed it was because they were uncomfortable around women. On the other hand, some of the smartest men in science did talk to her, did take her seriously.

Nancy decided Barbara must have done something wrong. Besides, the last decade had seen a rush of new civil rights laws that prohibited job discrimination. Nancy assumed they'd prevent it. Maybe Barbara's stories were true, but to Nancy they were like Barbara's science—something from a previous era. The old rules and assumptions did not apply anymore, or, at least, they would not apply to Nancy. Of that she felt confident.

Chapter 6

"Women, Please Apply"

Barbara McClintock had paid little mind to other people's stories; she insisted she had no heroes or role models and gave the impression she didn't need any. "Scientists should be dedicated but not consecrated," she told the *Atlantic*. Still, she recognized, somewhat discomfited, that she had a responsibility to be an example to those who came after her. She had been "caught," she once said, by her election to the National Academy: "I figured I couldn't let the women down." She would have to be a role model herself.

She was held up as an example, though it was not always convincing. In 1955, when Mary-Lou Pardue was a senior at the College of William and Mary, in Williamsburg, Virginia, her professors encouraged her to go to graduate school for biology. Look at Barbara McClintock, they told her. Mary-Lou was unmoved. It wasn't that she wasn't interested in biology—she was—and it wasn't that she didn't admire McClintock—she did. But even at twenty-one she was wise enough to understand the parts of McClintock's story her professors were leaving out: how universities refused to hire her, how few women were on any university faculties. Mary-Lou suspected that her professors were telling her about Barbara McClintock because Barbara McClintock was the only female scientist they could think of.

Mary-Lou had gone to college with a wide range of interests, engineering and Latin among them. She had started majoring in history—William and Mary's colonial-era campus was steeped in it, and its redbrick architecture and formal symmetry charmed her from the minute she saw it. She also hadn't had much choice; it was the only state school in Virginia that accepted women. Her father, a physicist and a dean at Virginia Tech, had told her to start at a public university and that she could go to a private school if she wanted to pursue a higher degree.

She'd switched her major from history to languages because it required her to learn ancient Greek, which she had always wanted to do. Then biology pulled her in. It had been the one class she didn't like in high school—it bored her, with its classifications and memorization. But she'd always been drawn to the everyday wonders of the natural world. She'd lived as a child in Lexington, Kentucky, dirtying her knees peering at insects and animals and trapping frogs in the woods. Her family had gone hiking in the Blue Ridge Mountains. The summer before she went to college, a friend's father who worked at the agricultural school at Virginia Tech had recruited her and four other girls as field hands for the university's experimental crops. They cross-pollinated corn by covering up tassels on some stalks and dabbing pollen on others. Mary-Lou had been fascinated by the hybridized corn she saw in the lab: rainbows of orange and purple and red, or yellow with squares of black kernels that could make an ear of corn look like a crossword puzzle. She changed her major by junior year. It was the year after Watson and Crick had discovered the double helix, and biology was fresh again with the promise of modern genetics. Mary-Lou was especially entranced by chromosomes, the dancing, shape-shifting strands of DNA that McClintock had studied.

But Mary-Lou could find few female role models in science. She knew one in the biology department, who looked to be breaking her neck to collect samples from local hospitals for her students to use in the lab class. That woman had a PhD, but it was obvious she was going to be teaching the lab class for the rest of her career. The other women Mary-Lou saw in the sciences were generally older and handling the heavy teaching loads, freeing the men, most of them younger and fresh from doctoral programs, to do the research that would get them the grants, faculty jobs, the respect of their peers. The men were the ones encouraging Mary-Lou, but the women set the example, and Mary-Lou saw no point in getting a PhD if theirs was going to be her life.

She thought maybe she'd travel. She was used to moving around. Her father had grown up in rural Scottsville, Kentucky, one of seven children in a family that owned the town's telephone company and grocery and feedstore. He was expected to quit school as a teenager and become a telephone lineman, but he made his parents a deal that if he worked in the grocery store, he could stay in high school. He did well enough that they allowed him to continue to the University of Kentucky, where his professors encouraged him to go into physics. He'd gotten his doctorate at Yale and returned to Kentucky to teach. Mary-Lou had been born there, but during summer vacation the year she was five her family drove from Kentucky to California, where her father did a year's sabbatical at Caltech. The next summer they drove back, stopping to see friends along the way. When Mary-Lou was ten, she and her mother and brother went with her father to Chicago and then Oak Ridge, Tennessee, as he worked on the Manhattan Project, the secret enterprise that developed the atomic bombs dropped over Hiroshima and Nagasaki at the end of World War II.

Mary-Lou's mother, too, had traveled as a young woman; she had grown up in Missouri, then worked as a teacher in Puerto Rico, traveling by boat on the rivers from Kansas City to the Great Lakes and out the St. Lawrence River to the Atlantic. Mary-Lou thought that sounded like fun: exploring, seeing what the world was like. She looked like someone you'd want on an adventure, with her wide, unflinching blue eyes and a smile that burst across her face like a firework. She wore her long brown hair pulled back in a casual coil, a thick fringe of bangs in the front. She had a sturdy confidence.

She applied for the graduate fellowships because her professors had asked her to, and she wanted to be polite. She won them all and turned them all down.

Her father, the dean of graduate studies at Virginia Tech, told her that Oak Ridge National Labs was hiring women to work as technicians in cancer research. Mary-Lou thought maybe he was trying to get her away from the medical student she was engaged to in Virginia. But she recalled that she'd liked living in Oak Ridge, and the research sounded interesting. The US government had seized sixty thousand acres of isolated farmland in eastern Tennessee at the height of the war to build the facilities that purified uranium and plutonium for the Manhattan Project, creating a "secret city" of seventy-five thousand people almost overnight. Now, a decade later, the Atomic Energy Commission was training the labs toward peacetime uses of radiation, which included making isotopes to treat cancer and tag biological samples. Mary-Lou knew this made it a good place to study biology; radioactivity was still one of the only tools scientists had to study living organisms, and the director of the lab was good at getting money and identifying talented young scientists to work there.

When she arrived, she found she no longer had a job; the woman she was supposed to replace was not leaving after all because her husband had failed his army physical. It turned out to be a stroke of fortune. She was put into the lab of a recently hired biologist, Jack von Borstel, and because he was young and new, Mary-Lou was his lab. She had to learn how to do everything, and he treated her like a colleague. He taught her the genetics of *Drosophila*. He sent her to New York to present a paper, as long as she promised to explore the city rather than spend time only with her friends from William and Mary. And he encouraged her to enter a competition for new National Science Foundation predoctoral grants that would pay for college graduates to take classes and do research in university labs across the country, which would result in a master's degree. Mary-Lou won on a proposal to study biology at Harvard. She hadn't expected she'd win, and she had met and married a young chemical engineer in Oak Ridge, having broken up with the medical student soon after leaving William and Mary. When she told von Borstel she was going to decline the fellowship, he suggested she switch her proposal to take classes at the University of Tennessee, just an hour away by car. She could continue to work in his lab and write her thesis under his direction. Mary-Lou liked the research she'd been doing in the lab and had assumed she'd have to give it up for marriage, so she jumped at the chance to continue.

Her life became a split existence. Her husband, Pete, was enrolled in the MIT Practice School, a kind of satellite internship for engineers run by the Massachusetts Institute of Technology. He was working at Union Carbide, which had taken over operation of Oak Ridge Labs, preparing to be sent off in three years to start a management career that could take them around the world. Their social circle was mostly through the Practice School. The wives were preparing to follow the husbands, set up homes, and be social assets wherever they landed. In the meantime Oak Ridge was a booming town of prefab houses and shiny supermarkets. The women organized parties and picnics, and the men kept boats in the marina and on weekends took their wives waterskiing along the lakefront. Mary-Lou liked this comfortable life enough, but she had a lot more fun with the people she knew at the labs.

At the end of the three years, Mary-Lou had earned a master's, but her professors at Tennessee told her she had done enough research for a PhD and encouraged her to write up a thesis. She declined. She loved the work and wanted to keep doing it. But she still saw almost no women around her with doctorates, and none of them was leading a lab. One of the few was Dr. Liane Russell, whose studies led to the warnings against X-rays for women who might be pregnant. Russell worked in the "Mouse House," the big mice genetics program, and was married to the director—she had once been his student. Otherwise, the women were research associates working in labs led by men. Mary-Lou expected she'd do the same, and all she needed for those jobs was a master's.

She had also begun to realize that she and Pete wanted different lives. Pete wanted a home base, which she had never needed. He wanted to have children—he would remarry and raise four. Mary-Lou was not ready to settle down. She wanted to continue her research. The marriage soon ended, inevitably and not acrimoniously.

Mary-Lou moved back to Virginia and took a job as a technician in a big university lab and started to make a list of the scientists she'd like to work for. It was 1961, and scientists were moving at a once-unthinkable pace to understand how DNA relayed its message through its single-stranded cousin, RNA, to the proteins that carry out the functions of the cell. At the top of her list were two men she'd heard speak at Oak Ridge. Seymour Benzer was a physicist who had been converted by the phage course at Cold Spring Harbor and gone on to break new ground in understanding the substructure of genes. He was isolating the molecular keys that allow the transfer of information from DNA to proteins. Joe Gall was studying DNA and RNA using multicellular organisms, known as eukaryotes. Mary-Lou knew almost nothing about techniques for working with protein and decided she could take what she learned in Benzer's lab to work with Gall, but not the reverse.

Benzer was at Purdue, in Indiana, and agreed to hire her as a technician. He set her to working with a postdoc on the structure of ribosomes, which synthesize proteins. Two women worked as technicians, and two other women worked as postdocs in what appeared to be a permanent role. Every day the postdocs—but only the men, who were all training for faculty jobs—went into the small lunchroom with Seymour to unwrap their sandwiches and talk about the day's work. When they were done, the technicians would go in to clean the tables.

One day Seymour told Mary-Lou she should go clean the tables.

"Seymour," she replied, "I didn't come here to clean tables."

He thought for a moment, then asked if maybe she'd like to come eat lunch with him and the others.

So she did, pretending she didn't understand the dirty jokes the men told in French, ignoring the other women in the lab who whispered about whether she was eating or sleeping with her new lunch companions.

Benzer was restless; he thought molecular biology was becoming too crowded. He was thinking of exploring what genetics could reveal about behavior; he had two daughters and was fascinated that they could be so different despite being raised the same way. He asked Mary-Lou to teach him what she knew about the genetics of *Drosophila*. They began dissecting the flies' brains (also cows' brains, which Benzer, an adventurous eater, would sometimes take home to cook for dinner).

Seymour soon decided that he wanted to go to Caltech and invited Mary-Lou to join him as a graduate student in his lab. Seymour was considered one of the leaders in his field, a cardinal in the Phage Church. By now Mary-Lou had enough experience in labs to understand how much an influential lab director could help your career—the relationships were so crucial that scientists drew family trees of who had mentored whom. And she was beginning to see the value of earning a doctorate. She knew more than some of the postdocs in Seymour's lab; they increasingly relied on her knowledge and her work, and she didn't like the feeling that they were taking advantage of her. But it was 1964 and there were still few women with faculty jobs, so Mary-Lou didn't think she'd require Seymour or anyone else to help her find one. She decided not to go with him to Caltech. She saw freedom in her choice: rather than plotting a career, she could follow the questions that most interested her. Now she wanted to understand better how genes work in larger, multicellular organisms.

One of the few people studying eukaryotes was the man who had been on her list a few years earlier: Joe Gall.

Gall had recently moved to Yale, which was finishing the new Kline Biology Tower on Science Hill. Designed by the celebrity architect Philip Johnson, it would reign as the tallest building in New Haven. Mary-Lou applied as a doctoral student at Yale and in 1965 became a graduate student in his lab.

Joe was five years older than Mary-Lou and had been the kind of kid she would have hung around with growing up. When he was fourteen, his parents moved from Washington to a five-hundred-acre farm near Charlottesville, Virginia. He roamed around collecting insects, taking them home and identifying their parts in a book, then mounting them with pins in old cigar boxes. His father, a labor lawyer, did some work for Bausch and Lomb and managed to bring home a microscope during World War II even though the company had suspended production. Joe gathered algae from the pond, lice from farm animals, the organs of rabbits and groundhogs—anything he could get his hands on—and smeared them on slides to study in a lab he set up in his bedroom. By the time he was a graduate student he had accumulated an encyclopedic knowledge of plants and animals, so he knew which one offered the best model for the biological question he wanted to answer. His natural habitat was over a microscope. He had dismantled the one his father gave him and used the parts to build a more powerful one with the help of an uncle who was a machinist—he still used that one in his lab.

Joe had an unhurried and unpretentious manner about him, which Mary-Lou appreciated. She'd seen other scientists try to show off their smarts when they answered questions by putting down the questioner. Joe never did. He encouraged independence, not competition, among his graduate students, helping each to find an interesting question to explore for a thesis. He himself often came into the lab on Saturdays, but discovering graduate students there, he'd say, "What are you doing here on a Saturday?" Mary-Lou had a tremor in one hand that got worse when she was anxious. Noticing it, Joe would join in a round of tennis to loosen everyone's nerves. Unlike other faculty members, who left it to postdocs and grad students to carry out experiments, he still did his own. While some professors put their names on grad students' papers just because they ran the lab, Joe did so only when he had contributed actual work.

Gall would also become famous for hiring and mentoring an unusual number of women. He had been the one who'd arranged, through Jim Watson, for Joan Steitz to switch from Harvard Medical School to a doctoral program in biology after Joan worked for Joe one summer at the University of Minnesota. The year Mary-Lou arrived, all seven graduate students in Gall's lab were women. They hadn't chosen him for his support of women; they wanted to be in his lab because of the scientific work he was doing. But they knew how lucky they were. They traded stories about how bad it was elsewhere. One woman had applied to Johns Hopkins and said she'd been told, "We already have a woman." The number was capped; they didn't want to waste a slot on a woman they assumed would quit science to have babies.

The women in Gall's lab became close to him, and to each other. They got a kick out of their unusual number; it made them bolder. Mary-Lou's first year at Yale, they piled into a car to drive to the annual meeting of the new American Society for Cell Biology in Philadelphia and slept eight to a hotel room, the most senior paying for the room with her travel allowance, the others sneaking in the back door with their sleeping bags. The next year the meeting was in Houston, and the women hadn't yet passed the qualifying exam to permit them a travel allowance. Mary-Lou convinced the department to pay for them anyway: "Fly now, pass later!" They agreed that they felt sorry for the speakers given the last slot at the meetings, when the audience had snuck out to cocktail hour or the plane home, so they resolved to pile into the front row at those sessions as a show of support. Soon they had a nickname, Gall's Girls.

"What does Joe Gall *have*?" male biologists at other universities asked, which both amused and annoyed the women: that their thoughtful, soft-spoken boss with his owlish eyes would be some kind of lothario, that a handful of women in biology would be so remarkable, that there must be some mystery to getting women to work in a lab.

Gall's reputation perplexed him. He shied away from the hero's cape, and he didn't like the mythology that he was somehow so naive that this had all happened by accident. But there was truth to it. He was not going out of his way to hire women—there was no mystery in it, he was simply hiring the best students. He hired men, too, he pointed out. The women just had more success, even if the pool was smaller—or maybe because it was.

If pushed to explain, Joe credited his mother. She had been the first in her family to go to college and majored in math, but when she graduated in 1920—the year after Barbara McClintock started at Cornell—she saw no jobs available to her in science. She became a homemaker and instilled her love of science in Joe. He recalled her making him an endless supply of butterfly nets from old coat hangers and helping him build the collections of critters he found in the yard. When he was ten, he begged her to buy him *A Boys' Book of Insects*. She agreed, but also picked up a copy of E. B. Wilson's *The Cell in Development and Heredity*, a classic early textbook of cell biology. Young Joe quickly moved on to read that, learning from it how to section the tissues that he studied under his bedroom microscope. His mother made him think it wasn't unusual for women to have an affinity for science. He got the impression his father, by contrast, was afraid of bugs.

Joe's scholarly demeanor could be deceiving; he was driven and intense. He had been the first to demonstrate that each chromosome was a single DNA double helix. He had been among the first to identify gene amplification, in which some genes leave their place in the chromosome to make hundreds or even a thousand extra copies of DNA outside it.

When Mary-Lou started in his lab in 1965, Gall had been working to develop a technique to locate specific gene sequences in cell tissue. He had the idea to use a radioactively labeled single strand of ribosomal RNA, mixed in solution, to locate a complementary single strand of ribosomal DNA in tissue. When the RNA matched with its complementary base pairs, it would act as a radioactive tag to pinpoint the genetic sequence.

He was working with cells from the ovaries of clawed frogs and had some results suggesting that he'd successfully hybridized the RNA and DNA. But no radioactive signal was coming through on the autoradiography images.

For her thesis Mary-Lou had proposed an elaborate question about gene amplification and how the extra DNA gets out of the chromosome: Was it a single strand or double? The problem proved too challenging—more than fifty years later the question remained unsolved. But she had become exquisitely proficient in purifying the single-stranded nucleic acids. And she had discovered that the fixative Joe had been using to immobilize the DNA on his slides was making them effectively too thick for the signal to come through. Joe asked her to work with him.

In late 1968 they successfully demonstrated the new technique, which would become known as in situ hybridization. The discovery, published the next year in the *Proceedings of the National Academy of Sciences*, was a breakthrough for biology and for medicine. Scientists had been able to map genes in the cell's nucleus, but that was a primitive sketch; this was seeing the leaves on the trees in a satellite image. In situ hybridization became the standard method for locating specific genes. It set the stage for the new field of genomics—mapping the entire set of genetic instructions to make an organism. In clinical settings it would create a diagnostic tool to detect the presence of bacteria or viruses in tissues, and, in prenatal testing, the rearrangement of chromosomes that indicate abnormal development.

By 1970, Mary-Lou had her doctorate and growing renown. She was now fifteen years out of college, a decade older than most of the others at the same level—including Nancy Hopkins, who received her doctorate from Harvard the next year. But Mary-Lou didn't regret any delay. Everything she had done had been valuable; she now had an education she calculated was worth several graduate degrees. She had been able to take risks, to pursue the questions that interested her, without worrying about whether the experiment would work or how it might affect her job prospects. In time, she would even say this was the reason for her success.

The recession that year had tightened the job market, and the male PhDs she knew were staying in the United States to do postdoctoral work, hoping it would put them closer to faculty positions at universities. Mary-Lou still didn't expect that she'd get a faculty job. She knew of only one woman on the biology faculty at Yale; most women with doctorates there and elsewhere were research associates, and that's what she expected to be. So she decided to do a postdoc in Europe, thinking that way she could also learn a language. But when she made her list this time, the four people she wanted to work with were all in Scotland. In the fall of 1970 she moved to Edinburgh on a fellowship from the American Cancer Society to work in the lab of the geneticist Max Birnstiel.

The following year Max informed her that he was moving to Zurich, and Mary-Lou thought she should think about finding her permanent job as a research associate. She wanted to work for a man who was so accomplished that he would not see her as a threat; she recognized she'd had this with Joe Gall, but didn't think she should work for her doctoral supervisor. She set her sights on Don Brown at the Carnegie Institution in Baltimore, a friend of Joe's who had been collaborating with her on some work in Scotland. She wrote Don saying that she expected there was competition to work for him, but she'd be happy to come the year after the next if that's when he would have an opening.

Don Brown wrote back that she was welcome to join his lab, but he thought she was selling herself short. With her experience she should be on the faculty somewhere, running her own lab, not working as a permanent postdoc for him. He told her to send him her résumé, he would share it with people he knew at other universities. He also sent her a letter to the editor he had clipped out from a February 1971 issue of *Science.*

The letter was from Boris Magasanik, the chairman of the biology department at MIT, and ran under the headline "Women, Please Apply."

"The role of women in society is being re-examined," the letter began. "The participation of women in many fields of endeavor is often restricted by their sex even when their qualifications are not in question."

"We recognize the existing obstacles," Magasanik wrote. "If women are to participate in science, we must actively support qualified women who aspire to do so. As a first step we are making the effort to locate qualified women for faculty positions in our department in the areas of neurobiology and development." MIT encouraged applicants, he wrote, and hoped to be a model for other universities.

Mary-Lou had seen the article that prompted Magasanik's letter. It had appeared in *Science* in September 1970, titled "Women in Academe" and written by Patricia Albjerg Graham, an associate professor of history and education at Barnard. Graham opened with a warning: American campuses, already struggling with "the state of siege mentality" brought on by the antiwar and Black Power movements, "were in for another round of crisis—this one dealing with the 'woman question.'" Women accounted for about 40 percent of the undergraduates on campus and earned about 10 percent of the doctorates—in both cases, the numbers were less than they were in the 1920s. Even in fields where women now earned a sizable proportion of the PhDs, they made up a far smaller share of the professors. Women constituted about 18 percent of the staff at colleges and universities, but that was mainly at small colleges or in the lower ranks—lecturers and other untenured teachers—at more prestigious schools. At the University of Chicago, just 2 percent of the tenured professors were women, down from 8 percent in 1900; the numbers were about the same at Stanford and Columbia, even though women were earning between 15 and 20 percent of the doctorates. The proportion of male faculty at Barnard—still a women's college—had been rising steadily since World War II. And at two of the Seven Sisters colleges, considered the women's Ivy League, men had replaced women as presidents. (One that had a female president—Radcliffe—had no faculty of its own.)

The discrimination of a decade earlier, when a "senior professor in a leading history department" confidently asserted that there would be no women in the department as long as he was around (and there hadn't been), was not as common, Graham wrote. Yet a new discrimination had replaced it, "now more subtle and less easily countered." Most of the blame she lay on stubborn cultural standards, the "internal ambivalences," among young women with ambition, "the fear that success in competitive achievement situations will lead to negative consequences, such as unpopularity and loss of femininity." The women knew it was unacceptable to earn more money than their husbands—and it was assumed women would marry. They saw few women in academic or professional roles, and the ones they did were spinsters or, by contrast, "the rare individuals who manage to marry a brilliant and successful husband, have five children, write intelligently on a variety of topics, and at the age of forty, be featured on the beauty pages of a woman's magazine." But most women on the precipice of decisions about how to live their lives recognized this model as "truly exceptional."

"Women's low expectations for themselves so infect the society," Graham wrote, "that both men and women refuse to think of women as generally likely to occupy important posts."

She presented a riddle making the rounds of cocktail parties: A man and his son are in a terrible accident; the man is killed. His son, critically injured, is rushed to the hospital, where the surgeon charges in and suddenly cries, "My God, that's my son!" Guests ventured complicated explanations, involving stepfathers or artificial insemination. "Almost invariably," Graham wrote, "the storyteller must supply the answer: 'The surgeon is his mother.' "

A striking number of the prominent and well-degreed women on American campuses had come from England and Europe, where women from the leisured class were permitted the money and time to work outside the home. "All of these women have direct experience with another culture, and presumably recognize a greater variety of options for women than the stereotype of middle America currently exemplified by Mrs. Nixon and Mrs. Agnew."

What was the solution? Maternity leave, for starters, which was almost unheard of. The ability to extend the time available to bid for tenure, which tended to expire around the peak in women's childbearing years. And American universities had to hire more women, though Graham cautioned that "just as tokenism has been rejected for the blacks so it must be rejected for the less militant feminine majority." To make it easier on the men on the faculty, she suggested starting by hiring women as junior faculty members: "Many mature male professors find it much easier to appoint young women to junior and subordinate positions (where they have little power) than to appoint women of their own age to positions truly equivalent to their own."

But if hiring posed a challenge, changing attitudes would be even harder, as the responses to Graham's article showed. An Ohio man mocked her "weird" idea for "complete equality," saying it was impossible because there wouldn't be enough women to work the day care centers where women in academia would "dump" their children. (He assumed only women would work in day care.) "If the facts were really known," he wrote, "they would probably show that the vast majority of women enjoy being women." A psychology professor at the University of Pennsylvania recognized her own struggles in what Graham had written. Day care would help, she wrote, but it would not change the assumptions that constantly undermined professional women with children: employers considered them uncommitted to their jobs;

teachers, guidance counselors, and judgmental neighbors considered them uncommitted to their children. "Until children are considered a family responsibility for two consenting adults, women cannot have equal opportunities in employment," the professor wrote. "Only when social roles require comparable efforts from professional men and women can equality of opportunity be said to exist."

Mary-Lou recalled the article even years later, though she would remember it for a problem that it didn't mention: women couldn't go out for drinks with men—they weren't invited, or if they were, they risked being seen as sex objects, not scientists—so missed out on the exchange of ideas and opportunities that were often discussed in more casual situations. Not that she wasn't familiar with the problems the article did mention; few women could fail to notice them. She just hadn't dwelled on them, or couldn't. "Girls in this society do not think of themselves as conquerors of the world," Graham had written. But they also didn't think of themselves as victims. It was the way you survived. You worked around the problems, otherwise they might overwhelm you or, at least, take you away from your science.

Mary-Lou still expected she would take the job as Don Brown's research associate when she finished her postdoc in 1972, but she recognized his kindness. In his letter he had told her to apply to MIT and to use his name—he was on the visiting committee for the biology department there, a sort of outside board of overseers. She wanted to keep him happy, so she did.

That summer, 1971, Mary-Lou was asked to teach a course on cytology at Cold Spring Harbor. She was sought out by Barbara McClintock, and they became friends and colleagues; Barbara would help teach the course over the next decade. Other invitations had followed, to give seminars around the country. Mary-Lou knew there was a push to hire women, and these talks were considered tryouts.

When she finished teaching the course at Cold Spring Harbor, she was headed to give a seminar at the Marine Biological Labs in Woods Hole, on Cape Cod. On her way she stopped in New Haven to see Joe Gall and his wife. Mary-Lou and Joe had continued their work together on in situ hybridization; the work had helped earn her the invitations to speak. She was surprised to find a letter waiting for her that had apparently been forwarded from several other addresses before ending up at Joe's.

It was from Boris Magasanik at MIT, following up on the application she had sent months earlier. The letter was short, and officious, and Mary-Lou read it aloud to Joe and his wife, Dolores: thank you for your application, we have a lot of interest from qualified candidates, only the best people will be considered, and we will not consider you.

"Read it straight, Mary-Lou, don't paraphrase," Dolores said.

"I'm reading it word for word," Mary-Lou replied.

She ripped up the letter. *To hell with them*, she thought. She was stung, having been on a high from all the requests to give job seminars. She also had a nice offer from the Carnegie Institution in Baltimore. Don Brown had reported back that she could have a job on the faculty there with a special appointment at Johns Hopkins, or vice versa if she liked; the chairman at Hopkins was out of the country and wanted her to come back over Christmas break to discuss it further.

Back in Edinburgh, there were more letters of interest, from Duke and Harvard. Mary-Lou was arranging a schedule of talks for Christmas when she got a call from someone identifying himself as Gene Brown from the biology department at MIT. He wanted to talk to her about a job.

"Who?" she asked. "*Where* did you say you were from?"

Another professor in the department, Maury Fox, had heard Mary-Lou's talk in Woods Hole and had urged Magasanik to hire her. Mary-Lou, determined to enjoy the moment, played it cool. She told Gene Brown she was scheduled to give a talk at Harvard in December, but she supposed she could come over to MIT after that.

Brown was persistent, and apologetic: "I understand there's a letter we might need to explain."

Mary-Lou told him the connection was bad; they could talk more about it when she got to Cambridge over Christmas.

Chapter 7

The Vow

The women's movement had lagged behind the civil rights and antiwar movements at universities, but by the late 1960s women were pushing for a place on campus, beginning with coeducation. Yale and Princeton began admitting women in 1968 and 1969, and Harvard and Radcliffe made the first awkward moves toward what would become an arranged marriage a few years later. The first two years of the new decade swept in revolutionary changes. "This is an untraditional generation," Gloria Steinem said in her commencement address at Vassar in May 1970, when ten men who had transferred the previous fall became the first male graduates of the former women's college. "You have made the campus part of the world."

It was "the year of Women's Liberation," as Steinem said, and she urged women to "live the revolution" by challenging and disrupting traditional roles and assumptions: "The first problem for all of us, men and women, is not to learn, but to unlearn." And more and more people began to argue that factors other than the desire to have children and stay home with them were keeping women out of scientific careers.

The growing outrage filled the pages of *Science*, with articles along the lines of Patricia Graham's about the status of women in various fields, and letters to the editor that objected to, among other things, job postings explicitly stating that women would be paid less than men.

Women began demanding that scientific groups do more to ensure fair treatment for women. Vera Kistiakowsky, a forty-two-year-old research associate in the physics department at MIT, started a women's committee within the American Physical Society in April 1971, having grown weary of male colleagues saying they couldn't hire women because there were no female physicists to hire. Her committee began compiling a list of qualified women to share with the men and also secured a $10,000 grant to survey women in physics about their challenges. The results were predictable and depressing: women entering doctoral programs were discouraged by their professors from finishing, and if they did finish, anti-nepotism rules made it hard for married women to find jobs. If they found jobs, they were often fired for having children or stuck at lower pay or in lesser positions than men—even, as one woman noted, when she had received the same doctorate from the same university under the same thesis adviser. Asked what she wanted, another woman wrote, "Less shitwork + more responsibility."

A similar caucus started within the American Society for Cell Biology after women in the Kline Biology Tower at Yale began discussing the lack of women on panels and podiums at scientific conferences. Two of them—Ginny Walbot, a graduate student, and Mary Clutter, a research associate—decided to put up signs in the women's bathroom at the society's 1971 annual meeting, inviting women to meet in the hotel bar. About thirty gathered, nervously hooting like they were at a pep rally. They agreed that one of the problems was in hiring, which was done largely by word of mouth; there were no job listings or other public advertisements. So the women passed a hat to raise money for a newsletter and told everyone to go back to their universities and collect intelligence about what positions might be open, which Walbot and Clutter then typed up, mimeographed, and sent out to all members of the society with female-sounding names. Soon meetings of the fledgling Women in Cell Biology were attracting three hundred women. The newsletter also included a roundup of sexist comments from speakers at meetings. The group began training women in mock interviews, helping them to come up with retorts to common questions such as "Do you plan to have children?" ("I haven't decided; how about you, do you have a family? Do you like your kids?")

The larger organizations were not always supportive; the cell biologists refused the women's request for money to keep publishing the newsletter. Other institutions, though, responded with changes that were small yet meaningful: in November 1971 the publisher of *American Men of Science*, a reference guide used by scientists, students, and reporters, agreed to change the title to *American Men and Women of Science*, since it had been listing women for much of its sixty-five-year history. (Dora Goldstein, who had been one of the first female graduates of Harvard Medical School and was an expert on alcoholism at Stanford, had complained about the name in a letter to *Science* two months earlier, citing it as one of the "trivial hindrances" that added up to a "widespread, low-key slighting of women, a refusal to be aware of what women can do and are doing.")

As more female students arrived on once-all-male campuses, the lack of female professors became more glaring. The all-male culture would be hard to crack, as suggested in several experiments that sent fictitious résumés to department chairmen. The résumés the researchers sent out were identical in almost all respects—both applicants were married with the same number of children and the same credentials—except one was a woman and one was a man. By significant numbers, employers found the male credentials more impressive or ranked the men higher than the women. One study of physical science departments in 1971 found the pattern especially striking at "higher-quality schools." Some chairmen at those schools offered unsolicited comments on the résumés of women: What would the applicant's husband do if she was hired? What would she do with her children? What was her personality like, and would she get along with the others in the department?

The civil rights legislation of the 1960s had established broad protections against discrimination in hiring and promotions, but those protections had not extended to women in academia. It had been a hangover from the anti-communist panic of the McCarthy era, when hundreds of professors were hauled before Congress or fired by their universities under suspicion of left-wing beliefs or affiliations. Congress wanted to show that the government would no longer meddle in university employment decisions, so it exempted higher education from the Equal Pay Act of 1963 and from Title VII, which prevented job discrimination on the basis of sex, race, ethnicity, or religion.

Now, nearly a decade later, it was clear that universities, for all their liberal-minded foundations, were not going to act without prodding to hire more women. And women were about to win a law designed specifically for them.

It began with Bernice Sandler, a psychologist known as Bunny, who had been teaching part-time at the University of Maryland in 1969 when she applied for one of seven full-time positions available and was rejected. A male friend on the faculty told her why: while her qualifications were excellent, "You come on too strong for a woman." Sandler agreed with him. She went home and cried and berated herself for all the times she had spoken up at faculty meetings. Her husband, a lawyer, asked whether there were any strong men in the department. Bunny replied that all of them were. "Then it is not you," he told her. "It's sex discrimination."

Bunny wasn't sure. She had bought into the image of feminists as abrasive, man-hating, and unfeminine, and she didn't want to become part of their cause. Many women felt the same way. Some had been turned off by newspaper reports of angry women burning their bras (the reports were later deemed false; the most anyone could verify was that some women had tossed their bras in trash cans on the Atlantic City boardwalk to protest the Miss America competition). Other women wanted to be judged by their merits rather than their gender— as lawyers and scientists, not "lady lawyers" or "women scientists." They thought of themselves as proof that hard work was enough to succeed in a male-dominated world. And some, including Polly Bunting, said they did not feel that they had been discriminated against. Maria Goeppert Mayer, the 1963 Nobel laureate hailed also as a San Diego mother, objected to being assigned to Vera Kistiakowsky's committee on women in physics, saying she had no expertise or interest in it.

Bunny Sandler soon changed her mind, after she was rejected for two more positions. One employer told her she wasn't a professional, "just a housewife who went back to school." Another spent an hour explaining that he didn't hire women because they didn't show up for work when their children were sick. Angry, and ever an academic, Sandler began reading to learn from the strategies of Black activists. She realized that the civil rights laws did not apply to women, but she also noticed a little-known and rarely enforced provision: in 1968, President Lyndon B. Johnson had extended an executive order prohibiting discrimination by federal contractors to include discrimination based on sex. Sandler was alone when she read this, but she shrieked at the power of her discovery; this meant universities, which received federal money, could not discriminate on the basis of sex.

She joined the Women's Equity Action League, or WEAL, a spin-off of NOW started by women who considered themselves less militant, which in January 1970 filed a class action lawsuit against all colleges and universities in the United States alleging an "industry-wide pattern" of discrimination against women. Sandler had received behind-the-scenes strategic advice from Vincent Macaluso, an assistant director of the office of Federal Contract Compliance in President Nixon's Department of Labor, who encouraged her to have women at universities across the country collect data about the numbers of men and women in various ranks in their departments. Sandler then compared those numbers to the number of PhDs being awarded to women in the fields. The results, again, were predictable and depressing: many departments had no women at all, even in fields where they earned a quarter of all doctorates. The higher the rank, the more prestigious the university, the fewer the women.

Representative Edith Green, an Oregon Democrat, was on the board of WEAL and led a House subcommittee on higher education. She had long wanted to hold hearings on discrimination against women on campus, but had been concerned she wouldn't find a constituency to testify. The information Sandler had gathered showed there was one.

Green began hearings in June 1970, declaring, "Let us not deceive ourselves. Our educational institutions have proven to be no bastions of democracy." The hearings lasted for seven days and resulted in thirteen hundred pages of testimony, which Green hired Sandler to compile. Green then found an ally in Representative Patsy Mink, a Hawaii Democrat and the first woman of color to serve in Congress, who had been rejected by multiple medical schools and then, after she graduated from law school, multiple law firms. Green and Mink wrote the legislation that would become known as Title IX. It prohibited sex discrimination in any institution of higher education that received federal funding, as almost all did. While it would become known mostly for expanding women's sports (though the word never appeared in the bill) over the next decades it would also bring changes in admissions and hiring and how colleges and universities dealt with accusations of rape and sexual misconduct.

Title IX attracted little notice when it was signed in June 1972, just one provision in a broad higher-education spending bill. The *Washington Post* story on the legislation included only one line, in the second-to-last paragraph, noting that "new bans on campus sex discrimination are also included."

But the Nixon administration had also started sending out teams to investigate universities, including Yale, Harvard, and the University of Michigan, demanding they produce affirmative action plans to hire and promote more women. The Department of Labor sent letters to departments with no or few female professors warning that they were at risk of losing federal funding if they did not hire women. Suddenly, universities found new urgency to do so—or at least to hire a token few. In some cases, they promoted women who had worked for years as research associates: Mary Helen Goldsmith was made a professor of biology at Yale, where her husband was the department head; Annamaria Torriani-Gorini, who had labored as a research associate for thirty years after receiving her doctorate in biology, was elevated to the faculty in Biology at MIT, as was Vera Kistiakowsky in Physics.

Lotte Bailyn was sitting in her office at the Sloan School of Management at MIT in 1972 when the dean walked into her office and told her he was hiring her to the tenure track—she would be the first woman there. He caught her off guard. She was forty-two years old, and it had been sixteen years since she'd received her doctorate from Radcliffe. She'd gotten used to the limitations that had surprised her when she had arrived in Cambridge. She had accepted that she would have to make her career out of a string of research projects. She had taken one at MIT in the fall of 1969 and turned it into a job there as a senior lecturer. MIT had been an adjustment after the formality of Harvard, where graduate students still wore coats and ties. At MIT they called her by her first name.

No one had asked her if she wanted the position—she never determined whether the dean had informed his faculty he was hiring her. And she had some reservations about taking a full-time job, with the demands of having to earn tenure. Her two boys were now fourteen and ten. Working as researcher, she could keep up with their school projects and music lessons, and the PTA expectations she had written about in her 1963 paper at the American Academy—making sure she brought the cookies to the meetings, which were always scheduled for the middle of the workday.

She had followed her mother into a career exploring people's relationship to work, though it had happened unintentionally. In 1969, Bud took the family on sabbatical to England. There Lotte met and began a project with Rhona and Bob Rapoport, sociologists who were studying dual-career couples, a configuration so new in public consciousness that the Rapoports had invented the term. Lotte used survey results she had gathered in that work to do a study on her own about the importance of husbands in a woman's ability to have a career. The survey had asked husbands whether they derived the most satisfaction from family life or from work and asked both husbands and wives their attitudes about women who sought careers.

Two patterns emerged in couples where both husband and wife described the marriage as "very happy." The first was among couples where both the husband and the wife wanted a traditional division of labor: he saw his career as most important, and she was against women having careers outside the home and saw her role as raising the children. The second was where the wife was trying to integrate raising a family with having a career and the husband said he derived the most satisfaction from his family. Those marriages, however, were happiest if he also had high ambitions for his career and a high income, and the wife had help with the children and household chores.

The marriages least likely to be happy were those where the woman was trying to integrate her desire for a career with having a family and the husband valued his career more than his family. (The research also suggested that few couples discussed these goals before they decided to get married—not surprisingly, this was to their detriment.)

Bud had always supported Lotte's career, and he'd had tremendous success in his own: his book *The Ideological Origins of the American Revolution* had won both the Pulitzer Prize and the Bancroft Prize in 1968. (The *New York Times Book Review* had lauded it as an instant classic upon publication the previous year: "One cannot claim to understand the American Revolution without having read this book.") He was on his way to being his generation's most influential scholar of American history.

Lotte took the job at MIT and continued her studies about attitudes toward work: she was working on one that used data collected from MIT alumni, and another about women who had succeeded in scientific careers. Inevitably she found that when you explored people's ideas about work, you ended up discussing their ideas about family. Her research led her to become one of the first to study another idea that was just poking up in the public consciousness, labeled with a phrase that Lotte herself disliked: *work-life balance.*

Mary-Lou had arrived in Cambridge to do her job talk at Harvard in December 1971. It went well, and her suitors there took her to the Harvard Faculty Club for dinner afterward. The club, an imposing brick mansion nestled on a street behind Harvard Yard, had only recently begun allowing women into the main dining room (and only because of a renovation, which required taking over the women's waiting area near the back door, where women had previously had to enter). Inside, the clubhouse was soothed by a gracious curved and crimson-carpeted staircase and dark wood paneling. Gilt-framed oil portraits told the story of three centuries of Men of Harvard. The leather chairs were deep, the conversation serious and subdued.

She did her job talk at MIT the following day, and her suitors there took her to its Faculty Club, which occupied the sixth floor of the Sloan School of Management. The building was on the far edge of campus toward the abandoned factories and vacant lots of industrial East Cambridge—it had once been the headquarters of Lever Brothers, the soap maker. The rooms were boxy and nondescript, like standard function rooms at a midrate hotel. Faculty members rarely dined there—they frequented the F&T Diner around the corner, with its pastrami and beef stew and the neon Bud signs in the windows—so MIT frequently rented out the rooms to local groups. The night Mary-Lou visited, the club had been rented out to a Cambridge bowling league, and drunk bowlers had commandeered the hallways. She thought the whole scene funny. It was far more relaxed than Harvard, and more her style.

She had been consulting with friends among the Gall's Girls about where they had received offers, where they were waiting to hear from, and where they wanted to go; they didn't want to compete against one another. Everyone got what she wanted: Susan Gerbi went to Brown, Ann Stuart to Harvard, and Mary-Lou agreed to start at MIT in the fall of 1972.

Mary-Lou was taking a risk because no one else in the department was doing the cell biology she did. (The field was so new that Joe Gall would later be considered one of its founders.) But she did so intentionally: she thought she could learn about new areas from her colleagues, which was exciting, and she knew she would still have collaborators at other institutions. She had never been to Boston before and liked the idea of living there. MIT offered her a position as an associate professor, on the tenure track. Her job offers elsewhere were one level below, for assistant professorships. She thought it was as nice an apology as MIT could offer for the form letter rejecting her first application.

She also liked that there were two other women in the department: Lisa Steiner, an immunologist who was the first woman hired onto the biology faculty, in 1967, and Annamaria Torriani-Gorini, the research scientist promoted to the faculty earlier that year. Every other place Mary-Lou had offers she would have been the first and only woman. She didn't worry about being the only woman; she'd been the only woman in plenty of rooms. It was more that she didn't want to think she'd been hired just because she was a woman.

A few months later, on a cool spring evening in 1972, Nancy Hopkins was walking with her friend David Botstein along a dimly lit stretch of Brattle Street in Cambridge, the old "Tory Row" of the Revolutionary War, with its grand homes and graceful elms. She had met David in the summer courses at Cold Spring Harbor, and he had quickly become a mentor and good friend. Back in Cambridge he sometimes played tennis with Brooke, and he and Nancy walked and talked science on many evenings when she was working in the labs at Harvard; by now they figured they had worn down every sidewalk in the city, mostly in the rain.

David was vocal and opinionated, but omnivorously curious, and Nancy liked that he was as interested in discussing other people's ideas as much as his own; not every scientist was. He had emigrated to the United States as a child with his parents, Polish Jews who had been completing their medical training in Switzerland during World War II. A product of Bronx Science and Harvard, he had been hired out of graduate school as an instructor at MIT and quickly turned that temporary job into an assistant professorship in the biology department. Now, sniffing back allergies, he was telling Nancy about a woman the department had just hired, Mary-Lou Pardue.

The whole thing was a travesty, he told her. Mary-Lou had applied before and MIT had no interest. Now they'd hired her under pressure. "They pulled her résumé out of the garbage just because they had to hire a woman."

Nancy nearly stopped in her tracks. She didn't know Mary-Lou personally, but she could still remember being in the Harvard library in 1969 and reading Mary-Lou and Joe Gall's paper about in situ hybridization. It had blown her away, the genius of the technique, and how it would change the field to be able to actually see the gene sequences.

David thought that others in the biology department just didn't understand Mary-Lou's work because it was too new. Nancy worried David himself didn't appreciate how important the work was, but she didn't say so. She also didn't let herself consider that Mary-Lou's work might not be appreciated because she was a woman.

As far as Nancy could see, civil rights laws and the women's movement were knocking down discrimination. The hurdle had been to find jobs, and now women were being hired. Nancy herself already had two potential jobs. The previous fall, Harvard Medical School had called to ask her to interview for a job in virology, which was expanding to do more research on cancer-causing viruses. Not long after David told her about Mary-Lou, Nancy was in the Biolabs at Harvard when she was called to the phone. This call was from Salvador Luria, who in 1969 had shared a Nobel Prize with Max Delbrück and Alfred Hershey for the work on bacteria they had started at Cold Spring Harbor Labs. Now, in 1972, Luria was at MIT and wanted to interview her about a job for a new center he was building, dedicated entirely to understanding the molecular biology of cancer.

Nancy recognized that they were interested in her partly because she was a woman. She remembered Jim telling her years earlier that Harvard would soon have to hire women; it was the first time she'd heard the phrase *affirmative action*. She didn't think it was a bad thing because she believed science was a meritocracy; it helped to have mentors, but the people she saw advancing were smart and worked hard. They'd earned the jobs they were getting. Her work with Mark and her publications had made her a nationally known commodity. Few researchers had done the work she had with bacterial viruses at Harvard, and now with animal viruses at Cold Spring Harbor. That work made her ideally qualified for jobs in cancer research.

The previous December, President Richard M. Nixon had signed the National Cancer Act of 1971, the stroke of his pen signaling the start of a War on Cancer. Families of cancer victims had lobbied for decades to mobilize the resources of American science toward finding a cure for cancer, billing it as a Manhattan Project for cancer, or now, with the public entranced by the Apollo 11 moon landing, a "moon shot for cancer." Cures were still far off. But with the Vietnam War dragging on and midterm elections approaching, Nixon had seen the popular appeal in fighting what was the second-leading cause of death in the United States and, as the legislation described it, "the disease which is the major health concern of Americans today." The act committed $1.5 billion to establish designated cancer centers at hospitals and universities across the country.

Jim told Nancy to interview for both jobs, but that she should take the one at MIT. (Barbara McClintock told her not to take a job at any university: "You won't be able to take the discrimination.") MIT had Luria, who had been Jim's mentor. Jim thought the War on Cancer was largely hype; he feared that hospitals that received the money would give patients false hope that a cure was around the corner. Yet MIT, he believed, had correctly understood that the first challenge was to understand the basic science of cancer, its fundamental cause. Luria also had David Baltimore, whose breakthrough on retroviruses at MIT had inspired Nancy's move into cancer research after graduate school. His discovery of reverse transcriptase had made Baltimore the leading virologist of his time, and everyone expected it would soon win him the Nobel Prize.

Luria had immediately secured $4.4 million under the National Cancer Act to renovate the old Brigham's chocolate factory near the Longfellow Bridge in East Cambridge to house the cancer center. Watson told Nancy that Luria and Baltimore had designed it brilliantly; they even made sure it had the newest and best water system, so they would have the ultrapure water that was best for growing the animal cells needed to study tumor viruses. Harvard, Watson told her, was stuck studying phage and hadn't grasped the importance of such detail.

In his letters of recommendation, Jim noted that cancer had been Nancy's "ultimate research objective" since she was a senior in college. The young woman he'd recommended eight years earlier as a "bright and cheerful" girl, he now described as one of the best PhDs Harvard had turned out in the previous decade; most important, he wrote, she had shown that she knew how to focus only on "very important problems" and could "stick with them no matter how many times new obstacles crop up."

"She is very bright, thinks quickly and is not at all sentimental in the way she judges ideas," Watson wrote Luria. Harvard did not like to hire postdocs right out of training, but Watson added, "If she were not a product of our department, I would strongly urge we make an equivalent offer."

But increasingly, the important problem occupying Nancy was outside the lab. She was now only a few months shy of turning thirty, her self-imposed deadline for leaving science to begin raising a family. After thirty she understood there were increased risks for mother and child; amniocentesis was still rare and dangerous, so little information was available to what were then considered older pregnant women. Nothing she had seen in any lab had changed what she had concluded in her first days in science: that you could not be a mother and a scientist, at least not the level of scientist she wanted to be. Science alone was already more than a full-time job. Even a puppy had proved to be too much for her and Brooke to care for while they both worked; they had fallen in love with one, adopted it, and named it Hector, only to decide it was better off living with Brooke's parents in Baltimore. Since Nancy and Brooke were earning only graduate students' wages, they would not have the money to hire help.

Nancy had been pushing aside doubts about marriage and family for a long time. In graduate school Brooke had given her an engagement ring that he and his mother had designed. They had consulted Nancy, and it was stunning, yet when she thought about the commitment the ring represented, she felt buried alive. She and Brooke had broken up and gotten together again while she was at Yale struggling to find her way in science. But she thought she had settled her worries; she had done the significant work she'd aspired to do with Mark and the repressor, and more than she could have imagined since then.

Now it felt as if she were in a love triangle between Brooke and science. There was no question she loved Brooke. Their mothers had thought the marriage was a bad idea, that Nancy was smarter than he was—and they had let their views be known. But Nancy had never agreed. She loved him for being her opposite, for bringing out her fun side, developing her in ways she considered herself deficient. Now he'd shown himself to be an accomplished scholar as well.

Brooke was tall and athletic, laughed loudly and moved with big motions, as if he had no fear. He loved to dance until he was drenched in sweat, cook big meals, listen to jazz. He debated philosophy and politics fiercely—he was a Democrat, rebelling against his conservative parents—and read voraciously, introducing Nancy to authors and filmmakers she had never heard of. He had a wide circle of friends, and many of them had become Nancy's.

Nancy and Brooke were both prone to worry and deep funks. He could be irritable, but with his huge brown eyes and the smile that slowly crept over his face, Nancy never stayed angry at him for long. They had started their long talks in college, over sandwiches on the banks of the Charles, and they could still talk for hours.

But Brooke hated science, wasn't interested in it. Nancy's science friends ignored him when he attempted to make conversation. He complained they couldn't talk about anything besides science. Nancy agreed, but she also didn't mind—she hung on their every word. Part of her envied the female scientists who worked in their husbands' labs. She wanted to succeed on her own and believed she could, but those were also the women who managed to have a family and stay in the field.

In the summer of 1972, she and Brooke traveled four days by train to California so she could take a course on tumor viruses David Baltimore was teaching at the Salk Institute in La Jolla. From Philadelphia Nancy wrote a note to a secretary at Cold Spring Harbor—she attached it to a form about grant money—reporting that their compartment was "very cozy," though she wondered where two beds would come from "or how they'll fit in—particularly one long enough for Brooke!"

"So far it's fun!" After the course they took a few days to drive north along the Pacific Coast Highway. They returned happy and relaxed.

By that fall, Brooke was increasingly depressed. He had been worrying about his job prospects. His lectures at Harvard drew crowds and standing ovations, which were almost unheard of. But he was supposed to be writing a book—it was about autobiography. His favorites were Proust and James Joyce. He had to finish it to be considered for tenure, yet he could barely write a word. He and Nancy had known from the beginning that Harvard rarely gave tenure to its assistant professors—they might go elsewhere and come back later, but the university liked to hire stars. History and conventional wisdom had always held that a junior faculty member at Harvard could get hired at any university in the country. Now, though, the economy was slowing and there were rumblings that the job market for English professors—even Harvard-trained ones—was drying up.

With the tenure clock running out on him, Brooke started seeing a psychiatrist to help him cope with the stress. Nancy did her own worrying. What if it took Brooke years to find a new job, to get settled, and they couldn't afford to have children right away? She pictured herself sitting at home in some university town far away, missing science. She had always known his career prospects were more important than hers, since he would be supporting the family; it never occurred to her that she could be the primary breadwinner. She tried to detach herself from science, told herself she loved Brooke, and thought about what her mother had told her about the joys and satisfactions of having children.

At night, Nancy had a recurring dream. She was a child playing carefree in rolling countryside on a glorious summer day. She dreamed she was wearing denim overalls, riding a milk truck, and happily bounding on and off as she helped the milkman make his deliveries. She had just placed the glass bottles on a back stoop when a woman burst through the screen door and glared at Nancy. "Are you a little girl or a little boy?" she demanded. Nancy could not answer; she did not know her answer. She woke in a panic, her breath short. She had seen science as the perfect refuge, all order and logic, free from human complications and emotion. She knew she had to give it up soon.

Job offers forced her hand. By early 1973, she had them from both MIT and Harvard Medical School. Nancy took the job at MIT. The cancer building would not be ready for nearly a year, so construction delays would give her a little more time. She applied for a grant to extend her work at Cold Spring Harbor and hoped that by the end she and Brooke would know where they were going.

But even if Brooke got a job that allowed them to stay in Boston, she still had to think about having children. She would be working with RNA viruses at MIT, and it was still unclear whether they caused cancer in humans, and whether they could be contagious; she knew Jim would not allow the RNA viruses in the labs at Cold Spring Harbor and he'd had all the cats there killed out of fear that the leukemia that could spread between them might jump to humans. No one knew what effect RNA viruses had on a fetus, and Nancy worried she might not even be able to get pregnant.

She had been looking for resolution for so long that she decided she had to make a clean break. Pulled by love, the promise of motherhood, the conventions of her generation, Nancy called MIT and told Luria that she was so sorry for the trouble, but she would not take the job after all.

That spring, while Brooke had time off from teaching, she rented a room for them in a house at Cold Spring Harbor, thinking that he might be able to write the book away from the pressures of Cambridge. She was finishing her postdoc, and in the evenings they could walk together out to the beach or into the pretty village.

The house sat on a hill along the main road through town. Their room faced the front, and in the middle of the night they could hear the grinding of the gears on trucks making their way up and down the hill. Brooke complained about it, couldn't sleep. Nancy tried to change the room but no other was available.

One afternoon she arrived back at the room and discovered Brooke's typewriter and clothing gone. She called him in Cambridge and got no answer at home or his office, which she knew was a bad sign. Nancy hated to fly, but the train would be too slow. She flew back to Boston and raced to their apartment on Prescott Street. Brooke's things were gone there, too. She sat on the floor, crying and eating canned pineapple with a plastic spoon—he had taken the silverware.

She found him at Eliot House, the undergraduate residence at Harvard, where he was a tutor. He cried, told her he still loved her, and returned to their apartment, then left again a few days later. He came home and left again, returned, left, the same heartbreak several times.

Brooke told Nancy that her job offers from MIT and Harvard had finally crushed his confidence, after she'd gotten them he had been unable to write a single word. She was brilliant, one success after another, wanted by prestigious places; he had failed, couldn't write, and wouldn't get tenure. He knew she would never be able to leave science, and he didn't want to ask her to.

Nancy felt guilty, in shock. She hadn't been thinking about career success, she had just wanted to do experiments; she hadn't cared about winning the Nobel Prize, she wanted to do the work worthy of it. She hadn't realized Brooke saw her this way. She had always thought marriage would get in the way of science; she hadn't considered that science got in the way of her marriage.

Now she hated science, blamed it for what had happened: she had cared more about her work than she had about Brooke; now she was paying the price. She began to see science as an uncontrollable obsession, and she wanted nothing to do with it. That summer she was supposed to be finishing her postdoc in Cold Spring Harbor, but she stayed in Cambridge. She took up smoking, an old habit from the lab that she had quit to try to get pregnant. Unable to eat, she lost twenty-five pounds and ended up in the Harvard infirmary, where she watched, alarmed, as they took away her shoelaces. Her single scientist friends visited and offered to have sex with her, thinking she was just lonely. She thanked them for their kindness but declined. She wasn't lonely; she missed Brooke.

A friend told her Brooke had a girlfriend, another tutor at Eliot House. Her friend introduced Nancy to the woman. Nancy asked her if she was dating Brooke. "Who told you that?" the woman replied. "That's not true." Brooke insisted there was no one else. He was renting a house that summer—"in the country," he'd told Nancy—where he would finish his book. He would be there alone. Ann, Nancy's sister, bet he was going with someone else. Nancy trusted Ann—Ann was so much more sophisticated—but didn't believe that Brooke would lie to her. She had to see it for herself.

She went to the garage where Brooke parked the Volvo they had shared. On the floor she found a hand-drawn map labeled *Peaks Island*, marked with an X. Peaks Island was in Maine, so Nancy drove to Portland and caught the passenger ferry there. The island was small, only a few stores, some bungalows along the rocky coast with dirt roads, tall woods, and barely any cars. As Nancy stood outside the hotel where she'd rented a room for the night, looking at her map to figure out where the *X* might be, a young man approached her, introduced himself as Sonny, and asked if she needed help finding something. Sonny was no more than twenty-one, and a total stranger. It was a measure of Nancy's desperation and the disorientation of the previous months that she told him the whole story and agreed to let him help her.

Sonny insisted they wait until early the next morning, when it would remain dark. Under a still-starry sky they crossed the island by foot, walking down the dirt roads and into the woods. Finally they crawled through some brush to the edge of a small yard with a house where the *X* had been marked on the map. As they watched through dense branches, Brooke came out onto the porch with a cup of coffee in his hand, stretched, and took in the new morning.

Nancy and Sonny emerged from the branches. Brooke startled and was speechless for a few moments. Then—looking resigned, Nancy thought—he invited them in. Through a doorway Nancy could see Cecelia, the tutor from Eliot House, still in bed.

"My job is done," Sonny said, and took his leave. Brooke reintroduced Nancy to Cecelia, and together they walked Nancy back to the ferry. Cecelia was pretty and young and smiled and chatted lightly along the walk. She was smitten. *They won't last*, Nancy thought.

The trip to Peaks Island proved to be Nancy's turning point. She realized Brooke had moved on. He had lied to her. She stopped blaming science and thought instead about how difficult it had been to leave the lab on time all those years, to get home to Brooke for dinner. Now she wouldn't have to get home for anyone; she could indulge her love of science. It left her less vulnerable. She went back to the lab; now it was no longer a question of whether she would work—like so many women she had to work.

That summer Nancy turned thirty. She resolved that she would not marry again, she would not have children. She thought of it as taking a vow. Years later she'd joke about it—she'd become a nun of science!

For now, she took her vow seriously. She called Salvador Luria and asked if the job at MIT was still available.

Part **2**

Chapter 8

"We Should Distance All Competitors"

MIT was barely a mile from Harvard along Massachusetts Avenue, but it was a universe apart.

Founded in 1861, it was the dream of William Barton Rogers, an ambitious geologist from Virginia who had for decades championed a new education for the mechanical age, one that would combine theory and practice, conveying to engineers, architects, and other "practical men" the principles of physics, chemistry, and mathematics as well as their application to real-world problems.

Rogers had married a Boston bride and fallen for what he called the city's "knowledge-seeking spirit." He'd had six enslaved people in his Virginia household. But as the still-young United States split over slavery, he put his faith in the North: "In the midst of the sad trials in which so many are called to participate, and in spite of the ever-present cares and claims of the war, we have daily proofs that New England, and especially Massachusetts, will continue with unabated zeal to urge forward the peaceful enterprises of education and humanity." He started the institute with money from the state under the new federal land grant program to establish universities in the service of a country that was newly industrializing. It was to be a place that respected and fostered "the dignity of useful work." And while MIT remained private, Rogers envisioned for it a public role, "to guide as well as to stimulate research and invention." Adopting the motto *Mens et Manus*—"mind and hand"—the new institute held its first classes two months before the end of the Civil War with just fifteen male students in a rented building downtown.

By 1891, enrollment topped one thousand, and the institute colloquially known as Boston Tech had spread out over ten buildings in the Back Bay and was helping to shape a booming nation. Early graduates included Arthur D. Little, who pioneered the field of industrial research and founded the consulting company that bore his name, and Pierre du Pont, who led his family's chemical company and later General Motors to become two of America's biggest manufacturers. Alexander Graham Bell did the early work that led to his development of the telephone in a fledgling physics lab on Boylston Street.

Other universities rushed to imitate the "scientific school," and Harvard tried three times to take it over. MIT, its finances wobbly, almost gave in. But by 1916 it had declared its intention to compete on its own terms—and on Harvard's turf—with a move to the Cambridge side of the Charles River. Built by an anonymous $2.5 million gift (later revealed to be from George Eastman, the founder of Eastman Kodak), the massive new campus rose like a citadel from newly filled tidal flats along Memorial Drive. Its neoclassical buildings spread out symmetrically from the Great Dome, held up by ten columns. The buildings were sheathed in white Indiana limestone, a departure from the usual university brick and a tribute to the purity and rationality of science. Names of great thinkers—Aristotle, Newton, da Vinci, Darwin—were etched on the facades around the quadrangle. The Infinite Corridor, said to be the longest hallway in the world, connected the campus from east to west, underscoring the seamless movement between departments and disciplines. As its mascot, MIT adopted the industrious, nocturnal beaver.

The institute grew quickly, and exponentially starting in World War II as money for military research and development flooded MIT labs, three times as much as for labs at Western Electric, General Electric, RCA, DuPont, and Westinghouse combined. MIT constructed new buildings to make room for the work of exploring what Vannevar Bush, the former dean of the MIT School of Engineering who became head of wartime research for President Franklin D. Roosevelt, deemed "the endless frontier" of science: a wind tunnel to test fighter planes, laboratories that developed the radar used by Allied troops, the navigation system for the Apollo moon landings, and the flight simulator that in turn led to the first real-time digital computer.

As World War II wound down, MIT worried about relying so much on money for military research. To round out its interests, it added two new schools—Humanities and Management—to the existing three: Architecture, Engineering, and Science. But thanks to the Cold War, military budgets kept growing. By the time of the Vietnam War, MIT was receiving more money from the Department of Defense than any other university. To meet the needs of a growing campus, the institute mushroomed into what had been industrial Cambridge, taking over buildings where factories once churned out candy and cookies, soap and shoe polish. (Where the Hood dairy company once made ice cream, metallurgists studied beryllium and zirconium to build nuclear reactors; an actual nuclear reactor, painted baby blue, was built down the street.) Compared to the elegant original design along the Charles, MIT's spread was utilitarian, function over form or fashion: machine shops alongside offices, classrooms on top of atomic-particle accelerators. "It is difficult for our students to entertain guests from other campuses without social embarrassment," a committee convened to evaluate campus facilities complained in 1949. Even the vast stretch of the Infinite Corridor, with its exposed ducts and battleship-colored walls, was more big-city high school than Ivy League, more rolled shirtsleeves than blue blazer. Over the years would come buildings from designer architects, including Eero Saarinen, Alvar Aalto, and I. M. Pei, an alumnus. Still, no matter how much MIT grew in size, influence, and prestige, the bridge over the Charles that emptied right at its doorstep would still be known as the Harvard Bridge.

But MIT celebrated its scrappiness, embraced its difference. No one tried to pretend all the buildings were pretty; after all, there was an advantage in being able to run a wire from one lab to another just by jamming a hole in the wall with a screwdriver. Noam Chomsky, the celebrated linguist and leftist who began his long career at MIT in 1955, cheerfully explained to Japanese visitors who puzzled at his cluttered quarters in a building nicknamed the Plywood Palace: "Our motto is physically shabby, intellectually first-class." Campus buildings were numbered, not named, as if to suggest science were above ego or influence. Students didn't major in chemistry or physics, they enrolled in Course 5 or Course 8. "Tech Is Hell" went one unofficial student motto; a later, darker one was "I Hate This Fucking Place," its acronym etched into class rings. But this was their paradise, too, and they loved it. Long before the world had hackers, MIT had its tradition of anonymous hacks: the campus would wake up to discover that overnight students had lifted a cow or a Model T Ford onto the roof of a dorm or transformed the Great Dome into a likeness of the Great Pumpkin. (One of the Model T hackers, James Killian, went on to become one of MIT's most celebrated presidents and science adviser to President Dwight D. Eisenhower.) The hackers used a freshman named Oliver Smoot Jr., from Bexar County, Texas, as a measuring stick to mark the length of the Harvard Bridge—364.4 Smoots, plus or minus an ear. They welded a streetcar to the tracks in broad daylight, tied up Harvard phone lines, changed the sign at the bridge to read TECHNOLOGY BRIDGE, and set up primer wire that would have detonated to spell MIT in fifteen-foot letters across the field at the Harvard-Yale game. (Cambridge police foiled that one, but doubling down, MIT students made The Game a centerpiece for many later hacks; one year they replaced the VE-RI-TAS of the Harvard crests on the scoreboard with HU-GE-EGO.) It was as if to remind MIT's fancy competitors: you can have your majestic libraries, your football stadiums, your daddy's money; we know how to *do* things. *Mens et Manus.*

Above all, MIT saw itself as a meritocracy. It didn't have a football team and it didn't set aside spots for legacy applicants. Children of the Establishment went to Harvard or Yale; MIT took first-generation kids and told them and all the other strivers that hard work was the only thing that mattered. You earned respect by taking on the problems that scared off lesser scholars and lesser institutions. And being accomplished enough to solve big problems, faculty members expected the privilege of sharing in the governance of the institution. They participated in standing committees that shaped university policy. There was no faculty senate, which kept relations with the administration less antagonistic. The president chaired faculty meetings, and his office was located on the same hallway as the lecture hall where freshmen took intro physics and chemistry.

When formerly all-male colleges began accepting women in the late 1960s and early 1970s, MIT liked to say that it had never been single sex. While that was technically true, the practice was more grudging. In its earliest days MIT allowed women only into its free evening lecture series. It rejected a request in 1867 from four Boston women who had attended the series on chemical manipulation and wanted to continue as paying students in the day classes. That same year, a professor suggested to William Barton Rogers that MIT could seize an advantage by admitting women if it expanded from a scientific trade school into a full-fledged university: "By taking the bold step of opening our doors freely to both sexes I believe we should distance all competitors."

Instead, the first woman entered through the side door on special conditions. Ellen Henrietta Swallow, a recent graduate of Vassar with an interest in chemistry, was steered to MIT by a Boston chemical company where she had sought an apprenticeship. She knew she had to be intrepid: "My life is to be one of active fighting," she wrote as she prepared to enter science. (She signed her letters "Keep thinking.") She applied with a recommendation from the astronomer Maria Mitchell, a friend of the institute's president. MIT admitted Swallow "in the nature of an experiment," as the minutes of the corporation noted in 1871, "it being understood that her admission did not establish a precedent for the general admission of females." Swallow was apparently the first woman at a "scientific school" in the United States, but MIT did anything but advertise this advance; it declined to charge her tuition so the record books would not officially count her as a student in the event (or likelihood) she failed.

She succeeded, graduating with a bachelor's in chemistry at age thirty-one in 1873, and remained at MIT for two years, teaching for free in the hopes it would earn her a doctorate. When the institute refused her, based on her sex, she married one of her professors. He had proposed to her in a lab, and for a honeymoon they took students on a tour of mines to collect samples of ore. After she married, Ellen Swallow Richards conducted a survey of contaminated water supplies in Massachusetts, work that led to the nation's first water-quality standards and sewage treatment plants. She founded the field of home economics, introducing the concepts of calories, protein, and carbohydrates into American households, created correspondence courses for women interested in science, and founded the organization that became the American Association of University Women.

In 1882, MIT allowed women as "regular students," and by 1895, they made up 6 percent of enrollment. Since there were no dormitories for women, Ellen Swallow Richards and others raised money to provide them with "suitable toilet rooms" and a lounge that was named after Margaret Cheney, a graduate who had died young. Beginning with women still bound by corsets and crinolines of full-length dresses, the Cheney Room became a haven on campus. It was re-created in a new building when MIT moved across the Charles because the institute still did not see fit to build dorms for women. While the number of women jumped during the two World Wars, it declined drastically when men returned from battle, as it did at other universities, and remained at about 2 percent into the 1960s.

There were no women on the faculty until Elspeth Rostow was hired in 1952 (told the Faculty Club was all-male, she suggested MIT build her a women's faculty club; instead it changed the policy to admit her). In 1945, the university finally opened a women's dorm in a Boston town house, but it had room for twenty students at most, and they had to travel a half hour by trolley and T or walk the mile and a quarter to campus, often across lashing winds on the Harvard Bridge. The town house was the former home of Katharine Dexter McCormick, a 1904 graduate who had married into a harvesting-machine fortune. She had donated it and set up a fund to pay for the dorm's residents to take taxis in bad weather.

Even here, women who were interested in science were seen as a strange breed, urged not to attract too much notice. Female students warned each other that a mistake by one could mean failure for all; they must not "get on the boys' nerves," as a handbook by MIT's Association of Women Students in the 1950s advised: "A helpless female is a nuisance, and her counterpart, 'one of the boys,' is resented."

Only about one in twenty women made it to graduation. The numbers were so dismal that in 1955 MIT's president, James Killian, appointed a committee to consider whether the institute should continue to accept female undergraduates at all.

The currents on campus ran against it. The housemother at the Boston dormitory argued that most of the young women there were "such immature lulus" that they would be better served by women's colleges such as Mount Holyoke or Wellesley and could come to MIT as graduate students. The director of MIT medical services, a psychiatrist, appreciated the "pleasure and ornamentation" women provided on campus, but recommended that their seats would be better filled by men, given that women would have to spend a minimum of four to fourteen years on "the business of raising a family."

"During this time, had a male student had her place, he could have been contributing profitably in his professional capacity," the psychiatrist wrote to President Killian. MIT could be a hostile environment for women, and they struggled to compete without abandoning their femininity. "With so much conflict at an emotional level, it becomes plain that their intellectual efficiency must almost inevitably become impaired."

Even the young men of the *Harvard Crimson* weighed in: "Few people are aware that M.I.T. is a coeducational institution. Indeed, to most Harvard students, the idea of a feminine mind concerning itself with electrochemical engineering or mining and metallurgy seems somewhat revolting."

President Killian had decided against including alumnae on his committee, and the man appointed to chair it, Professor Leicester Hamilton, apparently solicited no opinions from its members anyway. Instead, when Killian pressed him for a decision in the next year, he wrote a confidential memo recommending that the institute stop admitting women. Alumnae, however, had rallied, producing a survey that showed the success and contributions of female MIT graduates: 93 percent were employed in their fields after graduation, the majority working ten years, many returning to work after raising a family. Instead of giving up on women, they argued, MIT should devote more resources to making them feel welcome on campus.

Killian sided with them. "I do not see how the Institute, having admitted women for so long, can now change its policy," he wrote the committee chairman. "Nor do I feel that even if such a change were practical we should do it in view of the growing feeling that women should have access to our great universities."

The institute's chancellor, Julius Stratton, announced the decision at a tea for female students: "Women are here to stay and it is our hope to make them feel more a part of the MIT community."

Sputnik had launched and, with it, the calls for scientific woman-power. MIT moved to improve facilities and services for women—as a 1959 report called them, the "forgotten men" of MIT. That year the institute promoted Emily Wick, a research associate who had received her PhD in chemistry from MIT, to become its first female professor in science. Emily studied the chemistry of flavor. She would soon become one of the biggest champions for women on campus.

With a $1.5 million donation from Katharine McCormick, MIT unveiled plans to build the first women's dorm on campus, with room for 116 female undergraduates. McCormick Hall opened in 1963, named for McCormick's late husband, and slightly delayed because Katharine McCormick was training her efforts and money toward helping Margaret Sanger develop the birth control pill. It was gracious living: elegant reception rooms with nineteenth-century antiques, gilt-framed mirrors and ample fireplaces, a piano and a sewing room. Women in science were still unusual enough that *Time* and *Seventeen* magazines covered the opening. "The only girl in a class gets plenty of professorial attention," *Time* reported, "and a freshman reports that getting a date required only the merest smile." The new occupants of McCormick Hall were doing "as well or better than the boys," but they recognized that, as one told *Seventeen*, "She must also show that for all her brains, she is a woman."

As soon as McCormick opened, there were calls to build a second tower to double its capacity. The publicity around MIT's new commitment to coeducation had resulted in what Stratton, now the president of MIT, called "a rather spectacular increase in the number of applications for admission on the part of highly qualified young women." The number of women entering as freshmen in 1964 was double what it had been two years earlier. The total number of women on campus was relatively tiny—about 260 out of 7,000 students, with most women in Math, Chemistry, Biology, and Physics. The director of admissions anticipated their numbers would rise to 400 within five years, but, he predicted, "This rate of increase will not continue." Unlike men, women had to live on campus "for safety," unless they lived with their parents or "close relatives." The limited housing for women, the admissions director said, would "force us again to apply more rigorous standards in the selection of female than of male applicants."

There was still a towel-snapping culture on campus, and it remained a tough place for anyone who didn't fit the usual picture of a scientist, which was white and male. Through the 1960s, MIT had between two and five Black students in each freshman class, not unlike many other elite institutions. For a coed university, it had a relatively small proportion of female students. The antipathy toward women was openly allowed. The student newspaper, the *Tech*, ran a regular column titled Cherchez La Femme to assist male undergraduates in finding coeds at parties on other campuses in and around Boston—the assumption being that an interest in science had curdled women at their own school. "The story that most MIT coeds are dull and ugly is one of the first an entering freshman encounters," advised the *Social Beaver*, a student handbook in the early 1960s. "With exceptions, they are that and more." In the middle of that decade, a Lecture Series Committee started a long tradition of showing X-rated movies on Registration Day each semester in Kresge Auditorium, the largest gathering place on campus, with audiences of often two thousand men and women showing up to flout local anti-pornography laws.

But the women proved they could hold their own. In the class of 1964—the year Nancy Hopkins graduated from Radcliffe—the graduation rate of women matched that of men for the first time. That October, the Association of Women Students on campus organized the

symposium on women in science and engineering where Bruno Bettelheim and Erik Erikson spoke, and Alice Rossi argued her "immodest proposal" for equality in marriage and science.

There were still just ten women on the faculty. All but Emily Wick,
in the department of Nutrition and Food Science, were in the humanities, and only Emily had earned tenure. In 1965, she was promoted to
assistant dean of students with a special focus on "women students."
She made it her policy to know each one by name, and early on she
did. By 1966 the number of female students had surpassed expectations, hitting 401. A second tower had been built onto McCormick
Hall, doubling the number of rooms for women. MIT experimented
by allowing "girls" in the senior class to live off campus—provided
they were twenty-one and had permission from their parents—and
the next year extended the privilege to sophomores and juniors. This
made more room for women in the freshman class. The predictions
continued to be wrong: more and more female students were applying,
and many of them were more highly qualified than their male peers.
The "scientific school" was going to need more than a bigger dorm.

Chapter 9

Our Millie

In 1968, MIT hired a physicist, Mildred Dresselhaus, as its first female full professor and found in her its first female superstar. Millie—almost everyone called her Millie, or sometimes Our Millie—was smart, attractive, funny, and energetic. Starting with the push for affirmative action in the early 1970s and continuing for five decades, she would reign as the emblem for what all women could be at MIT, and in science.

Millie had been born in Brooklyn at the start of the Great Depression to Eastern European Jews, recently arrived and unemployed, who moved the family to the Bronx so that her older brother, a violin prodigy, could be closer to his teacher. Millie followed in his footsteps, learning to read music before she could read English and winning a scholarship to Greenwich House Music School in Manhattan. Her brother went to Bronx Science and left for college when Millie was twelve, earned his master's degree by eighteen, and worked on the Manhattan Project. By comparison Millie saw herself as the less intelligent little sister. She lied about her age to start working in a zipper factory—at age eight, according to family lore—to help her mother, the family's chief wage earner, who worked nights in an orphanage and took in sewing. But Millie was driven, and at the settlement house she learned about Hunter High School, one of New York's selective exam schools and the only one that accepted girls. Hunter rescued her from what she later called the "sordid mess"—the poverty and hopelessness—of her neighborhood. She had to teach herself math for the entrance exam; she discovered she liked it and in high school proved she was good at it. As her yearbook rhapsodized, "Any equation she can solve / every problem she can resolve / Mildred equals brains plus fun / in math and science she's second to none."

She went on to Hunter College intending to become a teacher, which looked to her like the best profession available to a girl with good grades and no money. But as a freshman, she took a physics class taught by Rosalyn Yalow, a future Nobel laureate, who encouraged Millie to keep going in physics and remained a mentor. On a bulletin board at Hunter, Millie saw a notice about a Fulbright scholarship, which she won and used to study at Cambridge University. She returned to the United States to earn her master's in particle physics at Radcliffe, where she wrote her exams alone in a room at Radcliffe while the men from her classes took theirs at Harvard.

She decided that the best doctoral program in physics was at the University of Chicago and began there in 1953 as the only woman among eleven students. Her thesis adviser told her he didn't believe women had a place in science; they were taking resources away from men who would use their degrees rather than just get married. He told her she couldn't work on particle physics, so Millie moved into the study of superconductivity. She worried her adviser was right, that she was not smart enough to keep up with the men. But she loved physics, so she figured she might as well stay as long as she was curious. She was soon mentored by another Nobel laureate, Enrico Fermi. They lived in the same neighborhood near campus on the South Side, and both rose early to get to the lab, Millie on foot, Fermi on his bike. Millie would wave to him across the street, and soon Fermi began hopping off his bike and crossing over to walk with her. He had a gift for explaining hard concepts in simple terms, and Millie credited him with teaching her to think like a physicist, to keep asking "What if . . . ?" and to stay interested in everything, because you never knew where the breakthrough would come from. Fermi had set off the first man-made nuclear reaction a decade earlier, building an atomic pile in a subterranean squash court beneath the University of Chicago football stands.

Millie repurposed the surplus equipment from the reactor to create superconducting wire and build microwave equipment for her thesis project.

Her thesis adviser showed so little interest in her that he didn't even ask what the project was until two weeks before it was due. But Millie said his neglect forced her to be independent. This was like her; it sometimes seemed she looked for adversity because she so relished the challenge of overcoming it.

In Chicago she met and married the man who would be her biggest booster, Gene Dresselhaus, and together they went to Cornell in 1958, Gene as an entry-level professor, Millie as a postdoc on a National Science Foundation fellowship. Her adviser liked her, but Cornell refused to hire her when her postdoc grant was up because it, like other top universities, had a rule against hiring spouses. Gene quit in protest. They could find only two places that would hire a husband and wife: IBM and MIT. MIT had never had an anti-nepotism rule and would even allow Gene and Millie to write papers together. Its willingness to open the door instilled in Millie an abiding belief that MIT was a meritocracy, where ability and hard work mattered most of all.

In 1960, she and Gene moved to start jobs at Lincoln Laboratory, a federal defense program that MIT operated in the nearby suburb of Lexington, near Hanscom Air Force Base. Millie was one of two women among a thousand employees. She was now an expert in superconductivity, but the director of Lincoln Labs told her the field was dead; the work at Lincoln Labs was on semiconductors, materials such as silicon and germanium that enabled the missile defense and communications systems for national security (and later, the efficiencies of increasingly smaller personal computers). Gene suggested Millie work on the magneto-optics of graphite, which at its most basic level was infinitesimally thin and strong. No one thought it would yield interesting results, and Millie liked that because it meant that if her work failed, it wouldn't be seen as her fault. She also liked that she had the field to herself because no one would notice if she had to take a couple of days off to care for a sick child. She and Gene soon had four of them.

The first, a girl, had been born while they were at Cornell; Millie pushed the baby in a pram to the lab, where secretaries helped keep an eye on her. Three boys followed while Millie was at Lincoln Labs. Legend had it that she left work pregnant on a Friday and returned to work Monday after giving birth. (She did not dispute this but added that it helped that one baby had arrived on a snow day, and another on a long weekend.) She took her fourth baby to work the day after he was born. Security at the labs, however, would not allow him onto government property because he did not have proper identification for clearance. A colleague who witnessed the ensuing exchange said it was the rare time he had seen Millie lose her temper.

She took a teaching job at MIT out of desperation, though she didn't admit that until much later in her life. In 1966, Lincoln Labs began cracking down on a government attendance rule that everyone had to be at work by eight o'clock in the morning. Millie had help at home— she hired babysitters and finally a long-term nanny from Harvard Wives, which had been started to find employment for women whose husbands were studying on the G.I. Bill—but babysitters and schools couldn't start early enough to allow Millie to get to work on time. The lab director scolded Millie and the other woman for arriving late; when the other woman got fed up and left for Tufts, Millie had to absorb more heat. Gene was angrier than Millie; he reminded her that she worked from home early in the morning, and again at night while he put the kids to bed, which by his calculation meant she worked far more hours than the men at Lincoln Labs. Millie feared she'd be fired if she complained that she was being unfairly punished. But when a colleague told her about a visiting professorship at MIT paid for by the Rockefeller family for a "distinguished woman scholar," she jumped at the chance. The job was for a year, and she figured that after that her children would be just enough older to make the mornings manageable.

MIT had been looking for someone to teach physics to engineering students. The students loved Millie—she had learned from Fermi to hand out concise, detailed notes at the end of her lectures so that students could listen rather than keep their heads down taking notes in class—and they petitioned the head of electrical engineering to hire her permanently. MIT did so the next year.

She brought the same ambition and discipline to raising four children as she did to her career. She sewed buttons on clothing during meetings. At home, she and Gene divided cooking and cleanup, discussing physics and children as they washed the nightly dishes. They taught the children chemistry over meal preparation and took them around the world on work trips that were sometimes several weeks long—Venezuela, Israel, Japan. The children were all trained in chamber music, like their mother. Her daughter, as a preteen, refused to put through a phone call from the Nixon White House—she instructed the increasingly insistent White House operator to call back Monday—because Millie had given specific instructions not to bother her while she was working.

Millie wore her hair in milkmaid braids on top of her head—she made it look efficient, not girlish—and an impish smile that made her look as if she was having the time of her life no matter what she was doing. She was brisk and no-nonsense, with a trace of outer-borough New York in her voice, but unfailingly kind and interested in young people. She welcomed students as family—to the degree that her daughter puzzled at an elementary-school assignment to do a family tree, wondering where so many "uncles" fit in. Millie paid special attention to students who were women or Black or had come from other countries—anyone she knew would find it hard to handle the intensity of MIT. She accompanied them to university mental health services if they needed it, found a suckling pig and roasted it for a student trying to keep up a Chinese New Year tradition, and invited students to parties at the Dresselhaus home in Arlington, where guests sat on the floor balancing plates of homemade stew on their laps. After dinner the children would play a chamber music concert, Millie often picking up her violin to join.

The Rockefeller-family professorship had been established in 1962, named for an only daughter who had never gone to college and shocked proper society by striking the word *obey* from her marriage vows. The professorship came with a mandate to encourage women in science and engineering. So when Millie took it in 1967, she consulted Emily Wick, the chemist who was now the dean of women students, to figure out the best ways to help. She and Emily began hanging around the Cheney Room so female students could know where to find them for advice. The young women were eager for an outlet. They complained that professors didn't take their questions seriously; afraid to seem stupid, the women stopped speaking up. There was usually only one woman in any given class, and no one would sit near her—like a dandelion sprayed with weed killer, as Emily said. Sometimes the tension in the classrooms was so overwhelming that women asked to take their exams in Emily's office.

Millie still felt insecure herself. She worried privately that MIT had given her the faculty job only because she was a woman. She decided that it discouraged women to tell them how hard it was to be a woman in science; it only made them feel more singled out. So she did not talk about her own challenges. Instead, she worked hard to give the impression of someone who let nothing dent her confidence. She liked to remind young women that she had taken off only five days to deliver her three sons and advised her graduate students with what one thought of as Millie's Quantum Theory of Babies: "You must have either two or four kids for a stable family life because kids are like electrons in an atomic orbital. They come with spin up and spin down, so you need to have a stable configuration. You can't have one or three kids because that would be unstable."

Emily liked to say that Millie picked up the ball on women at MIT and ran with it. But their efforts were swept along by the campus unrest of the late 1960s. Antiwar protests at MIT had not been as frequent or as violent as the ones at Berkeley or Wisconsin or Michigan. Students demonstrated against visits from recruiters for Dow Chemical, which was producing the napalm in the bombs ravaging Vietnam, but the protesters didn't block the entrance to the career placement office. They took over the student center for six days in November 1968 to provide sanctuary to a solider gone AWOL, but yielded to allow the junior prom to be held there; the dean of students affectionately but dismissively described the straggling protesters as "a small group of tired students, a few hippies, and several motorcyclists."

Nineteen sixty-nine, however, brought what the dean of student affairs called the "year of test and confrontation." On March 4, thousands of students and faculty walked out of their classes and labs to object to the institute's complicity in the Vietnam War—its work developing missile systems and counterinsurgency technology, but also that the Department of Defense was floating MIT's budget. Calling it a "research stoppage," they gathered in the space-age dome of Kresge Auditorium, where Noam Chomsky spoke about the social responsibilities of intellectuals, and Hans Bethe, an architect of the Manhattan Project, urged nuclear arms control. Harvard's George Wald railed against the destruction that technology had wrought and urged science back toward "human ideals." Faculty and students on at least thirty campuses joined in solidarity, and Wald's address, "A Generation in Search of a Future," was reprinted in full in the *Boston Globe*, and then publications across the country.

Eight months later, a coalition of local antiwar groups called for a week of "November actions," demanding an end to the war in Vietnam and all war-related research at universities. They trained their protests on MIT's Instrumentation Laboratory, which along with Lincoln Labs accounted for slightly more than half of MIT's total budget. The Instrumentation Lab, known locally after its founder as the Draper Lab, had been flying high just that July, having developed the navigation systems and flight simulators that guided Apollo 11 to land the first men on the moon. Now hundreds of demonstrators were at its gates off Main Street protesting the development of some of the same technology to guide ballistic missiles. Five and six people deep, they linked arms to surround the building, chanting and scuffling with employees who tried to enter, until a phalanx of police with riot gear and rifles dispersed them with dogs and tear gas.

For decades, MIT men and women had been proud of the institute's role in helping the country win World War II. But this war was different. Given MIT's reliance on money for government and war-related research, the growing dissent hit hard and prompted an existential crisis about how the university was organized, the work it did, and where it got its money. "We did not need a weatherman to tell us that the weather had changed," one dean wrote, an academic's paraphrase of Bob Dylan.

The civil rights movement, too, was pushing a reevaluation on campus. When Shirley Ann Jackson, the valedictorian of her public high school in Washington, DC, arrived at MIT in 1964, other girls in McCormick Hall shunned her, telling her to go away when she tried to join a study group, standing up to leave when she sat down to dinner. She was so miserable that she was looking elsewhere to do her PhD, but on the car ride to visit Penn, she heard the news on the radio that the Reverend Martin Luther King Jr. had been assassinated and decided that she had to stay at MIT. King's courage had inspired her: she would not allow those women who had denied her existence to win; she would help make MIT more welcoming to Black students. She joined others in forming a Black Students' Union, which pressured the university to increase the number of Black faculty and students on campus. The members fanned out on recruiting trips across the country and helped start a summer program to help incoming Black students who had been ill prepared by their high schools. In 1969, the freshman class included fifty-three Black students—still few, but a quantum leap from previous years.

By 1970, students at MIT were agitating for positions on committees that governed the institute. University officials were reexamining the expectations of serving in loco parentis. Students were adults, not children. Eighteen-year-olds were being drafted into combat; they were given the right to vote. *Girls* were now *women students*. And they were demanding to be scientists.

MIT now allowed coed dorms, yet the number of women on campus was still limited to the number of rooms available in McCormick Hall. Millie and Emily proposed to increase the number by making admissions gender-blind. Many administrators and professors resisted; they argued that women couldn't keep up in class or wouldn't be able to find jobs, and if they did, they would earn less than men—wouldn't this weaken alumni fundraising? But Emily and Millie won; women would now be evaluated by the same criteria as men. As they were, the share of women increased: in 1969, there were 73 women in the freshman class; four years later there were 122. Some argued that the increase resulted from lowering the standards for women; as Millie and Emily pointed out, if this was true, it was only because the standard for men had been lower all along. To meet the growing demand from women but to accommodate the men, MIT increased the size of the freshman class.

In January 1971, MIT introduced an Independent Activities Period in response to student calls for more flexibility in their coursework. The idea stuck, and the next January, Millie and Emily planned a workshop on women's issues and advertised it in the "Independent Activities" catalog as a session to discuss "issues of interest to women." They didn't expect many people would show up.

They held it in the Cheney Room, and when they arrived, they found they could barely get in; it was standing room only. Millie and Emily had forgotten to put "students" in the advertisement for the workshop; the hundred or so people who showed up were secretaries and research associates, junior faculty members, faculty wives, and some students, including two men. The undergraduate women seemed to be the happiest women on campus—it was largely graduate students and secretaries who complained. The crowd demanded a committee to come up with ways to improve the "climate" for women at MIT. Surveying the array of faces, Millie had a sudden realization: she was expected to be a leader for all women on campus, not just students.

MIT had a new president, Jerome Wiesner, and he asked Millie to form the committee the group had demanded, and to conduct a series of hearings about women around campus.

Wiesner, known as Jerry, had come to MIT in 1942 to work in the Radiation Lab and embraced a long tradition of university leaders who wove careers of science and public service, shuttling between Cambridge and Washington. He had worked on the Manhattan Project and served as President Kennedy's chief science adviser, helping secure a partial ban on tests of nuclear weapons, and recommending the phaseout of pesticides after Rachel Carson's *Silent Spring* raised alarms about DDT. His association with Democrats and his opposition to the antiballistic missile system earned him a place on Nixon's enemies list. For Wiesner's inauguration in 1971, Archibald MacLeish composed and read a poem hailing him as "a good man in a time when men are scarce":

"He saunters along to his place in the world's weather / lights his pipe, hitches his pants, / talks back to accepted opinions."

Wiesner's wife was one of the founders of the METCO program, started in 1966 to send students from Boston's most disadvantaged neighborhoods to suburban schools. Wiesner was determined to make MIT more welcoming to minority students and women.

Millie held the committee hearings and wrote up a dutiful report. Wiesner called her to his office and told her she had to do better, she had to rewrite it with the same data and rigor she would a scholarly article in a physics journal. It had to be something people would take seriously.

On the Role of Women Students at MIT, released that spring, noted that although MIT had been admitting women officially for ninety years, women made up just 10 percent of undergraduates and had received just 3 percent of graduate degrees over the previous five years. Less than 2 percent of all faculty members—tenured or not—were women. "Decisions on admissions, degrees and appointments may indeed be made with no deliberate effort to exclude women—at least at times—but policy must be judged by outcome, not by pronouncement. And here we find inadequate numbers of women at all levels, most significantly so at senior levels."

"Even if women were to be admitted in equal numbers, such a policy does not imply that they will be treated equally once they are here," Millie wrote. At hearings around campus, women had told stories of "open hostility, lavish or total lack of attention, demeaning and embarrassing comments or other subtle forms of sexual discrimination." Advisers tried to steer female undergraduates to "more feminine" fields. One professor liked to say that "his girls" were "nice distractions" for the men in class. Graduate students told of research directors who openly stated that they would not accept women as students, and male peers who denounced women for stealing their jobs.

Some women could persist undiscouraged. But most women felt beaten down, and conspicuous. With few women on the faculty and none in powerful roles, female students had few places to turn for support: "The woman's influence in the Institute is ineffectively diffuse and/or non-existent."

None of this shook Millie's faith in the meritocracy. As she saw it, there was a problem of supply: to get more women on the faculty, you needed to fill the pipeline with women as students. If that pipeline was slow, it was because not enough girls chose to go into science. The door was now wide-open; young women just needed confidence, encouragement, role models. But Millie underscored in her report that women did not want special treatment or lower standards; she urged the institute to hire "*qualified* women faculty." "Hopefully, as women fill more faculty positions, their presence will become less of a novelty and more natural and socially acceptable."

Millie did a back-of-the-envelope calculation: based on her experience, she determined that it took at least two women in a class before one of them had the courage to participate, and if you had twenty students in a class, you needed about 15 percent of women in each department to achieve a critical mass. Given the recent increase in applications from women, Millie thought that critical mass might even rise to 20 percent by the late 1970s. Either way, the men would see what women could achieve, and the women would, too. Those students would naturally, confidently follow their interests up the ranks— undergraduates would become graduate students and PhDs, then postdocs—and as they did, MIT would create not just the next generation of female faculty members but the next generation of leading scientists who just happened to be women. As Millie became the most visible role model on campus, and increasingly a leader in scientific societies around the country and the world, this became the prevailing principle for MIT's efforts to increase scientific womanpower.

The 1972 report concluded with the same sort of conciliatory note as previous manifestos on expanding opportunities for women: "If there is one point we wish to emphasize it is our conviction that these steps will improve the environment at MIT for *all* members of the

community. Remedying injustices against women enhances the quality of life for men as well."

Starting in the months after the report, the admissions department
added more pictures of female students to its catalogs and sent female
faculty on visits to high schools to encourage girls to come to MIT. The
Association of Women Students, which had gone dormant in recent
years, revived and trained its energies toward increasing the number of
female students: its members took turns sitting in the admissions office
to encourage visiting high school girls to apply and held telethons,
calling girls who had been accepted to encourage them to attend.
The institute also commissioned a map of campus bathrooms—and
increased the number for women—after female students complained
it could take twenty minutes out of a half-hour lunch break in an exam
just to reach one.

Emily left in 1973 to be dean of the faculty at Mount Holyoke.
Millie found a new partner in Sheila Widnall, who in 1963 had been
the first woman hired on the tenure track in engineering. Sheila had
arrived at MIT from Tacoma, Washington, in 1956 as one of 23 girls
in a freshman class of 936 and become an expert on the turbulence of
vehicles—cars, planes, and helicopters—at high speeds. She said her
self-confidence came from having a mother who was a juvenile proba-
tion officer and a father who rode bulls. She and Millie started a course
called What Is Engineering? to familiarize young women with the
manual skills that many boys had already learned from their fathers or
in shop classes that were still boys only; again, more women showed up
than they expected, and the course also became popular with male students from minority groups. Within five years, the number of women
studying engineering at MIT had increased about 250 percent.

In June 1972, President Nixon signed Title IX. MIT had already
begun trying to increase the number of female professors, giving tenure to junior faculty members and elevating women who had long
worked as lecturers or researchers in labs run by men. That year, Lotte
Bailyn was promoted from lecturer to associate professor at the Sloan
School, becoming the first woman on the tenure track. Mary-Lou Pardue arrived in Biology, welcomed by its chairman, Boris Magasanik,

who had written the letter to *Science* urging women to "please apply." And in 1973, Biology hired Nancy Hopkins to the cancer center.

MIT looked to be leading the way. That same year, Shirley Ann Jackson, who had been one of the founders of the Black Students' Union and counted Millie among her mentors, became the first Black woman to earn a doctorate at MIT. She was the second Black woman in the United States to earn a doctorate in physics.

Millie thought getting more women on the faculty might take fifteen or twenty years. But she felt certain it would happen. That was the way the meritocracy worked.

Chapter 10

The Best Home for a Feminist

When Nancy arrived as an assistant professor in 1973, MIT was marking the hundredth anniversary of its first female graduate, Ellen Swallow Richards, with a yearlong centennial celebration. President Wiesner used the occasion to kick off a fundraising campaign to endow a professorship in Richards's name "to strengthen the role of women in the Institute faculty." An exhibit at the MIT Museum showcased "100 Years of the 'New' Woman." Students organized a two-day conference on women in science and technology, and the alumnae association convened a symposium featuring a banquet speech by Katharine Graham, publisher of the *Washington Post*—which had just won the Pulitzer Prize for its coverage of Watergate—and the first female CEO on the Fortune 500.

Ellen Swallow Richards wouldn't have recognized the school she had known as Boston Tech. MIT had become a full-fledged university since its move across the Charles River at the start of the century, and now, emerging from the turbulence of the late 1960s, it was broadening its ambitions again. Writing his annual report in October 1973, Wiesner condemned Watergate and the American intervention in Vietnam as the arrogance of power. While MIT needed to champion "and perhaps defend" the role of technology, he wrote, it also needed to direct its research, resources, and influence toward "the changing issues of the times." That year, the university fully divested itself of the Draper Instrumentation Lab, signaling that it had heard the antiwar protesters' demands to replace war-related research with "socially constructive work." Within four years, the Department of Energy, the National Institutes of Health, and the National Science Foundation would replace the Department of Defense as the university's biggest sponsors. In World War II, MIT had helped the nation by exploring the structure of the atomic nucleus; now it was building housing for low-income elderly people in Cambridge, attempting to understand the basic mechanisms of cancer, and expanding opportunities for women and Black students.

"Now is the moment to try a little harder, to care a little more, to charge our intellectual and professional development, and that of our students, with an understanding and concern for the underlying question of human values," Wiesner wrote. The federal government had approved the institute's affirmative action plan the previous spring. Now Wiesner invoked his vision of "a truly human environment" where all students, faculty, and staff could thrive as individuals "and can fully participate in the life of an educational organization which takes them seriously as individuals and not solely as the fillers of job slots or as 'representatives' of 'the women's issue' or 'the black issue.'"

Affirmative action programs, he wrote, were necessary short-term steps. "They are predicated on what we know to have been years, decades, perhaps centuries of practice that tolerated bias, discrimination, and a treatment of people on the basis of stereotyped views and misguided convictions of what they want and what they are entitled to." These attitudes, he warned, "will not go away by virtue of a faculty vote or the forging of guidelines." But he believed that MIT was primed to tackle the challenge. He had hired an adviser on "women and work" and created an office of minority education. And he was confident that the nation's preeminent institute of technology knew something about how to innovate: "Social progress, like that in the sciences," Wiesner wrote, "is inevitably the result of many experiments."

The number of women at MIT was still relatively small, just 816 out of 7,850 students, but it had doubled in five years—women were now 11 percent of all undergraduates, up from 2 percent in 1960. As part of its centennial efforts, the alumnae association had sent out a brochure to 10,500 girls across the nation, the top scorers on the SAT, its title declaring MIT "A Place for Women." The brochure took care to show young women in active roles: playing sports and working on problem sets, as opposed to pictures in previous brochures of women watching their boyfriends playing guitar. The quotations across its pages reflected the steely optimism among the women on campus. "If you're looking for equality, this is the place," one said. "Nobody is going to give you special treatment. If you want something, go work for it."

"At least they can't call women here dumb broads," another student said. "And when there are more women at the institute, and/or when a few gentle reminders are given to the chauvinist offenders, the offenses will probably disappear. Women, if a professor makes a sexist remark, tell him—he probably knows not what he does."

Female students borrowed boats from the men's crew team to start a team of their own; they even used the men's hand-me-down sweat suits and finally, bravely, barged into the men's locker room, fed up with having to dash across four lanes of traffic on Memorial Drive to shower at McCormick Hall. Another group of students tracked down one of the alumnae portrayed in the museum's centennial exhibit, Florence Luscomb, who had been among the first female graduates in architecture and a prominent Boston suffragist. They found her at age eighty-six, living in a commune in North Cambridge, and invited her to what became regular dinners.

The men leading university departments—all the department heads were men—were eager to show that they, too, were on board. The report from Biology in 1972 noted that its hiring for the year had been influenced by its "intention to increase the proportion of women on the faculty," and that it had found Mary-Lou Pardue. At thirty-nine, she was a decade older than the typical man starting on the tenure track, but the department report trumpeted her as "a young woman, a developmental biologist of great promise." She had received her PhD from Yale, it noted, and "her work on the hybridization of DNA in cell preparations is an important breakthrough in cell biology."

The Center for Cancer Research, where Nancy arrived as a junior faculty member, was a centerpiece of the new energy at MIT. The Cancer Act of 1971 had directed $1.5 billion to new designated cancer centers across the country, but most of that had gone to hospitals to expand treatment facilities: Memorial Sloan Kettering in New York, MD Anderson in Houston, Dana-Farber in Boston, Roswell Park in Buffalo. MIT created the rare center not attached to a hospital. Free from the responsibilities of caring for patients, it could focus exclusively on understanding how cancer cells worked, what made them different from normal cells—and ultimately on curing rather than just treating the disease. The center occupied the entirety of a newly renovated former chocolate factory on the eastern edge of campus. Jim Watson spoke at its official dedication in early 1975 and warned against the "hoopla" surrounding the War on Cancer, but singled out MIT for its "unique" and "responsible" recognition that the most important hurdle was to understand the basic science of the disease: "The best thing I think our country can do now in the War Against Cancer is create just one more institute of high-quality cancer research equivalent to what MIT has started here today."

The center had been started and designed by Salvador Luria, one of the three founders of the Phage Church at Cold Spring Harbor, who with the others—Max Delbrück and Alfred Hershey—had shared the Nobel Prize in 1969 for decoding the genetic structure of viruses. Known as Salva, Luria had fled his native Italy after Mussolini banned Jews from universities at the beginning of World War II. Having escaped fascism, he embraced a belief that Americans had not just a right but a responsibility to engage in their democracy. He and his wife had hosted the Reverend Martin Luther King at their home in Lexington to raise money for the Southern Christian Leadership Conference and joined a local group pushing banks to lend to Black home buyers. He led a group of scientists that took out a series of full-page ads in the *New York Times* denouncing the war in Vietnam. He was so outspoken against the war that the National Institutes of Health blacklisted him in 1969, and when he won the Nobel that year, friends sent letters of congratulations that assumed he had won the Nobel Peace Prize. (In fact, his was in Medicine.)

Salva retained a thick Italian accent, a dry sense of humor, and a broad intellect—for years he taught a literature seminar in addition to his duties in Biology. He lavished his Italian flirtatiousness on women. But he also supported them professionally: in the late 1940s he had insisted that Polly Bunting present a paper on bacterial genetics at Cold Spring Harbor despite her protestations that she was busy with three young children and pregnant with her fourth; the paper led to her getting lab space to resume her career at Yale. Salva's wife, Zella, was an ardent feminist and a professor of psychology at Tufts who studied how children construct gender identities, and when she proposed in the late 1960s that she and Salva divide the housework—alternating weeks to do the cooking, cleaning, and bill paying—Salva went along, first grudgingly, then boastfully, stopping at specialty food shops to buy dinner on his way home. He told the *Times* that when the Swedish Academy called to inform him that he had won the Nobel, he was washing dishes.

While Salva was the center's founder and director, David Baltimore was its star. Bearded and serious, Baltimore was known as a single-minded genius—as Salva wrote of him: "brilliant, hard-driving, arrogant, insensitive, even ruthless." He meant it nicely, or nicely enough, as he had hired Baltimore at MIT and built the cancer center around him. Baltimore's discovery of reverse transcriptase in 1970 had been earth-shattering—"Central Dogma Reversed," an editorial in *Nature* blared—reviving the field of basic cancer research by opening the possibility of finally understanding how viruses caused cancer. Virologists seemed in easy reach of identifying and understanding the function of oncogenes, the genes that, under certain circumstances, caused healthy cells to turn cancerous.

Baltimore's advance had inspired Nancy to change fields when she went to Cold Spring Harbor, shifting her research from phage to cancer viruses. She thought he was a genius; he seemed to know everything. While Jim was always popping into the lab to offer his latest idea or gossip, Baltimore was quiet and listened more than he talked. Yet when he spoke, people leaned in to hear. He had an unusual ability to sum up a new set of facts, to file them in perfect order in his mind, to recall the information and immediately grasp why a new discovery mattered. He always knew what the next question needed to be.

Like Salva, Baltimore believed that scientists had a responsibility to speak out on the political issues of the day. Even as he was closing in on the discovery of reverse transcriptase, he ordered work in the lab to cease on the day the United States invaded Cambodia and again weeks later when Ohio national guardsmen killed four students in the antiwar protests at Kent State University.

The cancer center had just six professors to start: two studying the immunology of cancer and four, including Nancy and Baltimore, studying cancer viruses. Nancy was caught up in the excitement, proud to be part of the center. And she was confident. She had built up a strong reputation, doing highly visible work even as a student. The best molecular biologists at Harvard—Jim, Mark, Guido Guidotti—had recognized her talent and mentored her, and some of the most highly regarded scientists in her field called her to discuss her work. Baltimore had chosen her because of her experience in phage, the simple bacterial viruses; her training would allow her to apply the powerful techniques of molecular genetics to the viruses that cause cancer in animals. As Salva's work had shown, the techniques and principles of phage were fundamental to understanding virology; Baltimore recognized that he himself lacked that knowledge. An MIT photographer arriving to document the new center captured the fresh excitement of the moment in a picture taken early on of Salva and Baltimore standing on either side regarding Nancy. The lab shelves around them were still empty, and Nancy had dressed up in a skirt and a blouse, a silk scarf knotted at her neck, a jacket draped impractically over her shoulders. The Mary Tyler Moore of the microscope.

The center was part of the biology department but operated largely as an independent unit. Salva and Baltimore had made sure it had the latest equipment, and facilities for working with tissue cultures that few other universities had. Everyone felt the urgency of the mission; the virology labs were on the fifth floor of the building, and around MIT a legend developed that the lights there never went out. Nancy was working fifteen-hour days, seven days a week, like everyone else, but she felt she especially had to make up for lost time. She had changed fields again; at Cold Spring Harbor she had been studying DNA tumor viruses; now she was working with RNA viruses and would have to build a body of work to establish a reputation in this new line of work. The department evaluated junior faculty after three years to determine whether they could continue on the track to tenure; after that they had another three years to win it. She still had a lot to accomplish.

Nancy's relationship with Brooke was behind her, but her personal life had cost her time, too. Three months after she arrived at MIT, her mother became sick. Budgie had once been infinitely energetic, spending weekdays running between the New York City public schools where she ran art programs, setting out on weekends to new museums or city neighborhoods or Cambridge to visit Nancy. Now, at sixty-three, she complained that she was so exhausted she could barely walk. Nancy feared it was her fault, that in her months of grief over the divorce she had literally worried her mother sick. On a visit to New York, Nancy encouraged Budgie to take a walk and became alarmed at how she dragged behind, how her ankles had swollen. A doctor diagnosed colon cancer, but too late. The surgeons found that the cancer had spread too far to operate. For the first two months of 1974, Nancy cycled between Cambridge and New York, keeping vigil by her mother's bedside in Ann's apartment, writing grants and setting up her new lab, meeting with the lawyer negotiating the details of her divorce.

She and Brooke signed the divorce agreement on February 12. Two weeks later, Nancy was back in New York; the nurse she and Ann had hired told them their mother was near death. Nancy imagined that death would be beautiful, gauzy like in a movie; her mother would look ageless and sigh gently as she took her last breath. Instead, Nancy could barely watch as Budgie writhed in pain and gasped for air.

Budgie died on March 3, 1974. The next morning Nancy took the train back to Boston and boarded the T at South Station. A few stops later, by coincidence, Brooke got on.

"My mother died yesterday," she told him.

"Oh, I'm sorry," he said. "How are you otherwise?"

They had spent thirteen years together. Now, within months, their relationship had been reduced to pleasantries on public transportation. Soon after this, Ann moved to Singapore, where her husband had a new job. Somewhat suddenly, Nancy found herself alone. The people she had been closest to were gone or out of reach.

Salva had tried to make her feel at home, and even protect her. He had given her the year off from teaching to care for her mother and recover from her divorce. As she and Brooke haggled over the settlement—she discovered he had trust funds that could have covered the cost of hiring someone to care for children—Salva insisted Nancy's lawyer was too much a gentleman and had her replace him with his own more hard-charging one. Salva also arranged for her to meet with his financial adviser, who took Nancy out to lunch and told her he had seen too many single women lose all the money they had; the adviser instructed her to invest the small amount of money she'd inherited from her parents in Treasury bonds and, to save her bank fees, escorted her to the federal building in downtown Boston to buy the bonds directly.

Salva was generous with young people, excited about the new professors he'd recruited and their science, and eager to help them flourish and get tenure. He'd regularly pop into the offices of the junior faculty members to say, "Write another grant!" or "Are you writing a paper?" He was looking for other bright young minds for the cancer center, and when he asked Nancy for recommendations, she immediately mentioned Phil Sharp, who was a staff scientist and one of her favorite people to talk science with at Cold Spring Harbor. Later that day Salva ran into Nancy in the hallway, and his eyes twinkled. "I'm on my way to call Phil Sharp," he said. "Will that make you happy?" It had, and now she and Phil had offices on either end of a shared space, with a secretary in between whose salary they paid jointly from their grants. Two of her other closest friends and occasional tennis partners—David Botstein and Ira Herskowitz—were working in the Biology department, across the street.

Yet for all the good intentions, and all Nancy's early renown, she was still the only woman on the faculty at the cancer center—when Baltimore, as expected, won the Nobel Prize in 1975, there were more Nobel laureates than women on the faculty in Biology, and that would remain the case as its size increased over the next years. No one—not Salva, not Nancy herself—had anticipated how difficult her position would be.

Nancy had known she would lose some status once she was out of Jim's immediate orbit. Still, she had assumed that people at MIT would recognize and respect her experience the way others did—the lab heads at Harvard and the National Institutes of Health, the high-level scientists from other universities who treated her like a peer. But the postdocs in the fifth-floor lab knew nothing of her work in phage; they were all studying virology. She was barely older than they were and looked the same age. The other women in the center were mostly technicians, and she could tell by the way the technicians talked to her that they thought she was one of them. Deliverymen asked her to sign for packages, assuming she was a secretary. The secretary she shared with Phil did his work first, and Nancy's only if she had time, leaving Nancy to wait for the secretary to go home at night so she could type her own papers and grant proposals.

The physical layout aggravated the problem. A traditional lab was self-contained and assigned to an individual professor as the lab's principal investigator, with a series of rooms with equipment and benches for postdocs and grad students and an office for the professor. The fifth floor of the cancer center was an open plan—a contiguous lab space with wide arches instead of doors between each professor's area and a large area in the middle where everyone shared benches and equipment. The offices were around the perimeter: Nancy and Phil on one side, and on the other, Baltimore and Bob Weinberg, who had started as a junior faculty member the same year as Nancy.

Baltimore had gotten the idea of the common space from the Salk Institute in California, where he worked before Salva recruited him to the faculty at MIT. Salva had argued against it at first but gave in to Baltimore's vision. Baltimore thought the open space would spark creativity and collaboration among colleagues leaning over equipment together, that science moved faster as a shared enterprise. The high-light of the week was the floor meeting, when faculty members, gradu-ate students, and postdocs from all the labs gathered over bag lunches in a conference room to take turns presenting their work, giving stu-dents the chance to have their work helped along by the top scientists in the field.

Nancy knew it was the way science should work in an ideal world. But for all the hopes that it would foster collegiality, the setup also bred fierce competition. Without walls, there were few boundaries, and without boundaries Nancy felt the students and postdocs didn't respect her. Students raided her incubator filled with the cells she had cultured, took her reagents without asking. She found herself wait-ing to use a microscope she had paid for with her own grant. The lab heads each had distinct lines of inquiry, but postdocs from the other labs took results she presented at the floor meeting and used them to design their own experiments in Nancy's line. She had always loved to talk about science, the ideas for new experiments bouncing in her brain. Now for the first time in her career she decided she'd better keep her mouth shut for fear that someone would make off with her best idea.

The competition was almost inevitable, given Baltimore's outsize influence and the ambition of the students who wanted to work with him. Everyone felt pressure. The postdocs wanted splashy results and Baltimore's attention, to publish their own papers so they could find jobs as heads of their own labs. Nancy and other junior faculty mem-bers had to worry about getting tenure—securing grants and graduate students and postdocs to help carry out experiments, making discov-eries that were interesting to journal editors and would be recognized as significant advances by leaders in the field, those who would write letters of recommendation for tenure. No one got tenure because of a group project; the whole system was built on individual credit. Yet the open space encouraged the impression that everything was up for grabs.

Nancy thought she'd only make it worse if she spoke up or told her graduate students to say no when others came looking for cells, or even if she said, "Have them come ask me." She didn't want to be known as a bitch, or a difficult woman. She wasn't sure if Baltimore knew what was going on; he was traveling frequently and taking an increasing role in national science policy, especially after he won the Nobel.

She asked David Botstein for advice; he told her people took his ideas, too. She talked to Jim, who laughed in recognition. "It's those hungry postdocs," he said. "Watch out for the hungry postdocs." Finally, she went to Salva. He listened carefully, then told her she should stop going to the floor meetings and not talk to anyone about her results. Nancy knew this would isolate her, but she agreed it was the only solution. When the others on the floor gathered in the conference room, she stayed by her bench and pretended not to notice she was the only one not joining. She began talking more to her colleagues outside MIT about her work.

She had to keep her eye on tenure. "Survival is my highest ambition at the moment," she wrote to Jim. "That is looking reasonably good, I think."

Many of the men on the faculty were still her friends; she could talk with them the way she had at Cold Spring Harbor, they invited her to their homes for dinner. But even as her work went well, she began to feel a chill from others on the fifth floor. Botstein told her he'd heard that Baltimore was upset with her. In early 1976, the chairman of Biology, Boris Magasanik, came to see Nancy in her office in the cancer center. Nancy liked Boris, found him charming, a great storyteller. Now, his brows were knitting and he got right to his point: "You're never going to be brought up for tenure. David Baltimore does not like you." The tenure clock ran for six years, and at midpoint the department solicited letters from prominent researchers in the field to determine whether assistant professors had done the work likely to earn them tenure. Boris told her she would be reviewed, but the letters would not matter. No matter how good they were, he said, the department would not nominate her for promotion: "We can't have this lack of collegiality." *Lack of collegiality.* He enunciated every syllable.

As soon as he left, Nancy picked up the phone to call Jim, who was now running Cold Spring Harbor full-time. "I think I've just been fired," she told him. She needed to find another job. Where did he think she should apply?

Jim cut her off. "How is your science going?"

Well, she told him. It was; she had made the first recombinants between mouse retroviruses, taking traits from two different viruses to show that they could make a novel one. She had mapped the genes of the viruses that cause certain leukemias in mice.

In that case, Jim told her, she didn't have to worry. The department would solicit letters, and the letters would testify that her work was first-rate. "Pay no attention to them, Nancy. Just keep working. When the time for tenure comes, if your letters are good enough and MIT rejects you, I will personally sue them for six million dollars. Just tell them that."

She told them nothing. Instead, she blamed herself. For not being aggressive enough to compete on the fifth floor. For sharing her worries with Botstein and Jim; she should have known their chatter would get back to Baltimore. Baltimore had reason to be mad at her, she thought: she had accepted the offer at MIT, then turned it down only to call back and ask if she could still have it; she'd taken time to recover from her mother's death and her divorce; now she was complicating Baltimore's collaborative vision for the virology lab. How else could she have expected him to react?

She got back to work and trusted that would be enough.

Millie Dresselhaus had gotten a grant from the Carnegie Foundation to continue her work helping women on campus, and she used part of the money to start monthly lunches for the new women on the faculty. They talked about how to help female students—they worked to get the women better locker rooms—and Millie and Sheila Widnall coached the junior faculty on how to win tenure, since the new women reported they were getting little mentoring from their department heads. With Sheila playing the role of applicant and Millie the department head, they relayed strategies for writing successful grant proposals, quantified how many papers you had to publish for tenure, and explained the importance of letters of recommendation.

The women at the lunches wasted little time on grievance. Across the country, feminists were rallying for the Equal Rights Amendment, which Congress had released to the states for ratification in 1972; across town, a coalition of women's groups had taken over a Harvard-owned building for ten days, christening it the Liberated Women's Center and demanding the city develop affordable housing and childcare programs. But the women at the lunches were not radical that way. They were grateful to MIT for the opportunity it had granted them; they felt they had finally joined the club, as one said, and wanted to help make it a better place for women. Sheila told President Wiesner when he joined them on one occasion, "You're lucky to have us."

Nancy went to one lunch to be polite, but she couldn't wait to leave and never went again. She understood that Millie wanted to help young women, but she didn't think it was the help she needed. She thought her problems were specific to the fifth floor, and competition. She didn't think she needed help getting tenure; she felt confident that with hard work she'd earn it—in this, she shared Millie's basic faith in a meritocracy. And Jim had told her that the science was what mattered, to worry about the quality of her experiments and results, not the politics of the institution.

Mary-Lou had invited Nancy to the one lunch she attended. Mary-Lou embraced the chance to talk to other women outside Biology. Typically no more than a dozen were at any lunch, but she marveled at their intellect, how briskly they identified a problem and went to work on the answer, harnessing the new power of Title IX to help students get more women's teams or space on campus. She had never before been around so many women doing such interesting work, and she loved hearing about it: Judith Wechsler in art history; Molly Potter, the psychologist who had pioneered the study of rapid serial visual presentation to understand how quickly the human mind processes information; Judith Jarvis Thomson, the moral philosopher who developed the thought experiment known as the trolley problem and whose writing on abortion had shifted the philosophical debate from the rights of the fetus to the rights of the pregnant woman before the Supreme Court's 1973 decision in *Roe v. Wade.*

The Biology department was small enough that most days everyone ate together in the lunchroom, and Mary-Lou made friends among her colleagues there, recruiting them to join her on weekends on hikes exploring New Hampshire's White Mountains.

Most of the Biology faculty were studying bacteria or viruses, nothing larger than a single cell. She was one of the few working with eukaryotes—the multicellular organisms—which was the next step toward understanding human development. Graduate students typically gravitated toward the labs of professors with full tenure and bigger labs, but Mary-Lou was pushing into a new area, and they eagerly signed up to join her.

But like Nancy, she found herself navigating new conflicts, largely around questions of credit.

Mary-Lou had come to MIT with the idea of using in situ hybridization—the technique she had discovered with Joe Gall—to observe gene expression in a developing organism. She was using larval salivary chromosomes that had an unusually large amount of DNA; under certain stimuli those chromosomes developed regions with high gene transcription, known as puffs. With her first graduate student, she began using heat shock and other stimuli to observe the sequence of gene transcription in the puffs. Another student, Allan Spradling, asked to join them; he was assigned to the lab of a fully tenured professor, Sheldon Penman, but was growing bored with his work there measuring the life cycle of DNA. Sheldon willingly agreed that Allan could join Mary-Lou on her project. Soon Allan and Mary-Lou were producing exciting results, showing that heat shock dramatically altered the pattern of transcription, shutting off many genes but also producing new proteins.

Sheldon began discouraging Allan from the work with Mary-Lou, telling him the project was not interesting, that it was a waste of time. Allan reported this to Mary-Lou, and they agreed that he'd remain with her but keep a hand in projects in Sheldon's lab to keep the peace—they joked that Allan was her stealth graduate student. As they prepared to publish their first results, she sought advice from another professor on how to share credit. Normally the person who does the work is listed as the first name on the paper and the lab head appears as last author—a mark of seniority and an indication of whose lab sponsored the work. The professor advised that when the work was split between two lab heads, they should take turns as last author. As this was Mary-Lou's first project, and she needed the publication to help get tenure, they decided she would appear as last author this time. She'd give Sheldon credit on the next paper, if there was a next paper.

She was surprised to see Sheldon outside her door a few days later. "This little project Allan is doing with you," he began. "Since neither of us has done any work on it, I was thinking we should not put our names on it." Mary-Lou chose her words carefully, telling Sheldon that he was welcome to leave his name off if he did not think he had done any work on the project. "But I've done quite a lot of work," she said, her tone level, "and I intend to have my name on that paper."

She walked off irritated but didn't think she needed to tell anyone. Allan wrote the manuscript and asked Sheldon if his secretary could type it, and Sheldon agreed. A few days later, Allan came running down the hall to Mary-Lou's office with the typed paper in his hand and pointed to the list of authors. Sheldon's name was not only included, he was listed as last author.

Mary-Lou confronted Sheldon. She told him she would not allow the paper to be published until her name was restored as lead author. He backed down; he didn't have much time to argue because *Cell* wanted their paper, and they knew a lab at Harvard was about to publish a paper in the *Proceedings of the National Academy of Sciences* about similar work. As humiliating as it might be to appear below a junior faculty member on the list of authors, getting beaten by Harvard was worse.

Small conflicts continued. A new postdoc from Berkeley arrived in Mary-Lou's lab and reported that Sheldon had discouraged her from signing on there. The postdoc had won a prestigious fellowship, and Sheldon told her she'd have trouble using it in the lab of an untested junior faculty member. When Allan finished his doctorate, he asked

Mary-Lou to sponsor him as a postdoc, and she raised grant money to pay for him, but Sheldon insisted that Allan would be his postdoc, that he would pay.

Allan's mother was a biologist, so he was sensitive to the way that women were treated in science, and he thought all this was sexism. Mary-Lou wasn't sure, but regardless, she had little time to make a fuss; she had to work around any problem. Allan's mother advised her to keep a record of who did what work on the experiments and mail it to herself so she'd have postmarked evidence if she had trouble establishing credit when she came up for tenure. Mary-Lou took the money she had raised to pay for Allan's postdoc and spent it on a new photo microscope they could use to publish micrographs. They nicknamed the microscope the Sheldon Penman Memorial Fellowship.

———

Most women given the new opportunities in science wanted to fit in, and who could blame them? Calling attention to being a first—one of the few, an exception—was to invite suspicion. While universities dutifully filed plans to make their faculties more diverse, critics of affirmative action had doomed it from the start—and even some supporters weren't helping. On June 28, 1974, the front page of the *New York Times* asserted that the program "by which the Federal Government is compelling colleges and universities to hire more women and blacks is lowering standards and undermining faculty." The article—on the page with news of Watergate, Nixon's visit to Moscow, and the ongoing energy crisis—continued, "Moreover, it is charged that new minority and women appointees may be paid more than white male faculty members at the same level and that some do not have the proper qualifications for the tenured and untenured positions to which they are appointed."

The article was based on a newly released report by the Carnegie Commission on Higher Education. The author, Richard Lester, had considerable stature: he had been vice-chairman of Kennedy's Commission on the Status of Women and was an economics professor and former dean of the faculty at Princeton. He supported efforts to diversify faculties. His quarrel, he said, was with the process. The competition for "the limited number of minority academicians" had "at times driven up salaries 'well above those for whites with equivalent or better qualifications.'"

Lester had based his conclusions not on hiring data, but on meetings he'd conducted with administrators at twenty leading institutions, which went unnamed. (About fifteen hundred colleges had federal contracts, requiring them to file affirmative action plans.) The *Times* article noted, toward the bottom, that "the charge that women and minorities are not prepared [to be] as potentially excellent educators as white males cannot be substantiated." Lester himself cautioned that abandoning affirmative action would be "premature." But the damage was done. Other publications piled on, including *Time* magazine, under the headline "Affirmative Action: The Negative Side." Its article reported how "Washington bureaucrats"—rarely a term of affection—had temporarily withheld $28 million in federal research funds from twenty leading universities as a penalty for not yet tendering their affirmative action plans.

In public, *Time* said, "college administrators and faculty members ritually endorse affirmative action." In private, they complained it was "a 'painful experience' that has accomplished little for minority groups while doing violence to a long tradition of academic independence and excellence." Setting hiring goals—or "quotas" as they would come to be called—might make sense when looking for "typists, bricklayers or punch-press operators," Lester said, but they did not make sense when "choosing a medieval historian."

His report, though, was projecting what might happen, not accounting for any changes that had already taken place; affirmative action plans had been in effect for less than a year at most campuses, which had not been enough time to hire many new faculty members. As *Time* noted in its last paragraph, surveys had shown that the percentage of Black and female professors had barely budged since 1968. Black professors made up less than 3 percent of all faculty, and women, 20 percent.

Lester's solution was not new; rather than "shoehorning more women and minorities into professorships," as *Time* put it, he wanted to increase the number of Blacks and women going into PhD programs and have them work their way up. But even this approach was already under fire; a week before the *Times* report, a thirty-five-year-old white man named Allan Bakke had sued the University of California, alleging that he was the victim of reverse discrimination; that California's practice of reserving a percentage of seats for "qualified" minority applicants had resulted in him being rejected twice from medical school. Hearing the case four years later, the US Supreme Court struck down the use of racial quotas but upheld affirmative action. Yet affirmative action was on the defensive, as were women, Blacks, and Hispanics—any group assumed to be admitted under lower standards. They were on campus, but they didn't belong there. Discrimination changed shape, from outright bans to more quietly held biases. Less overt, it was harder to identify, and harder to confront.

Women at MIT were provided a vent in the president's new adviser on women and work, Mary Rowe. Mary had an office on the Infinite Corridor, close to the president's suite, and had strategically placed a bowl of peanuts to encourage visitors. Some of what she heard was blatant: Women were not allowed to work in the campus observatory at night, told, "You're likely to get raped up there." Calendars featuring topless women hung on the doors of engineering labs. One young woman reported that a professor had invited her to his home and tried to have sex with her, offering that she could use his wife's diaphragm to avoid getting pregnant.

The most commonly reported problem, however, was not sexual assault or the rank discrimination that civil rights statutes had made illegal. It was what Mary called "the minutiae of sexism," the slights against women that were so casual that in isolation they weren't "actionable."

"Most are such petty incidents that they may not even be identified, much less protested," she wrote in a presentation for the American Association of University Women in 1974. These were the invitations to seminars and meetings that were not extended, the pages that were not typed, the professor who refused to learn his female students' names or vowed that if a woman was given tenure, he'd make her life so miserable she'd quit: "It is her work which by mistake is not properly acknowledged, not reviewed, not responded to, not published, her opinion which is not asked for."

Mary titled her presentation "The Saturn's Rings Phenomenon." The slights to women were so small and scattered that a woman might brush them away if she noticed them at all. Taken together, like the dust and ice around Saturn, "they constitute formidable barriers." Mary described it as "harassment"—making her one of the first people to use the term to describe endemic and persistent sexism; until five years later, when a young legal scholar named Catharine A. MacKinnon wrote a book about "sexual harassment," most women could hardly label what was happening to them. But whatever they called it, the problem reinforced itself. Made to feel invisible, women doubted themselves, which in turn made them seem too insecure to be promoted.

When a man was confronted about his behavior, Mary noted, he would often say, "I harass everyone." But much of it took sexist form: dirty jokes, or dismissing a woman's comment with "You sound like a housewife." A woman might be more sensitive to the slights, but it was hard not to take it personally when she was the only woman in the department, or a junior faculty member. Because she stood out, she attracted more attention. Because she had less status, she was an easy target.

Many women decided that the best strategy was to put on blinders. "Women need either to cope with slights against themselves or develop a considerable shield—a 'denial' of such experience—both of which processes take considerable energy," Mary wrote. She could understand. Asking women to confront the problem "is also in some cosmic way unjust; to require the victim to begin redress of grievance *itself* constitutes 'unequal opportunity.'"

Chapter 11

Liberated Lifestyles

As Nancy struggled to sort out the source of her problems on the fifth floor, one of the people who offered advice was Alice Huang, a microbiologist who had recently joined the faculty at Harvard Medical School. Alice was also David Baltimore's wife. She had emigrated from China at age nine, graduated from Wellesley and Johns Hopkins, and met Baltimore when she won a postdoctoral fellowship and was thinking of going to work with Jonas Salk, who had discovered the first vaccine for polio and started the Salk Institute in California. Baltimore was already working there, and a mutual friend suggested that Alice talk to him, then told Baltimore that Alice was a good scientist, and beautiful. "Take her," the friend said, so Baltimore did. ("Don't work for Jonas Salk," he'd told her, with characteristic self-confidence. "Work with me and bring your own money.")

Alice had followed Baltimore to MIT as his postdoc when he started on the faculty there in 1968. She would have been content to continue that arrangement—she had the money she needed to do research, and none of the responsibility of running a lab. But she had watched and learned from the experience of Annamaria Torriani-Gorini, a biologist who had spent decades as a research associate at MIT, then found herself unable to get grants of her own when the man she worked for decamped to Columbia—all the grant money went through the lab heads. Alice had joined other female postdocs in lobbying MIT to make Annamaria a professor in her own right. Then, resolving not to tie her own future to someone else's star, Alice started looking for a job running her own lab and found one at Harvard Medical School.

Zella Luria, Salva's wife, had organized a small group of older women to mentor Alice and two other women who were starting as science professors in Boston, meeting monthly in the older women's homes in the suburbs over after-dinner coffee. Alice enjoyed the company of the older women—Zella, Annamaria, and Ruth Hubbard, from the Harvard Biolabs. But she heard their stories the way Nancy had Barbara McClintock's; they were from a different time, before doors had opened for women. Her own generation, she thought, would be one of transition, of careful navigation: the women had to work twice as hard as men to show they deserved the jobs that had opened up to them, and they had to be reasonable, not too aggressive—not too male. Alice was willing to work hard, and she didn't want anyone to think she was seeking special treatment, because she didn't think she needed it.

She was shaken, then, when she was applying for a grant shortly after she arrived at Harvard in 1971 and discovered she was earning less than male professors at the same level. While she wasn't seeking special treatment, she did want equal treatment, and salary was the clearest measure of worth. She wrote a letter to her supervisor, demanding that he give her a raise. A secretary in the typing pool at the medical school, a woman, saw Alice's draft and told her not to send it—"You'll be seen as an angry woman." Then she nudged Alice's supervisor to increase her salary.

The incident had inspired Alice to consider the treatment of other women in her field. How many hadn't been fortunate enough to benefit from the intervention of a secretary who knew how to walk the line between assertive and angry? Alice did a study based on a survey of the membership of the American Society for Microbiology. Zella introduced her to Lotte Bailyn, who was now on the tenure track at the Sloan School at MIT and establishing herself as an authority on women's experience at work, and Lotte helped analyze the data.

Alice's report, written with three other women, was published in *Science* in February 1974.

Women were about a quarter of the society's membership. They reported that they worked as many hours as the men did, published as many papers, stayed at their jobs the same length of time, and shared the same motivations—men and women alike worked because they needed the money and because they loved the work and the sense of professional accomplishment. Yet on measure after measure the women had lower status. They earned less than men with the same qualifications, with the most educated women suffering the widest wage gap. Women had more trouble finding jobs, it took them longer to become professors, and they were absent among department heads and other jobs in top administration (which paid more).

Administrators sometimes argued that they paid men higher salaries because men were expected to support families, but the study found that men were paid more than women whether they had children or not. Women were less likely to be asked to speak or consult outside their institutions, to write a review or a chapter or serve on an editorial board—all signs of professional respect.

The study belied the bold promises a decade earlier to women who had hoped to combine family and career. The married women with doctorates reported the most dissatisfaction of anyone in the survey. They were more likely than their married male peers to have been discouraged from pursuing advanced degrees, less likely to have role models, and more likely to mention "bias." While most men in microbiology were married, less than half the women were. Most of the women had no children, but the opposite was true for men. Most women said they could move only if their husbands found good jobs; most of the men said they would move regardless of whether their wives found a job they liked. Not surprisingly, women were twice as likely as men to say that their life and career had not lived up to what they envisioned when they finished their training.

"If women were to receive continued encouragement, scientific contact, and professional recognition at each stage of their professional lives, they would undoubtedly become more visible," Alice and her coauthors wrote. "The lack of encouragement and self-confidence leading to isolation, which then leads to lack of recognition, is a vicious circle that must be broken for the woman professional."

The authors concluded, "This study of a select group of scientists probably has general applicability to all women professionals in their role vis a vis men."

The results did resonate, as more women began speaking up in a slew of new reports on "the status of women" at leading universities. At Harvard, which had no women among its full professors, a survey by graduate students in 1971 found that the majority of Harvard faculty believed that women had to choose between an academic career or marriage and family. "As long as such a choice is required, academic women will be, and will be regarded as, 'exceptional,'" the report concluded, "a tag that few of us, male or female, can carry for long, and that inevitably means discouragement and frustration for married women, and even for single women, attempting to pursue careers." By the middle of the decade, many scientific associations had done their own reports; one written in 1976 counted seventeen others that had been conducted in the previous four years.

That 1976 study came out of a meeting of thirty women—Black, Puerto Rican, Native and Mexican American—at an estate turned conference center in the horse-country town of Warrenton, Virginia, in December 1975. These were women who found no other harbor: organizations and programs to improve opportunities for women tended to include mostly white women, while those for minorities were overwhelmingly male. The women were, as one of the organizers wrote, "the most underrepresented and probably overselected group in the scientific disciplines." Their numbers were infinitesimal. The National Science Foundation counted about 245,000 scientists with doctorates in 1973. Of those, 1,611 were Black men and 249 were Black women; 106 were American Indian and just 3 of those were women. Black women were 6 percent of the population and one-tenth of 1 percent of the scientists. White men, meanwhile, were 41.5 percent of the population and 90 percent of the scientists.

The meeting produced a landmark report titled *The Double Bind: The Price of Being a Minority Woman in Science*. It echoed some problems white women had reported. They had trouble getting grants for research, were not invited to be on study committees to evaluate journal articles or grant proposals, or were given dead-end jobs because, they were told, "the men won't work for you." While men were seen as having innate abilities for jobs in management, women had to show they had extra credentials and were expected to plan receptions and do other tasks traditionally done by faculty wives or secretaries. One woman recounted being asked to relinquish lab space to a newly hired man. Another told how she and her husband had interviewed at the same universities; her husband was offered more senior jobs even though she had more publications and more significant research. The women felt conflicted about their choices to marry or pursue a career. They tended to credit luck over determination or hard work for much of their accomplishments.

But the women of color also had to overcome educational disadvantages built up over generations. Their secondary schools had lacked labs, advanced math classes, and guidance counselors. Often a sympathetic teacher had urged them on, but others had suffered teachers with "a patronizing, missionary attitude." Men "frequently showed surprise" that women of color "were able to perform well." Being doubly unusual in the scientific establishment, the women had been subject to excessively insulting treatment at conferences: "One young black scientist told of having been questioned by hotel detectives who had assumed she was a prostitute because she was going in and out of rooms for meetings and receptions, as were her colleagues who just happened to be white males."

In situations like that, "it becomes difficult if not impossible to determine which 'ism' is in force," the report noted. Although, it added, "In such a case, it does not matter whether one is being hit with the club of sexism or racism—they both hurt. And this is the nature and essence of the double bind."

The women felt additional isolation within their own groups: the Mexican American women felt that Black women had an advantage because Black people had been organizing for longer against discrimination; Native American women complained that programs to increase opportunity were all aimed at urban areas. The conference had not included Asian American women; they were not considered "minority" because they were represented among doctoral scientists at the same level they were in the population—under half a percent in each. Still, the report noted that Asian women in science suffered an "appallingly high" rate of involuntary unemployment. Twenty-three percent of Asian women with doctorates in chemistry and 21 percent of those with doctorates in physics said they could not find jobs.

Like many white women, the women gathered in Warrenton spoke up only reluctantly. They did not necessarily consider themselves feminists, and they did not want "to concede their own specialness," as the report said. One woman said she let no one know she had even attended the meeting.

———————

Nancy felt awkward talking to Alice, given that one of Nancy's main problems was Alice's husband.

Alice wanted to help. She was four years older than Nancy and realized how much she herself had been helped by other women, from the secretaries at the medical school to the women in Zella's group. She was sympathetic to the problems of competition and agreed with Salva that it was best for Nancy not to talk to anyone; Alice told her she didn't talk to Baltimore about her own work. She tried to warn Nancy that women inevitably faced discrimination as they came up for promotions or more powerful positions. "It can happen to anybody," Alice told her. "No matter how good you are."

Alice doubted Nancy heard her. She thought Nancy a bit full of herself, though she understood why: she had seen the letters when Nancy was being considered for the position at Harvard Medical School and knew she came highly recommended. Alice had run into other women she considered "queen bees," older women who believed that they had made it without help, and therefore others could, too. She figured that Nancy—with her job at a prestigious institution, Jim as her mentor, and no children or marriage to tie her down—was on her way to becoming a queen bee.

Nancy had heard Alice. She just hadn't decided what to make of the advice.

Nancy could see that life in the lab was different for women, especially single women. At a reception soon after she started at MIT, a senior professor in Biology, Gobind Khorana, told her that his wife stood by the door each morning to hand him money to buy his lunch. How could he concentrate on his science, he asked Nancy, if he had to worry about remembering his lunch money? (Khorana liked to retreat to a rented cottage to write without the distraction of a telephone or radio, which meant that his wife also had to drive an hour to relay the news that he had won the Nobel Prize.)

One evening Nancy waited to type a manuscript. Her secretary—the one she shared with Phil Sharp, Nancy's friend from the Cold Spring Harbor days—was rushing to finish up some typing for him and apologized for making Nancy wait: "You deserve better than this, Nancy."

Seeing an opening, Nancy asked the secretary why she did Phil's work first.

"Because he's a man. It's just the way it is." The secretary showed Nancy a note Phil's wife had written, referring to him as a Superman. "You need a wife like that, Nancy."

Nancy didn't disagree that this was the way life was for women. She had grown up in a world where women were not supposed to be professionals. When she'd started at Harvard, no women were on the faculty. That was discrimination: not being able to get a job. Nancy had the job. She didn't think discrimination described the problems she was having on the fifth floor. Taking materials, jockeying over experiments, rushing to be the first, that was competition, inherent to science. She had seen it with Mark and Wally Gilbert at Harvard, read about it in Jim's account in *The Double Helix*. It could ruin friendships—among men. She had always known she would have a hard time being aggressive enough to compete like that; she had been raised to be nice, polite. But she thought her hard work would speak for itself, would demand to be recognized. She couldn't say for sure that what she was facing on the fifth floor had anything to do with her being a woman; she was the only woman on the faculty in the building, so she had no control group.

She had made friends in Cambridge with a translator and children's book author who had been a few years ahead of her at Spence,

Sheila LaFarge. Out to dinner one evening, Sheila fixed on the idea that Nancy should write about being a woman in science for the *Radcliffe Quarterly*, the alumnae magazine. She introduced Nancy to a book editor who was on the *Quarterly*'s advisory board.

Nancy's essay appeared in the June 1976 *Quarterly*, in an issue devoted to what the cover called "liberated lifestyles," which included contributions from alumnae on a commuter marriage, "Why I Am a Redstocking," and commune living, by a member of the Boston women's collective that had published *Our Bodies, Ourselves* in its third edition that year. The variety suggested how many women were struggling to sort out what had been won and lost in liberation. A mother of five and grandmother of four from the class of 1943 worried about the "landslide" in birth rates among highly educated women. A Radcliffe senior lamented the persistence of the age-old inquiry "So when are you going to get married?"

Nancy's was a report on the status of her own thinking about being a woman in science. She was thirty-three and had now been working on the fifth floor for three years, and she had landed on two explanations for why science was so hard for women: children and competition. The headline was "The High Price of Success in Science: A Woman Scientist Disputes the Notion That a Woman Can Be a Successful Wife and Mother as Well as a Successful Scientist."

She was responding, she wrote, to the "flood of women's literature, much of which I find to be very unrealistic." And she was disinclined to complain: "I do not mean to imply that being a scientist is harder than any other job—for example, lawyer, doctor, coal miner, politician. I think most of my colleagues think that research is to some extent a luxury—and a privilege. I merely want to describe the way it is."

A scientific career proceeded along two parallel tracks, she wrote: real science and professional science. Real science was doing experiments, making discoveries, the work she'd fallen in love with back in the Watson lab, the long and winding conversations about science along Bungtown Road. "A unique kind of exhilaration and excitement," as she described it. "What is known is useful and sometimes beautiful, but only what is unknown is of interest. It is this open-minded quality of science—the excitement of the search, the constant, but usually unsatisfied, striving toward the solution of some very difficult problem—that is addictive."

Professional science was figuring out how to get paid for making discoveries: writing grants to buy equipment and pay grad students and postdocs, publishing papers and speaking at scientific conferences to get "exposure" in the field, winning tenure, teaching, managing a lab. It was a job "constructed by and for men (a certain type of man)," she wrote. Someone with a wife at home to manage life outside the lab, and with supreme self-confidence.

Real science and professional science were each a full-time job, with intense pressures and long hours. And because discovery did not arrive according to any schedule, those hours were unpredictable. That made children—a third full-time job—nearly impossible. It was no wonder so many women dropped out early in their careers. "There are exceptions," she wrote, women who managed to have children and a career as a professor, "but the chance circumstances that surround them are sufficiently unusual and, most of all, too unprogrammable and unpredictable to be taken as general solutions."

Nancy expressed sympathy for men, who faced the stress of professional science and providing for a family. Those pressures, she wrote, fostered an attitude of "success at any price" and led some people to "pinch" results, she wrote. "Ninety-five percent or more of the scientists I've known have at times been distressed to the point of incapacitation by feeling that their work has been usurped." She had yet to see a woman accused of this behavior, maybe because there were so few of them in the lab. But Nancy thought it was more because women could not compete like that: like her, they were bred to be nice. "A woman who chooses to conquer this world at its higher echelons usually requires a major overhaul of self and world views."

And society expected women to have children. A career in science, Nancy wrote, "requires attitudes and behavior that are alien to most women's upbringing at present." Those who chose to remain single and have a career were "placing themselves out on some totally undefined social limb."

While she described a system that implicitly discriminated, she mentioned discrimination only in passing—so much had already been written, she wrote, "that I will say nothing about it except that though the problem is getting better, it still needs to get a lot better." She didn't think discrimination explained what she was seeing in the lab. She mentioned it because she knew so many women were talking about it. She thought she had to.

She trained her anger on the promises that had been made to women of her generation: "the Mary-Bunting-superwoman-brainwash (successful career woman, mother of several children, husband optional)." Nancy had been shocked, she wrote, at how many of her Radcliffe classmates at their tenth reunion had deemed themselves failures because they had not lived up to Bunting's "myth." One had described Bunting's exhortations about combining work and family— with little or no practical discussion of who would watch the children— as "the most pernicious influence in Western civilization." Still, Nancy wrote, "the superwoman-soft-sell of the 60s cannot compare with the disregard for reality and the unreasonable expectations being raised by the bionic-woman-hard-sell of the 70s."

Some of the other essays in the *Quarterly* argued for solutions: the development of quality childcare, encouraging men to take a more equal role in raising children and to think of the work-family balance as their challenge, too. Nancy's essay offered only hard-won advice to women who wanted to pursue dreams of science: "If they are to be successful, there are personal sacrifices they must be prepared to make."

The article prompted outraged letters in the next *Quarterly* from women who complained that Nancy had, as one physicist wrote, "done a disservice to women and science." Another letter—from Ursula Goodenough, a junior professor in biology at Harvard and a friend of Alice Huang's from Zella Luria's after-dinner women's group—noted that the biographical blurb that accompanied Nancy's article defined her in terms of the three Nobel Prize winners who had hired her: Jim Watson, Salva Luria, and David Baltimore. Goodenough insisted that the article described only the life of molecular biologists who were "vaguely amusing" in their cultish self-regard. "I would hope that any readers of Nancy's article would bear in mind, before handing it over to putative women scientists for consideration, that Nancy is describing an extremely rarified and competitive scientific field," Goodenough wrote, "and that neither 'real' nor 'professional' science *has* to be like that, even at those prestige academic institutions she mentions." (Goodenough did not get tenure at Harvard; she moved to Washington University in St. Louis and also raised five children.)

The criticism stung. But another biologist at Harvard—Nancy Kleckner, who had been four years behind Nancy at Radcliffe and a postdoc of Botstein's at MIT—declared the article "terrific," saying Nancy was "absolutely right in all counts. Her having written it all down so well will save my having to do it sometime in the future." (Kleckner became the first woman tenured in biochemistry at Harvard—only the fifth woman tenured in the sciences—in 1984.)

Nancy remained proud enough of her assessment that she sent the article to Jim and to Barbara McClintock. Barbara wrote back that it was "a truly brilliant exposition of the present situation." Men, Barbara wrote, had been recommending it to her. "Your description of the nature of real and professional science hits them straight on with the agony of recognition showing in their expressions."

"As for women, God help them," Barbara added. "Their rewards must depend upon pleasurable mental activities overpowering cultural patterns of put-down. Successful competition with men is just out of the question. This applies even when the woman is intellectually superior to most of her colleagues. In short, if you can't have fun stay out of the laboratory!"

Nancy still believed she could compete with the men. And she was having fun, at least with the "real" science part. She was studying host range, trying to determine what causes a virus to grow better on one cell type than another, which determined how quickly or whether the virus causes cancer. Host range was typically determined by how the proteins on the surface of the virus fit those on the surface of the cell. Nancy had discovered that it could also be determined by a protein deep inside the virus particles, known as P30. This surprising discovery led to an advance in her professional science, too—and just in time. Soon after the *Radcliffe Quarterly* article, she ran into Boris Magasanik, the chairman of Biology, in the large courtyard outside the cancer center. Boris had insisted earlier that year that Nancy would not be considered for tenure because David Baltimore didn't like her, but now he looked happy to see her, even relieved. He smiled and put his arm around her in a half embrace. The letters the department had solicited for her interim review were very good, Boris told Nancy, patting her arm: "Maybe everything will be all right after all."

She was promoted to the next level, associate professor without tenure, effective July 1977. It coincided with the appointment of a new chairman of Biology, Gene Brown. That fall he appeared in the teaching lab where Nancy was working with undergraduates and told her that if her letters the next year were strong, he would see that she got tenure, or he would resign.

These were good men, Nancy thought. They would take care of her. They would be fair.

Mary-Lou earned tenure the next year. It was no surprise: she was well-known nationally, on the boards of four different journals, and on a study section to evaluate grant applications at the National Institutes of Health. Still, when the news echoed back to the Kline Biology Tower at Yale, Joe Gall came running into his lab waving a letter in front of a young Australian postdoc named Elizabeth Blackburn: "Mary-Lou Pardue has gotten tenure at MIT!" Working in Gall's lab, Blackburn had not yet had to think about the problems for women in science, and she had only just begun the work on the sticky ends of chromosomes known as telomeres for which she would later win the Nobel Prize. But from the jubilation and relief in Gall's voice, she understood that getting tenure at a place as competitive as MIT was a big deal, and especially for a woman.

Nancy took heart from it, too. She saw it as proof that the world worked the way it was supposed to: Mary-Lou had done excellent science and had been rewarded.

The situation on the fifth floor was no easier, but Nancy continued to ascribe any difficulties to the pressure, and competition. As new
problems arose, she explained them by their circumstances.

In the fall of 1977, one of Phil Sharp's postdocs, Sue Berget, came to
see Nancy in her office when Phil was away. Phil and Sue had recently
published a landmark result for the cancer center: the discovery of
so-called split genes. Until then, molecular biologists had understood
the DNA within genes to be a long, continuous stretch that copied its
code onto equally long stretches of messenger RNA, which then made
proteins. This was the way it worked in bacteria, so it was assumed
that this was the way it would work for all organisms. Looking at elec-
tron micrographs of the genes of adenovirus—the virus that causes
the common cold—Sue had seen loose bits of RNA flapping at one
end, unconnected to any DNA. From this she and Phil had designed
an experiment that found large segments of DNA that looped off as
messenger RNA was being made. This discovery of what would later
be called introns changed the fundamental understanding of where
and how genetic activity occurs.

Racing against researchers at Cold Spring Harbor, Phil and Sue
had operated in such secrecy that Sue had not told even her husband
when she and Phil confirmed their results. Both teams published their
results in August 1977, and the discovery was big enough that the haggling over credit began almost immediately. Cold Spring Harbor had
published four papers by ten authors, and already people were debating which names would be on the Nobel Prize.

Yet when Sue went on the job market that fall, she received no offers.
She was at a loss to explain; she had given seminars and had interviews
at Harvard, Stanford, Columbia, and the University of California at
Berkeley and Los Angeles and thought they went well. A friend had
offered to make a call to someone he knew at one of the schools and
reported back to Sue that the school thought Phil's letter of recommendation was weak. Phil had praised the work Sue did before split genes
but mentioned nothing about her work on their big discovery. The
effect was lethal, suggesting she had been no more than a technician.

Sue understood the sensitivities around credit; Phil didn't have tenure yet. But she also believed her work had earned her a fulsome recommendation. Another postdoc, a man, was out on the job market at the same time, and she guessed that Phil favored him over her. She needed advice, and maybe help, and she came to see Nancy because Nancy was the only woman on the faculty in the cancer center.

Nancy was alarmed, for Sue but also for Phil. She and Phil had been friends since their days working together at Cold Spring Harbor, talking science on the steps of the cottage he'd shared with his wife and their infant daughter. Nancy and her friend David Botstein went to see Phil and told him that shorting Sue on credit would make him look bad. He was going to win the Nobel Prize, they told him, and he did not want to look petty.

By his own recollection, Phil had been raised with traditional attitudes about women. He grew up on a small family farm nestled on a bend of the Licking River in the hills of northern Kentucky, where his father gave him a plot of tobacco to work as a way to earn college tuition. His sisters got no land and did not go to college. He'd stayed close to home, graduating from Union College, a United Methodist college in Barbourville, because any place bigger would have intimidated him. He'd gone on to graduate school in chemistry, but in 1968 had read the record of the Cold Spring Harbor symposium on the genetic code and, like Nancy, been captivated by the study of DNA.

Phil was normally full of energy, inquisitive and chatty, but in his office with Nancy and Botstein he said little, mostly listened. Later, Sue worked up the courage to see him herself, and he showed her the letter he had written. He offered to write a stronger letter, and Sue got job offers at Rice and Carnegie Mellon, excellent institutions but not as prominent as the places where she had initially interviewed. She took the one at Rice.

Nancy had heard nothing further from Phil or Sue and didn't think much more about it. She believed Sue deserved better. It was unusual and even unheard of that a postdoc who had played such a major role in such an important discovery would not get a job at another top institution. But she did not know Sue as well as she knew Phil, and she assumed Phil, for whatever reason, did not think highly of her. Nancy did not think—at least, not then—that it could have anything to do with Sue's being a woman.

Nancy was increasingly preoccupied with her tenure case, which came up the next year. To prepare any case, Biology typically solicited more than a dozen letters from the most prominent people in the field, outside MIT, testifying that the candidate had made significant contributions and would continue to. And as Jim predicted, the letters for Nancy came in strong—a friend already tenured told her they were the strongest letters the department had seen in ten years. Nancy had published seven papers the previous year, which was a lot—at the women's lunches, Millie and Mary Rowe told tenure-track candidates that they had to publish four a year.

But getting tenure still required a series of cage matches. The tenure committee in the department recommended candidates and solicited the outside letters. Then it sent names to the department for a vote, and the department sent the names it approved to the Science Council, which was made up of all the heads of departments and centers in the School of Science and the school's dean. The Science Council voted and sent its recommendation for the final vote by the Academic Council—the university's president and other top officials. Science Council debates were known to be savage; each chairman wanted to prove his department top dog and regularly challenged and even derided the nominations from other departments. Biology prided itself on never losing a case at Science Council. So the debates within the department were fierce, and the tenure committee worked hard to build a failproof case.

Mary-Lou led the tenure committee that year. Two other people were up with Nancy: Paul Gottlieb, an immunologist, and Bob Weinberg, who had done his undergraduate and doctoral degrees at MIT and worked as a research associate in Baltimore's lab for a year before starting on the faculty at the cancer center. Weinberg and Nancy were both studying cancer viruses in mice. A few years later, Weinberg would discover the first human oncogene. But in the fall of 1978, Nancy's work was farther along—she had mapped the genes of mouse retroviruses and identified those that allowed the viruses to grow especially well on certain types of cells, explaining novel mechanisms for the different host ranges of viruses. The outside letters evaluating Weinberg were strong, but Mary-Lou could see that the ones for Nancy were stronger.

Typically, the committee would write a letter for each candidate to send to the Science Council, but in this case they decided to write one letter nominating all three. The trick, as Mary-Lou saw it, was to capture the extraordinary praise for Nancy's work in a way that did not make Weinberg's look lesser by comparison. The committee worked the language carefully, with one paragraph describing each candidate. They ordered the paragraphs according to who they thought most deserved tenure: Nancy first, Bob second, Paul third.

Mary-Lou returned from a trip shortly after the committee had finished its work to find a final copy of the letter on her desk. It was the same letter, except the order of the paragraphs had been changed. This new letter recommended Bob first and Nancy second. Mary-Lou was furious, but she didn't know where to direct her anger. It seemed so petty, who would do that? She asked Maury Fox, who had helped recruit her to the department in 1972. He sympathized, vaguely. But he didn't say much or know who had changed it.

Mary-Lou went to Mary Rowe, the president's adviser on women and work. Mary had already heard about the letter in Biology. Often, the women who came to complain to her were secretaries—they saw a lot—and in this case, David Baltimore's secretary had come to complain about the letter. Baltimore had asked her to change the order of the names, and the secretary thought it was wrong, but did not want to confront him.

Mary Rowe and Mary-Lou both liked the secretary and didn't want to get her in trouble. They knew that they would have a fight if they asked Baltimore to change the names back. So they, too, decided not to confront him.

The debate at the department meeting ran long. It ended with a decision to recommend both Nancy and Weinberg, but not Paul Gottlieb. Mary-Lou was more upset about the letter than about losing a candidate. She had been in enough meetings evaluating tenure cases and grant applications to know that women were judged more harshly, they had to rise to a higher standard. Paul could get a job anywhere, but if the Science Council didn't give Nancy tenure, she'd be forever marked by her rejection from MIT.

David Botstein came to Nancy's office following the debate and reported that Baltimore had been open about his opposition to her: "She's the smartest person we've hired," Botstein recalled him saying, "but I don't want her here."

"I've never seen him so irrational," Botstein told Nancy. But Nancy was thrilled. David Baltimore thought she was smart! She knew he didn't like her. He could have killed her case by saying she was not smart enough. Instead, he had done the right thing.

She did not hear about the letter the tenure committee had written—it would be nearly four decades before she did. For now, she knew nothing more than what Botstein had told her, and soon the Science Council voted to give her tenure. To Nancy, it was more proof: her hard work had paid off.

The morning after the Academic Council officially voted to give her tenure, she was sitting in her office when she looked up and saw David Baltimore leaning through the frame of the door and smiling. He was the first person to congratulate her.

Chapter 12

Kendall Square

MIT had moved to Cambridge in the early twentieth century to be closer to the industries that had built up around the canals and railroads near the Charles River, the companies that made meat, furniture, machine parts, or candy and shipped them off through Boston Harbor and the Atlantic Ocean to the world. Depending on the day and the breeze, the air on campus would swell with the stench of animal renderings or fragrance of Charleston Chews and Squirrel Nuts.

But by the 1960s, the migration of manufacturing toward cheaper labor in the South and other countries had especially desolated once-teeming East Cambridge; a dozen companies left each year. City officials hoped NASA would build Mission Control to replace what they considered the "urban blight" of Kendall Square. When Mission Control went to Houston instead, the Cambridge Redevelopment Authority razed more than fifty buildings to create the space agency's Electronics Research Center. Only about half the new buildings were finished when the federal government closed the center in 1969. Cantabrigians would debate endlessly whether the development had fallen victim to NASA budget cuts or Nixon's revenge on the Kennedys. Whatever the truth, by the time the MIT cancer center moved into the former chocolate factory in 1973, the neighborhood around it looked less like a university or factory town and more like the cratered surface of the moon. It was mostly parking lots. The F&T Diner persisted, as did the stifling smell of boiling rubber from the old Boston Woven Hose plant across Main Street, but the crowds had gone from the Kendall Square T stop. It became unsafe at night; after others were mugged, Nancy began driving to work instead.

Still, while it was the ugly edge of campus, by 1980 the work being done in the cancer center was becoming more central to MIT—the new MIT that was looking for fresh sources of money to compensate for a decade of dwindling federal research dollars. The growing importance of biology would change the university and the city. In time, it would also alter the course of Nancy's career.

The shift had started with an epic town-gown fight, the kind of spectacle endemic to university towns but especially Cambridge, where the rare citizen without an opinion about the Red Sox could be counted on to offer one about the politics of the presidential race, rent control, or plans to redirect traffic in Harvard Square. (It was a native son and local hero, House Speaker Tip O'Neill, who taught the rest of the country that all politics is local.) The city pressed working-class neighborhoods up against old money and the ivory tower; well-heeled college students next to multigeneration families in triple-deckers; one hundred thousand people and an abundance of class resentment into less than seven square miles.

This town-gown fight was about recombinant DNA. By the early 1970s, discoveries about the genetic structure of bacterial and animal cells had progressed so much that biochemists could manipulate fragments of DNA from different organisms—taking a bit from bacteria and their viruses and inserting it into DNA from a monkey virus—to create wholly new organisms, or recombinants, that did not exist in nature. The new technologies offered enormous promise for curing diseases caused by defective or missing genes. They also exposed previously unimaginable dangers and ethical concerns: that someone by design or by accident would create and unleash new organisms that were toxic or cancerous or drug resistant.

In 1974, a group of influential scientists had met in David Baltimore's office at the cancer center and agreed on a two-year moratorium on most recombinant DNA work in the United States, to allow them time to develop rules that tried to anticipate all the ways the new technology might be used or misused. In heated discussions over four days at the Asilomar Conference Center on California's Monterey Bay in 1976, 150 scientists, lawyers, and government health officials hammered out standards establishing four levels of containment for labs working with recombinant DNA—levels P1 to P4—aimed at ensuring that no dangerous pathogens could escape.

The National Institutes of Health were close to issuing final approval of those standards in 1976 when Mark Ptashne, Nancy's former boss, filed a proposal to build a P3 containment lab on the third floor of the Harvard Biolabs. He hit up against the city's mayor, a colorful populist named Alfred E. Vellucci.

Vellucci had grown up in East Cambridge and dropped out of school in sixth grade to deliver telegrams to Harvard students on his bicycle; his first job following his driver's license was selling Milky Way bars from a truck, and by the time he ran for his first office on the school committee, he was managing a sandwich shop. He had a local's dropped *r*'s and antipathy toward universities, which didn't pay taxes and were expected to deliver something in return—charitable payments, community services, jobs for residents—but never did enough. "That guy John Harvard never did anything for Cambridge except give the city six lousy books on Protestant theology," the mayor once said. As a city councilman in the 1960s, he had suggested using federal urban renewal funds to raze the Harvard houses along the Charles and proposed placing a parking garage under Harvard Yard, and he briefly convinced the council to rename the area around it Christopher Columbus Square. He reserved special animus for the editors inside the goofily Gothic *Harvard Lampoon* building; he said it looked like a witch on a broomstick and attempted to pass a city ordinance to declare it a public urinal.

Mark's proposal had divided the Biolabs faculty before it divided the city. George Wald, a Nobel laureate and an organizer of the March 4, 1969, antiwar protests at MIT, argued that recombinant DNA experiments should be done in deserts, like the Manhattan Project, or at least not in a densely populated city. Wald encouraged Vellucci, who had never heard of DNA and did not want to be pushed around by, as he said, "those people in white coats."

The mayor ordered a public hearing in June 1976, shortly before the nation celebrated its Bicentennial. While Mark's proposal was for Harvard, any action the city might take would more acutely affect MIT, and specifically the cancer center, because it had already built its labs. City Hall was halfway between MIT and Harvard along Massachusetts Avenue, and even the second-floor gallery of the council chambers was packed that night.

"Refrain from using the alphabet," Mayor Vellucci instructed Mark and the other scientists sitting at the hearing table across from the council on the dais. "Most of us in this room including myself are lay-people. We don't understand your alphabet. So you will spell it out for us so we know what you're talking about."

Scientists testified to the potential of recombinant DNA to cure cancer; city council members worried it could unleash contagion into the sewers. One had recently seen the movie *The Andromeda Strain*, based on the Michael Crichton thriller about a mutating microbe kill-ing off an entire town. Vellucci worried aloud about the new science creating incurable diseases or real-life Frankensteins. He proposed extending the moratorium on DNA work another two years.

"This is a deadly serious matter, sir," he said, peering through black thick-framed glasses at Mark. Then, belatedly noticing Maxine Singer, a biochemist from the National Institutes of Health who was sitting next to Mark at the table, he added, "Ma'am."

The council imposed a three-month moratorium instead and established a Cambridge Experimentation Review Board to consider whether the city should allow recombinant DNA work in the labs. The board—nine citizens including a nun and a nurse—met twice weekly for the rest of the year. Phil Sharp toured them around the labs on the fifth floor of the cancer center to show the safety precautions set up after the Asilomar agreement. The board conducted a mock trial on recombinant DNA with biologists as their witnesses, and scientists on either side of the debate set up booths at street fairs to try to sway the public. By early 1977, what might have resulted in a Not in My Backyard stalemate had produced instead the nation's first municipal bio-safety ordinance, modeled largely on the National Institutes of Health guidelines.

Recombinant DNA would quickly give rise to an entire new indus-try. There was no biotechnology industry to speak of in 1977, but the technology for the first time had allowed scientists to mass-produce proteins that could treat or cure diseases once considered to be death sentences. Scientists and venture capitalists had started Genentech in San Francisco in 1976 using technology licensed from the University of California and Stanford and had by 1978 used recombinant DNA to produce insulin and human growth hormone. That same year, two venture capitalists —one had graduated from MIT—approached Phil Sharp and Wally Gilbert, still at Harvard, with a proposal to form the company they named Biogen. They set up in Kendall Square and started by developing protein antigens and alpha interferon to fight hepatitis infections.

Biologists were snobbish toward industry; the best minds had always aspired to academia. Some predicted the industry would never grow beyond the two companies, one on each coast. In fact, a prolifer-ation of companies divided and replicated, and as they did, biologists shed their snobbery, and Kendall Square would become the biotech-nology capital of the world. It offered land that was lonely for develop-ers, and empty warehouses with high ceilings and strong floors, perfect for installing the heavy equipment required for labs and manufactur-ing facilities. And it had the regulations that had been hammered out in the summer and fall of 1976, as Biogen noted in its letter applying for the first municipal license to do recombinant DNA. Everywhere else the process was uncertain; companies would have to convince municipal officials that they were not poisoning the water system, and the public would likely resist. In Cambridge, everyone already knew the rules of the game. Mark Ptashne encountered resistance even a few miles away when he and colleagues looked for headquarters for their new company, Genetics Institute, in Boston and then in Somerville. It ended up in Cambridge. At the ribbon cutting ceremony for Bio-gen headquarters on Binney Street in 1982, the industry was already a small cluster, and Mayor Vellucci declared he had no fear of recombinant DNA as long as it paid taxes.

Universities, and especially MIT, would become deeply entwined with the new industry. MIT had entered the 1980s under the pressures of double-digit inflation and with a new president, Paul Gray, warning that the decade would be one of "diminished affluence and falling expectations." Students were flocking to subjects that seemed to guarantee more lucrative career prospects; enrollment in the Department of Electrical Engineering and Computer Science had doubled in less than ten years, with one in every three undergraduates declaring it their major. A steady decline in federal research dollars was putting what Gray called "unrelenting pressure" on faculty members to apply for more grants, and on the university to get more money from private industry. By 1982, industry sponsorship accounted for $20 million in MIT's research budget—10 percent of the total, and triple what it had been five years earlier.

With Exxon now paying for combustion research at MIT and DuPont for molecular genetics at Harvard Medical School, faculty members worried about conflicts of interest, that companies and scientists would be less inclined to tackle big questions and more inclined toward developing products that could make money. President Gray, an avuncular engineer who spent his entire career at MIT, played down the concerns. MIT was founded to help power the nation's first industrial revolution, now it would help drive a second one, which would be led by information technology and biotechnology.

By 1982, another new building was rising across from the cancer center on Main Street. Built on land once designated for NASA, it was the dream of Jack Whitehead, who had made his fortune selling bioanalytical equipment and approached David Baltimore with the idea of creating the "Taj Mahal of biomedical research." MIT proclaimed that the new Whitehead Institute would change the face of life sciences, although in the Biology department not everyone agreed this was a good thing. Whitehead contributed $135 million to build and fund his institute—a sum greater than the entire endowment of Brown University—and it was to be self-governing (with Whitehead and his children serving on the board). It would hire faculty and own the patents on any discoveries made there but split the revenues with the university.

President Gray, facing a budget deficit, noted that Whitehead's gift would pay for research the university could otherwise not afford. "MIT is not a stranger to inventing new ways to support intellectual and educational objectives," he wrote in his report to the campus that September. Even before ground had been broken on the Whitehead site, MIT and other institutions had successfully lobbied Congress to pass the Bayh-Dole Act, allowing universities to patent and profit from discoveries made in federally funded research. Technology had created the new entrepreneurial university.

———————————

As more biotechnology companies sprouted, Nancy noted that she couldn't think of any women involved. Phil had just started Biogen, and he told her that investors—businessmen—would not work with women. Nancy and Phil still had offices side by side, she considered him a good friend, and she knew he was right. She recognized that many men still couldn't see women as professionals. She wasn't surprised to be excluded—she understood the industry was off-limits to women. She accepted it as the way things were.

By some appearances, women had made revolutionary advances. In 1979, they made up half of all college students in the United States for the first time, a proportion that would continue to grow. On television, an advertisement for Enjoli, "the eight-hour perfume for the twenty-four-hour woman," featured a glossy blonde boasting that she could "bring home the bacon, fry it up in a pan," suggesting that the suburban despair Betty Friedan had described fifteen years earlier could be cured by giving mothers briefcases and commutes; with the right fragrance they'd still find time to fold the laundry, make the breakfast, dole out kisses to the kids, and deliver in the bedroom.

Left unmentioned was that the "bacon" was still sliced thin: women were earning sixty-two cents for every dollar earned by men in similar positions, two pennies more than they took home before the Equal Pay Act was signed in 1963. The same year as the Enjoli ad, 1979, the Equal Rights Amendment was declared dead, after it fell three states short of ratification.

In 1980 Jane Fonda produced and starred in the hit movie *9 to 5*, based on stories she'd heard from an association of female office workers in Boston. The movie was a comedy and a fantasy: the working women on-screen strung up the boss who had stolen their ideas and their dignity, then exacted their revenge in raises, promotions, job sharing, and on-site day care centers, innovations that were still rare for real women. The next year, when Sandra Day O'Connor became the first female Supreme Court justice in US history, it shocked the existing order so much that nine out of ten American televisions tuned in to coverage of her confirmation hearings.

MIT had elected Sheila Widnall as the first woman to chair the faculty in 1979. That year, David Botstein asked if Nancy wanted to teach the undergraduate course on genetics with him. This large lecture class was known as tough to teach: the material was dense and not intuitive, and students frequently took out their frustrations on the professors in scathing teaching evaluations. Some of the most prominent members of the department had tried to teach it and given up. But Nancy was known as a good lecturer who could distill complicated concepts. Genetics was a required course for biology students, so there was stature in teaching it. She said yes immediately.

Botstein said he would ask the approval of Gene Brown, the chairman of Biology, then returned a few days later to tell her that Gene had rejected the idea; he agreed that Nancy was an excellent teacher but said that undergraduates would not trust scientific information delivered by a woman. It was one thing to teach a lab course, an informal setting where she was helping students one-on-one interpret their results. In a large lecture they would not take her seriously. They'd challenge her authority, maybe try to humiliate her.

Again, Nancy accepted it was the way things were. Mentally, she thanked Gene for sparing her embarrassment, a debacle in front of the students; she knew how much he cared about good teaching, and he'd always looked out for her. She thought he was right about how the students perceived women. There were still few of them on the faculty.

Nancy had continued to feel the isolation and the stress of the fifth floor. But she thought she'd found a solution: a lab space of her own. Finding it wasn't simple. Gene had offered her a lab being vacated by a retiring faculty member in the Biology building, across the street from the cancer center, but no one in that building was doing virology, so Nancy would have to raise grants to buy a lab's worth of specialized equipment, and she wasn't sure she could do that single-handedly. Then he suggested she move to the third floor of the cancer center. Salva thought it was a bad idea: the floor had been created as teaching labs, so hers would be the only research lab, and she and her students would have to go to other floors to use the equipment they needed. Phil warned her she was isolating herself even more. But Nancy was finding it increasingly difficult to concentrate on her science when she was always looking over her shoulder, worried about who was try- ing to take her cells, her equipment, her results. She'd begun getting migraines. She wanted out. She decided to move to the third floor.

"Perhaps it is premature to say, but it is just possible that a small miracle is occurring here," she wrote Jim Watson. "I am happy to report that for the first time I feel the potential loss of a crushing 5 year burden." As she stepped off the elevator onto the third floor the first day in her new lab, she could feel every muscle in her body soften, a feeling of lightness return.

She could not entirely escape the tensions of the cancer center. In 1980, Gene confessed that the department had voted to hire Susumu Tonegawa, an immunologist, and that Nancy hadn't been invited to the meeting, even though Tonegawa would be the only other faculty mem- ber on the third floor. Salva was having trouble convincing David Bal- timore to hire Tonegawa and worried that if Nancy was in the meeting, Baltimore would never go along. Nancy burst into tears when Gene told her, the only time she had ever cried in front of anyone at work. When she had won tenure, a colleague had said, "Of course you got tenure, everyone likes you," as if tenure were a matter of being likable, as if all the back-and-forth about her supposed lack of collegiality had never been an issue. Now she was reminded all over again that the most important person in her field did not like her, so much that he could not tolerate her presence in a meeting, and the whole department knew. It was mortifying.

Still, in her new space she could get back to science the way she thought it should be: real science, the flutter of curiosity, the thrill of figuring things out. She was continuing the work she had done on the fifth floor, mapping the genes of retroviruses that caused cancer in mice. For all her troubles in the building, she had developed a wide circle of colleagues and collaborators in her field outside MIT, at the National Institutes of Health and Sloan Kettering. One, Ed Scolnick, became a close friend after his wife, a Radcliffe graduate, read Nancy's essay in the *Quarterly* and recommended it to him. They were both studying tumor viruses—Ed at the National Institutes of Health— and he was interested in getting more women into science, so invited Nancy to come speak about her work. Two renowned virologists at the National Institutes, Wally Rowe and Janet Hartley, sought her out to study the interactions between viruses and host cells. They had a lab full of mice, which allowed Nancy to extend her work in the animals without having to pay for them or keep them.

It was now widely accepted that different types of viruses could cause different types of cancer. The mystery scientists were trying to figure out was why, and how. Much depended on how well and how fast the viruses grew.

Work on chicken retroviruses had suggested that the viruses turned a normal gene into a cancer-causing one by essentially moving in next door and inserting their own genes, turning a proto-oncogene into an oncogene that then causes cells to become cancerous. The key seemed to be in what were known as promoters—on-off signals that live next to genes and control their expression.

But by the early 1980s scientists working with monkey viruses had discovered another level of complexity: so-called enhancers, which were sequences of DNA that determined the level of gene expression— whether the signal was weak or strong—and could be situated far from the gene they activated. That raised the possibility that an enhancer might be responsible for that overexpression of the normal gene.

Nancy ventured further: Could the enhancer be responsible for determining which type of cells a virus grew best in and, therefore, what kind of cancer the virus caused? Wally Rowe guessed no, because this had not proven the case in experiments done on chickens. But they were working with viruses in mice, which were different and had more genes in common with humans. They decided to test Nancy's hypothesis.

A postdoc in Nancy's lab took two viruses that caused different types of leukemia and swapped out the ends—known as the long terminal repeat, or LTR, the sequence where promoters resided. In one they replaced the sequence in a virus that caused T-cell lymphoma with the ends of a virus that caused erythroleukemia, a red-blood-cell leukemia; in the other they did the opposite. Wally and Janet injected the new viruses into mice. The results confirmed Nancy's hunch: the different sequences resulted in two different cancers.

The experiment provided the first evidence that enhancers controlled more than simply whether a mouse retrovirus caused cancer; they could also determine what was known as tissue specificity.

Nancy thought of this, with the modesty befitting a Spence girl, as a "nice result": not only did the experiment prove her hunch, but all the steps had worked along the way, which didn't always happen. She worried over the paper announcing the results—she was always a worrier. Wally Rowe urged her to assert more boldly that this was a significant advance, and Phil helped come up with a new title that made that clearer. "We are now firmly out on a limb," she wrote to Jim. "Wally Gilbert said not to worry, now is the time for hand waving. Well, we are waving our hands and, I suspect, about to make grand fools of ourselves. (I do not sleep too well anymore.)"

The paper, with Nancy as senior author, was published in the *Proceedings of the National Academy of Sciences* in July 1983. Some months later, Phil stopped her in the hallway, his face screwed up in confusion: "I thought you discovered tissue specificity of the enhancer." He knew the work; the title he'd provided for the paper described the role of the enhancer in "disease specificity." He was just back from a talk by Bill Haseltine, a former postdoc of Baltimore's, now a junior faculty member at Harvard Medical School. "He said he discovered it."

Nancy had known Haseltine since his days as a graduate student of Jim Watson's in the Biolabs, and on the fifth floor. She thought it was a little odd that he had moved into the same approach her lab had been using in mouse viruses; typically researchers continued in the same line of work they had pursued as postdocs, and like many others in Baltimore's lab, Haseltine had worked on replication of retroviruses. It was his right to do the same work she was doing, maybe vaguely flattering. She was surprised by what Phil told her, but she didn't spend much time thinking about it. Maybe he had misunderstood.

Then she got a call from the office of George Khoury, the chief of molecular virology at the National Cancer Institute, letting her know that he would be in Boston and would like to see her. Khoury had been one of the first to discover enhancers and was considered a leader in the field. Nancy didn't know him, but knew he was a friend of Wally Rowe's, and Nancy presumed he wanted to talk science. Maybe share an interesting result, collaborate. She was excited to meet him.

Khoury was about her age, handsome and lively, with a thick mustache, and when he arrived in her office, his jacket and pants were rumpled, which Nancy took as the mark of a true scientist. He was friendly, but had a busy day scheduled, so got to his point quickly: he knew that Bill Haseltine was claiming credit for her discovery on the enhancer. It was unprofessional, Khoury was offended by it, and he wanted to tell Nancy how sorry he was.

Nancy barely knew what to say. She was stunned that he had made an appointment to say this, that he was so decent. She was struck that his decency was so rare. She knew science was competitive. Now that there was real money to be made, biology had become even more so; journal editors complained that authors waited to secure patents before publishing results. Nancy wanted other researchers to repeat and extend her results, it validated her and advanced science. She did discuss results with plenty of people in the field—they saw each other at Cold Spring Harbor or in the hotel bar at meetings, they picked up the phone to talk regularly. She had talked to some of Haseltine's postdocs about her work at meetings at Cold Spring Harbor.

Haseltine was just across the river, practically next door. He had known of her work, but never approached her to discuss it; he did not consider her part of the circle of scientists he talked to. That was the problem: he had not treated her like a colleague. It was no accident, Nancy thought, that he had come out of Baltimore's lab. Once again, it was as if she were invisible in her own institution.

She never saw or spoke with George Khoury again. He had lymphoma, which she learned only later, and died in 1987 at age forty-three. But his brief visit had planted the seed of another breakthrough. For a decade Nancy had thought the source of her problems at MIT was the fifth floor. But the problems had followed her. The struggles over ideas and credit affected more than whether she got tenure, more than her reputation in the department. A scientist's discoveries were her currency and her reputation in the field as well. Who got credit—or took it—could influence what grants she won, what talks she was invited to give, and, Nancy suspected, what companies someone would be invited to join or start. The person who got credit for the first big result in a paper would then be cited in the papers that followed. Nancy's discoveries, it seemed, did not belong to her. They were there for the taking.

Now she thought the problem might be the cancer center, or MIT. Maybe it was the field. And maybe it was that she was a woman, though she was reluctant to think this way. She considered that she might have to leave—MIT, the field, or both—but the prospect exhausted her, and she was already tired. She had decided against moving to the Biology building because it would have been tough to secure grants to furnish a new lab there, and she knew it would be even more difficult to win grants in a new field or at a new institution. She knew how lucky she was to be at MIT, to have the resources, the prestige, the remarkable students so eager to learn and work in her lab.

She knew, too, that men stole credit from other men. Botstein told her it happened to him all the time. She suspected it was easier to steal from women, when there were so few of them in the field. But how could she ever prove it?

Some time after Khoury's visit, a postdoc who worked in Susumu Tonegawa's lab casually referred to Nancy's discovery as Haseltine's. Nancy still didn't correct him. It seemed petty to do so. Again—for now—she kept going.

Chapter 13

"This Slow and Gentle Robbery"

Nancy was still working late hours at the lab, but now in her late thirties she was building a life outside as well. She was now a full professor at MIT. She'd made money on the Treasury bills that Salva's moneyman had helped her buy, and in 1980 she used that along with the raise that came with her promotion to buy her first home, moving from the rental she had shared with Brooke on Prescott Street to a condo in a newly renovated Victorian on Chauncy Street, on the opposite side of Harvard Square. She'd never wanted to own anything, and the place was more expensive than she had ever imagined she could afford—Salva's moneyman wrung his hands that she'd lose everything she'd earned—but developers were buying up apartments across the city, and she worried about being priced out. It was her first grown-up apartment, and she thought it was the nicest one she'd ever seen, with its cathedral ceilings, and a wall of windows that flooded it with sunshine. Her study was like a tree house peeking over Cambridge; her bedroom overlooked a garden that burst with fruit tree blossoms in the spring.

In August 1982, she met a man named Arthur Merrill in the library at MIT. He was a vice president at Citicorp in New York, a Harvard man, divorced, and two decades older than Nancy. He was auditing a course on biotechnology. They'd had one date before he'd left Cambridge; Nancy took him to Pizzeria Regina in Harvard Square, and he told her he'd never eaten pizza before. The next week he flew her to New York and arranged a stretch limousine to sweep her from LaGuardia to lunch in the Pool Room at the Four Seasons, the most elegant meal she'd ever eaten. They had a drink at the top of the World Trade Center before the limousine returned her to the airport. The whirlwind courtship began a long-distance relationship that lasted several years.

She didn't tell many people in her science life, not at first. She had always worried that people looked at single women and wondered about their sex lives. She tried to make herself seem asexual, dressing conservatively and not showing up to work parties with any boyfriends.

She liked to think it made men take her more seriously. She was finding out it didn't always work.

In June of 1983, she was invited to a dinner party at the home of Benjamin Lewin, who had founded the journal *Cell* and quickly built it into one of the most prestigious platforms for scientists looking to promote big discoveries. Lewin was trained as a molecular biologist himself and understood how to exploit the insecurities and competition among scientists. He tortured them with his edits or, worse, his rejection. Nancy thought he enjoyed his power a little too much, but she liked him and was charmed by his wife, who like Benjamin was a Brit.

She knew that a friend from graduate school who was in from out of town would join them for dinner. When she arrived at the Lewins' home in suburban Wellesley, she was delighted to also find one of her colleagues from MIT, who was there without his wife. He worked in the Biology building across the street, and Nancy hadn't seen him in a while. She was eager to hear about his latest work.

They continued talking after he offered her a ride home, and as they sat in his car in front of her apartment on Chauncy Street, she was so engrossed in the conversation that later she would barely remember how he ended up coming upstairs.

As soon as they were inside the door, he grabbed her, tried to kiss her and force her toward the couch. Nancy tried to push him off, toward and out the door. He wasn't tall or heavy and she wouldn't have thought that he was especially strong, but she now realized he was stronger than she was.

He's going to rape me, she thought. He must have had too much to drink. But not so much that he couldn't drive or carry on their conversation. He started to sound deranged, told her that sometimes he could barely control his urges: when he saw women on the street, he just wanted to jump on them. *He's lost his mind*, Nancy thought.

He pushed her onto the couch, and she finally struggled out from under him. But every time she tried to stand up, thinking he had calmed down, he grabbed her again.

This went on for hours. Nancy thought about Gene Brown, the department chairman, how embarrassed she'd be if he found out. She blamed herself, thinking she shouldn't have let this man upstairs. But most of her friends were men, she'd let many of them into her home, and none of them had ever done anything like this.

Finally he left, worn down by time, alcohol, or frustration.

The next morning Nancy woke up with a fever blister swelling her lip, and chills, as if she had the flu. She stayed home a couple of days and ended up at the doctor when whatever infection she had migrated into her ear. She did not even think of reporting what had happened to anyone at MIT. She assumed that anyone she told would blame her for inviting him up in the first place.

Many months later she ran into her friend from graduate school at a seminar at Cold Spring Harbor. They were standing on the stoop of Blackford Hall when her friend mentioned how much fun dinner at the Lewins' had been.

Nancy considered her response a long moment, then said, "It didn't end so well for me."

Her friend answered immediately, "I bet I know what happened."

He told her that the colleague who had assaulted Nancy had tried to sexually assault another woman he knew, a biologist at a different institution. The circumstances were similar: they had been driving to a conference together and the woman thought they were talking science, until he suddenly stopped the car and forced himself on her.

Nancy's friend had recalled this as he watched her and her MIT colleague talking so intently that night at the Lewins'. He had meant to warn Nancy not to get into a car with him but had forgotten. By the time he got back to his hotel room and telephoned the house to ask to speak to Nancy, she and her colleague had already left.

Nancy hadn't thought about whether her colleague would attack anyone else. Now she thought of course he would—he had—and she had to tell someone at MIT.

Maury Fox, who had recently replaced Gene Brown as the chairman of Biology, was as much a feminist as you could find in the department. One of his sisters was Evelyn Fox Keller, who had earned her doctorate in physics and become a feminist historian and critic of science; his other sister was Frances Fox Piven, a political scientist and activist who advocated mass disruption to force social change. Nancy began carefully as she told Maury what had happened, not using a name. She told him she would not have mentioned it except that she was worried that this man might try to assault a student. Maury listened, captivated. Who was it? he wanted to know. He didn't look surprised when Nancy told him.

Nancy didn't follow up to see whether Maury told anyone else or addressed the incident with her colleague. She figured she had done what she could. Whenever she saw the colleague in the hallways, she walked the other way and retreated down the nearest stairs.

The early 1980s saw the emergence of what was called feminist science, with scholars arguing that science was a world constructed by men, based on values that had traditionally excluded women: it was objective and controlling and moved according to facts; women were emotional, unpredictable, they responded to feelings. The bias was baked in. Science could not be entirely objective because its practitioners were influenced by social context, just as anyone else was. A feminist science would recognize the interplay between science and values and environment. It would move away from biological determinism, including the idea that women were not innately suited to scientific enterprise.

But feminist science was often characterized as a feminine way of doing science, which only reinforced the notion that biology was destiny. In 1983, Maury's sister Evelyn Fox Keller had published a biography of Barbara McClintock, titled *A Feeling for the Organism*, which defined Barbara's unappreciated genius as her ability to recognize the deep complexities of the plant systems she was examining. The book burnished Barbara's reputation as the mystic of genetics—alone and intimate with her plant samples, as if she were communing with them. Even as Evelyn insisted otherwise, the book was taken as evidence that Barbara's, too, was a feminine science and turned her into a feminist icon. Three months after it appeared, she won the Nobel Prize.

Barbara recoiled at the crush of television cameras that descended on Cold Spring Harbor for the Nobel announcement—"Oh, dear," she was heard to mutter when she heard the news she had won—and rejected the romanticization of her work. Nancy, too, hated the idea that women did science differently; she saw no evidence it was true, and she suspected that "different," when it came to women, would always mean "lesser." A feminist science could only segregate women more, make it harder for them to compete on the same field.

In 1984, MIT established a new department of Women's Studies in the School of Humanities, inaugurating it with a series of events headlined by Evelyn and Shirley Malcom, who a decade earlier had been the lead writer on the report about the "double bind" of being Black and female in science. The new department was led by Ruth Perry, a professor of eighteenth-century literature and the kind of feminist Nancy ran from. They were the same age, and completely different. While Nancy was taking math classes at Radcliffe, Ruth had been president of the folk song club at Cornell, performing with her band to raise money for a voter registration drive in Tennessee and coal miners striking in Kentucky. Since then Ruth had organized the Cambridge food co-op, no-nukes protests, and demonstrations against apartheid and the Iran-Contra arms sales, all while writing books about the early English feminism of Mary Astell and Jane Austen. Ruth wore chunky turquoise jewelry over her tunics and taught exuberantly, guiding students through the dances of Austen's England and re-creating an eighteenth-century feast with salmagundi and tansy pudding for a conference titled "Science, Myth and Knowledge."

More than a hundred universities around the country were offering degrees in women's studies, but the discipline was still widely mocked—"a problematic addition to America's colleges," as the *New York Times Magazine* wrote in April of that year. "Given the suspicion among many who hold power in the academy that women's studies is an accomplice of affirmative action, some young women are avoiding contact with the field, lest they come to be seen as infected."

At MIT, even the humanities had always struggled for status—the club sport at a university where science and engineering were varsity— and Ruth knew that Women's Studies would never be taken seriously if it did not offer any courses in the School of Science. She created an advisory board of scientists and engineers for her new department, and she wanted Nancy on it. Ruth knew that Nancy was no feminist, but there were few full, tenured female professors to choose from. She considered Nancy educable and saw a glimmer of a feisty spirit—Ruth had heard Nancy say that she'd never give any money to Harvard since it couldn't seem to find any women worthy of its faculty.

Nancy kept insisting she had no time to meet with Ruth. But she could see that Ruth wasn't going to give up, so she invited her to Chauncy Street for tea, where Ruth convinced her to be on the board. Ruth then pushed her to teach a class in Women's Studies. Another professor on the board—a man, an electrical engineer—scolded Nancy for resisting. "If we're ever going to change the culture, everyone has to get involved," he told her. "You have to put your oar in the water."

Nancy did think it was important to teach basic molecular biology to non-biology students; the revolution in genetics had made it essential for everyone to understand it. Arthur encouraged this idea; he was fascinated by the field and recognized that the students were in the generation that would be making the most decisions around genetics, especially in their decisions around having children. Maury Fox, too, liked it. So Nancy—despite adamantly declaring herself not a feminist—agreed to design and teach the first women's studies course in the School of Science at MIT, a class in reproductive biology that started in the spring of 1987. She taught the fundamental molecular biology and genetics behind new reproductive technologies, and Ruth recruited a feminist philosopher to lecture on the ethical issues they raised.

The class was a huge success: it earned credibility for Ruth's new department, and for Nancy, the highest ratings of any course in Biology. The two became friends—it helped that Ruth was a great cook, and that Nancy still ate most of her meals at the F&T. Ruth set about trying to educate Nancy, arguing how gender had influenced the construction of knowledge and the understanding of science, how history, literature, and psychology would be different if women asked the questions, did the research, interpreted the results. She gave Nancy frequent readings. While Nancy found the readings more interesting than she had Millie Dresselhaus's lunches for women faculty, they didn't resonate any deeper with her experience. Instead, it was a book she discovered on her own that changed her perspective.

The book, *Rosalind Franklin & DNA*, was the complementary strand to *The Double Helix*, the story of the woman who had produced the X-ray photograph that had helped push Watson and Crick over the finish line to be the first to decode the structure of DNA. The author, Anne Sayre, was a friend of Franklin's—they'd met through Sayre's husband, who like Franklin was a crystallographer. The book aimed to establish Franklin as "one of the world's great experimental scientists," a role denied her not only by her early death, Sayre ventured, but because of her gender.

In Watson's telling, Rosalind had been "Rosy," and an assistant to Maurice Wilkins, who had shown Watson photograph 51. Rosy was stubborn, uptight, and resolutely unfeminine—as Sayre said, "something of a termagant." She refused to wear lipstick and took not even the slightest interest in her clothing; Watson mused as he watched her lecture that she might be pretty if she took off her glasses and did something with her hair.

In fact, Sayre noted, Rosalind had been hired at the same level as Maurice Wilkins. She "invariably" wore lipstick and never glasses—"she had the eyesight of an eagle." Minor errors, but to Sayre they exposed the vulnerability of Watson's more consequential claim: that Franklin had not realized that her photo was as important as it was, that the structure it showed was a helix. Sayre cited the notes of a lecture that Watson heard Franklin deliver, where Franklin concluded that the photograph showed a "big helix in several chains."

"She underlined it," Sayre wrote.

Sayre's book had come out a decade earlier, in 1975, but Nancy had not read it at the time. She figured she knew the story of DNA; she'd been among the first to read Jim's version, had heard it from his very mouth.

Reading the book now, she thought, *This is my life.*

Rosalind had arrived in England from a lab she had loved in Paris and found herself unwelcome and in an intractable personality conflict with Wilkins. While Watson and Crick volleyed ideas over drinks at the Eagle pub, she and Wilkins apparently worked together for two years "never to have achieved so much as a simple conversation."

Nancy had been a graduate student when she read Jim's account. Now she had more than twenty years of experience doing experiments. She understood how hard it was to collect data, the painstaking work required to get an experiment right and then to repeat it. She could see how much work Franklin had had to do to get even a single photograph; it took her nine months just to create the apparatus she used to capture the photo Wilkins showed Watson. She had taken fifty photographs before she got one that made the helix clear.

In Sayre's telling, Nancy saw a woman who was more sympathetic—and familiar. Franklin was intense, uncompromising, determined—"because there was something important to be determined about." She had a temper, and a disdain for weakness in others and herself: in her twenties, she had once knelt on a sewing needle, driving it deep into her knee joint, then walked "alone, and for a fair distance" to a hospital, where the doctor who examined her insisted it was impossible that anyone could have walked at all with a needle angled the way it was. Franklin laughed at him.

She loved science for its rationality, its reliance on objective evidence, logic, and proof. To Franklin, science was a calling, something she had to do, not merely something she wanted to do. She scorned "triflers" in the field. And while she loved children and wanted them, she believed she had to make a choice between science and raising a family. "There's no use in doing *two* things badly," she told another female scientist. Having chosen science, Franklin assumed that her sacrifice would signal "a single-minded and single-hearted commitment," that she would be seen not as a "woman scientist" but as a scientist pure and simple: "Nothing but sheer ability would count."

She suspected that being a "girl" disadvantaged her in science. But like Barbara McClintock, like Nancy, she resisted being called a feminist because she was not an activist. "For herself she asked no favors, privileges, or special, softer standards of judgment because she was a woman, which meant that, to her, equality was taken for granted," Sayre wrote. "To have it raised, then, as a separate point, was maddeningly illogical."

Jim became the villain in the Rosalind Franklin story. But Sayre did not entirely blame him. She recognized that Franklin's photograph on its own did not lead Watson and Crick to their famous discovery; it was one piece of the puzzle. What offended Sayre was that Watson's telling had robbed Rosalind of her true personality. He had warped her to create a familiar stereotype, casting her as the "wicked fairy" in a cautionary tale about a bright woman who dares to think there is a place for her in science.

It was not an unusual or unacceptable view, Sayre wrote: "The intelligent use of an intelligent woman's intellectual gifts and aptitudes is an affront to the natural order of things." In this world, men were naturally geniuses, while women could only hope to work hard, and risk being ignored, minimized, or ripped off. "This slow and gentle robbery," as Sayre called it, continued past the theft of Rosalind's photograph. Museum exhibits and encyclopedia entries about DNA failed to mention her. Linus Pauling, another of the competitors in the race to understand DNA, had written a commemorative essay giving the credit for Franklin's photograph and analysis to Wilkins.

Nancy, too, could not blame Jim. He had been honest about how he saw Franklin—he'd only said things other men thought about women. He'd admitted he took the photograph without Rosalind's knowledge—without his book, the world might not have known its role in the discovery of the double helix. (The *New York Times* obituary when Franklin died in 1958—a decade before *The Double Helix*—was just four paragraphs and did not even mention her study of DNA; it described her as a researcher who worked on tobacco mosaic virus, "one of a select band of pioneers unraveling the structure of nucleo-proteins in relation to virus diseases and genetics.") Jim had admitted he was callow and insensitive during his days in England. And how could Nancy blame Jim? He had given her a life in science— her life.

Still, Sayre's account of Franklin's life left Nancy deeply discouraged. She had wanted to believe that it didn't matter if you were a woman, that as long as you did a great experiment, you could win a Nobel Prize. Here she had been confronted with clear evidence to the contrary. Maybe only death had denied Rosalind Franklin her Nobel. But Franklin had gone to her grave not knowing that Watson and Crick had seen her photograph, the essential role it played. It was tragic, Nancy thought. It affirmed what she had begun to suspect: that because women had no legitimate status in science, men could and would continue to take from them.

Nancy had bought the idea of the scientist as great thinker, the solitary genius. She was beginning to see how wrong she had been. But even if that was a fiction, the system was built around it. She wondered how many women would ever be allowed to play the role of genius.

Chapter 14

"Fodder"

Boris Magasanik, the former chairman of Biology, called Mary-Lou from Washington in April 1982 to share big news: she'd just been elected to the National Academy of Sciences, the nation's most prestigious scientific association. The academy, headquartered in a massive marble temple not far from the Lincoln Memorial, had been founded during the Civil War to designate leaders of the field to advise policy makers on science, and its members now served the National Research Council and other government agencies. It was an elite group, with membership reserved for about one-half of 1 percent of all American scientists, and nominations and elections were highly secretive—no one was supposed to know who put names into nomination, and nominees were not supposed to know they'd been elected until the academy made its formal announcement. Boris, who'd long ago declined to even interview Mary-Lou for a job, couldn't contain his excitement.

Mary-Lou's professional stature was growing: by the time she was formally inaugurated into the academy the following year, she had also been elected president of the Genetics Society of America. In 1985, she was elected to the American Academy of Arts and Sciences, and the year after that as president of the American Society for Cell Biology, the first of three of Gall's Girls to serve in the role. She had been teaching the summer course in molecular cytogenetics at Cold Spring Harbor for a decade. At MIT, she continued studying how cells respond to stressors in their environment and had also moved into work on transposable genes and telomeres, the protective ends of chromosomes that help prevent damage to DNA.

She turned fifty the year she was president of the Genetics Society and that December represented the United States at the International Congress of Genetics meeting in New Delhi. After the meeting, she fulfilled a longtime dream to trek the Annapurna Range of the Himalayas, one of the steepest hikes in the world. Annamaria Torriani-Gorini, a friend from the Biology faculty at MIT, joined her on the trip to Nepal, and on their way home, Mary-Lou stopped into an artist's stall in a crowded market in Kathmandu where a young man was painting thangkas of Buddhist deities. They got to talking, and Mary-Lou told him she was a geneticist. The young man's face lit up in recognition. "There's an old lady who is a geneticist who won the Nobel Prize this year," he said.

She had to laugh: the lady was Barbara McClintock, the woman Mary-Lou's college professors had tried to hold up as a role model for her thirty years earlier. Now she was only the eighth woman to win a Nobel Prize in science, noteworthy enough to be news in Kathmandu. Still, women on campuses had few role models in the sciences, especially at the most elite levels. Forty years earlier Barbara had been the third woman elected to the National Academy. The year Mary-Lou was elected it had about twelve hundred members and less than 5 percent of them were women.

At MIT, the dean of the School of Science noted when he stepped down in 1981 that there were nineteen women on the faculty, a striking increase from 1967, when he'd started and there'd been only one, and she didn't have tenure. The new president, Paul Gray, had first arrived at MIT as an undergraduate when women made up less than 2 percent of the student body. As chancellor he had worked with Shirley Ann Jackson and the Black Students' Union to increase diversity on campus, and now in his new role he committed to increasing the number of female students. By 1985, women made up 22 percent of students, up from 17 percent at the start of the decade. They were 27 percent of the incoming class. And women were declaring their presence and their

Nancy Hopkins, photographed by Brooke, with the dog they adopted then sent to live with Brooke's parents.
Courtesy of Nancy Hopkins

Brooke Hopkins photographed in Europe during his Oxford fellowship the year after he and Nancy graduated from Harvard.
Courtesy of Nancy Hopkins

Mark Ptashne, at right in sunglasses, at the Cold Spring Harbor Symposium on Quantitative Biology in 1968, the year after his discovery that the repressor bound to DNA.
Courtesy of Cold Spring Harbor Laboratory Archives

Jim Watson presents Nancy with a wedding gift, the stuffed leather pig she had spotted in a shop window in Harvard Square.
Courtesy of Nancy Hopkins

Barbara McClintock in her lab at Cold Spring Harbor in 1971, the year Nancy started her postdoc there.
Courtesy of Cold Spring Harbor Laboratory Archives

Mary-Lou Pardue at the 1977 Cold Spring Harbor symposium.
Courtesy of Cold Spring Harbor Laboratory Archives

Salvador Luria, Nancy Hopkins, and David Baltimore in the new Center for Cancer Research at MIT, 1973.
Courtesy of MIT Museum

Nancy, the only female scientist at a dinner table at the 1979 Cold Spring Harbor symposium. Ed Scolnick is second from left, in glasses; to his left is Michael Bishop, who would share a Nobel Prize in 1989 for his codiscovery of the oncogene; and Wally Rowe is to Nancy's left. Jim Watson is at the head of the table, smiling, with his wife, Liz, two to his left. Third from her left, with beard and glasses, is Phil Sharp, behind him is Fritz Lippman, the 1953 Nobel laureate in medicine.
Courtesy of Cold Spring Harbor Laboratory Archives

Our Millie: Millie Dresselhaus grades papers at her kitchen table, in a photograph taken by her husband, Gene.
Courtesy of the Dresselhaus family

Millie in Japan with the other participants at the Japan–USA Seminar on the New Materials Science of Carbon and Graphite in September 1970, a trip sponsored by the National Science Foundation.
Courtesy of the Dresselhaus family

Nancy Hopkins at the 1982 Cold Spring Harbor Symposium on Quantitative Biology.
Courtesy of Cold Spring Harbor Laboratory Archives

Eric Lander takes the microphone at the 1986 symposium, in an early discussion of the undertaking that became the Human Genome Project. (Nancy is in the foreground.)
Courtesy of Cold Spring Harbor Laboratory Archives

Chuck Vest arrived on his first day as president at MIT in 1990 to discover that student "hackers" had disguised the door to his office as a bulletin board.
Courtesy of MIT Museum

Bob Birgeneau was named dean of the School of Science the following year.
Photograph by Donna Coveney/MIT

Lotte Bailyn at the Sloan School in 1999.
Photograph by Len Rubenstein

The photograph of the Women of Science that ran in the *Chronicle of Higher Education* in 1999. From left: Sylvia Ceyer, Paola Rizzoli, Penny Chisholm, Nancy Hopkins, Leigh Royden, JoAnne Stubbe, and Mary-Lou Pardue.
Copyright Rick Friedman

ambition in what had turned out to be a go-go decade: undergraduates rushed a new sorority—MIT's first—and alumnae offered a new Independent Activities course titled Get the Job You Want in Industry: A Woman's Guerrilla Guide to the Pin-Striped World. In 1985, Millie Dresselhaus became the first woman named as an Institute Professor, the highest and most prestigious rank among faculty, reserved for just twelve professors at a time.

Still, that same year, the provost warned in his annual report, "The Institute's recent performance in hiring women and underrepresented minorities, especially for the faculty, is not what it should be at one of the nation's leading universities." While MIT still advertised itself to applicants as "A Place for Women," female students were surprised by what they sometimes found once on campus. Every spring, residents of Senior House, the rowdiest of the dorms, hosted a weekend bacchanal known as Steer Roast, spit-cooking an entire cow over a firepit they constructed beneath the Doric columns in the dorm's courtyard, a snowball's throw from the austere Italianate villa that served as the university president's home. Hundreds of students and alumni joined the celebration, watching over the weekend's awarding of the Virgin Killer prize to the Senior House resident who had slept with the most women in the previous year. On the other side of Mass Ave., the student-run Lecture Series Committee kept up the tradition of showing pornographic movies in Kresge Auditorium, the largest gathering place on campus, on Reg Day at the opening of the semester. It was near impossible for women to ignore the posters all over campus advertising the movies, or the crowds that watched, especially given that the only all-women's dorm on campus was just across the lawn. In scientific fields that involved field research in exotic locations—oceanography, say, or geophysics—men still upheld a tradition of sending postcards of topless, flower-draped native women to their colleagues back in university offices, where women then had to try to ignore them piling up in the mail tray or pinned to the wall.

Many women played along. Others seethed silently. But in 1983, a group of female graduate students in the department of Electrical Engineering and Computer Science worked up the courage to write a report detailing the demeaning treatment they had endured. Computer science had become one of the most popular and competitive departments on campus, as the development of large integrated circuits created the possibility of computers that were smaller, faster, and more universally available. MIT was leading the way. It had recently kicked off Project Athena, a five-year, $50 million project paid for by IBM and the Digital Equipment Corporation to install a campuswide computer network, one of the first anywhere.

Men vastly outnumbered women in the department. Many doubted the women's professional commitment and qualifications and assumed they were open to or actively seeking dates. The women complained of being ignored and interrupted in meetings; men explained concepts to them as if they were high school students instead of colleagues at one of the nation's best engineering schools. ("Oh, Jane," they reported men saying when a woman tried to offer her technical opinion.) Women could not win: people in the department equated aggressiveness with competence, yet women were expected to act "quiet and feminine" and were "socially ostracized" if they did not. All this meant that the energy women wanted to spend on their professional work was instead wasted managing awkward passes and walking a tightrope of acceptable behavior. Some reported they stopped going to the "bull sessions" where the department tossed around challenges—only to feel more ostracized.

The women noted that the atmosphere at MIT wasn't necessarily worse than in computer science departments at other universities. But it was enough of a "locker room" that prospective female students were choosing other schools over MIT. The report provided a catharsis for the department; in meetings and lunches to discuss it, many men said they had not realized their behavior upset the women. They wanted to do better.

In 1987, Mary-Lou was teaching the signature course required of all first-year graduate students in Biology, known as Methods and Logic. Graduate classes were considered the most prestigious to teach, and desirable, not least because they gave professors access to students who might want to work in their labs; graduate students were the department's most coveted resources because MIT had a federal training grant that covered the cost of them for professors. Mary-Lou was the only woman in the department teaching the course. She was also trying—unsuccessfully, she felt—to help a bright graduate student stuck in a conflict with her thesis adviser.

The graduate student had arrived in the department in 1983. Mary-Lou was on her dissertation committee and remembered her as a standout from the oral preliminary exams that graduate students take at the end of their first year. By 1987 the student had finished the project the committee had agreed to for her PhD thesis—a relatively quick finish—and had published two papers about it, one of them in *Cell*. She had lined up a plum postdoc position in the lab of a highly regarded professor at Harvard Medical School. She was supposed to spend her last few months at MIT tying up loose ends in the lab and writing her dissertation, based on the papers she'd published about the project, and defending the dissertation before her committee.

But her graduate adviser had assigned her another project in his lab and told her that she couldn't graduate until she finished it. The adviser, Frank Solomon, took a particular interest in this new project; it pitted him against a competitor at another institution, and he wanted to win.

The project required methods outside the student's area of expertise, requiring her to do immunoelectron microscopy, so she told herself that at least she was learning a lot, and she was interested. But after several months, her data were incomplete and the project was clearly not going to conclude for many more months. The student cared deeply about the project, but she was already late to start her postdoc. While her new boss at Harvard Medical School was understanding, she knew he couldn't be patient forever.

She explained to Frank that she was worried about further delaying the start of her postdoc and asked again if she could defend her dissertation as originally agreed, complete her PhD, and be on her way to her new job. She offered to return to MIT on nights and weekends during her postdoc to finish Frank's project, and to train another student to take it over. But Frank wouldn't budge. He told her the new project was her responsibility, and he wouldn't let her defend her dissertation until it was done.

The graduate student had chosen Frank as her adviser because she liked him, as many students did. He prided himself on being the first in his family to graduate from high school and had gone on to a history degree at Harvard before turning to science in graduate school at Brandeis. Now he ran a lab in the cancer center and had become a fixture of life in the Biology department, taking over the teaching of the literature seminar that Salva Luria had taught. Frank was known as a friend and counselor to graduate students and spent hours listening to their problems, personal or scientific. He talked to them like a peer, had a good sense of humor. He made science seem fun, but also engaged students to think about its higher purpose in society. The graduate student council had given him its teaching award three times already.

But the graduate student, like others in Frank's lab, had come to believe there were two sides to his personality—as one of her peers described them, Personality A and Personality B. One was thoughtful, kind, and supportive. The other was arrogant and contemptuous—he'd once told the student that many scientists working at other universities were just "fodder"; she had never heard that word used to describe humans before. She recalled a more senior student confiding, "There is one side of Frank you just love, and another side that kicks your guts out." Lately, the graduate student's interactions with Frank ended with her crying in the stairwell.

Unsure of what to do, she went to the director of graduate studies for the department and told him her dilemma; she was already late to get to Harvard, and she felt Frank was now holding her graduation hostage to the completion of a new project. The director smiled a sage smile. Graduate school, he told her, is like the old feudal system. The adviser is the lord and the graduate students are the serfs.

"And where are the police?" the student asked. The director's smile faded. He reminded her, as others had, that as her PhD adviser Frank would have influence over her career for many years; people would call on him as a reference whenever she was up for jobs or honors. She said she would take her chances. If she couldn't have a successful career in biology without Frank's recommendation, she would leave science.

The student's father was an accomplished scientist at another university, and during weekly phone calls home, he told her not to give up: "You're just in a rat's nest right now. Don't let them get you down. You'll move on to a different place and find people who treat each other better." He encouraged her to write a chronology of her conversations and experiences with Frank, seal it in an envelope, and mail it to herself so she'd have a dated, postmarked version in case Frank ever claimed she was trying to rewrite history.

She did so and also took a copy of the letter to Mary-Lou and asked her to keep it in her files. The student was happy to have a woman she could talk to. There were few in the department—six out of fifty-six faculty members—and the numbers alarmed female students. At department seminars the doctoral students sat in the balcony above the faculty members in the auditorium, and looking at the sea of bald heads below, they joked that you had to lose your hair to win tenure.

The graduate student told Mary-Lou that she didn't want to get Frank in trouble, she just wanted to be allowed to finish her dissertation and get on with her career. Mary-Lou thought back to her own early battles over credit on her heat shock experiments and told the student that she'd once had to put a "letter like that" in a senior colleague's files. She'd never had to use it and hoped that the student would not, either. She agreed to safeguard the letter.

Mary-Lou liked Frank; the two of them had organized a party for Nancy when she got tenure, and Frank had organized a small dinner for Mary-Lou when she got into the National Academy. Now Frank was tenured himself. They had taught Methods and Logic together. But Mary-Lou was also wary of him. She noticed that he seemed to like to talk about other people's mistakes and misfortunes, and she assumed that if he was gossiping to her about others, he would just as easily talk about her behind her back.

Mary-Lou also thought Frank had given the graduate student a lousy problem to try to solve, one based on poor reasoning. The student had worked hard to get the experiments right, and Mary-Lou thought she had done enough to show that Frank's hypothesis was invalid. Mary-Lou went to Richard Hynes, a cell biologist who was also on the student's dissertation committee. "I don't think she needs to do any more," she told him. Richard was sympathetic, but he was also close to Frank and reluctant to overrule him. Richard offered a compromise, which became the solution: Frank wanted the student to stay another year; instead she would be required to stay for one more semester and give the experiment one more try. If it didn't work, she could be on her way.

The student went back to Mary-Lou several weeks later. Frank had asked her to write an abstract and present a poster about the new project at the American Society for Cell Biology meeting that December. She'd written the abstract and sent him a copy, but he did not think it was forceful enough—he had scolded her before that she waffled too much in her writing, that she needed to be more declarative. She stood next to him at his computer as he rewrote it. They submitted it last-minute, but Frank justified it by saying that presenting a work in progress at the conference would be a great way to get feedback from others. The student presented the poster at the conference, yet when she returned to MIT, Frank told her that the data presented had been premature. Now, he said, she needed to stay until she finished the project.

Mary-Lou was fed up. She went to see Mary Rowe, who had been hired as the president's adviser on women and was now in the MIT ombudsman's office. Mary-Lou urged Mary to intervene to let the student leave. Mary hesitated—Frank was such a good guy, she told Mary-Lou, he was always trying to help others. But Mary agreed. After another meeting in Biology, the student was given two weeks to write her dissertation—a truncated timeline—and was scheduled to defend it the next month, in early 1988.

Frank refused to allow an outside reviewer to sit in on the student's defense, which made her fear she would not pass, that Frank would try to sink her, and no one would be there to stop him. Mary-Lou assured her that it was in Frank's interest to have his students pass—though he might make it difficult.

Standing up in the front of the room at her defense, the student could see Mary-Lou among the other faculty members, smiling her encouragement. She also saw Frank sitting next to the director of graduate studies—she had recently discovered they were good friends. The director posed a tough question about the material, and Frank gave him a playful punch in the arm, as if to say, "You got her." The director of graduate studies asked how she knew the epitope they were following with an antibody was still attached to the tubulin transgene she had put it on. "How do you know the transgene wasn't just rearranged?" It was a valid question, and the student had anticipated it. She told him that would have resulted in an altered pattern on the Southern blot, which probed the integrity of the transgene at the level of DNA. "The Southern blot would have shown if it was rearranged, but it wasn't." Again the director's face fell. The Southern blot was the right answer.

Mary-Lou waited for the student as everyone filed out and told her, "Frank behaved terribly in there."

"I know," the student replied. But she had survived.

After defending the dissertation and submitting the final version, the student took a copy of the chronology she had written and delivered it to Mary Rowe. She told Mary she wasn't asking for anything—she was on her way out—but she predicted that someday another student, likely a woman, would have a similar problem with Frank and would not be lucky enough to have a supporter like Mary-Lou on her committee. "If such a student comes to you, you're going to think she's crazy, and you might not believe her," the student said. "I'm telling you what happened to me so that when some future student comes to see you and she is at the end of her rope, you will at least listen with an open mind and maybe recognize a pattern."

Mary listened kindly but told the student that she imagined that Frank would have his own version of events: if he was there, he'd smile and say, "Let me tell you a story." The student was disappointed in Mary's response—Frank seemed to have charmed everyone—but reassured herself that she had at least left a record of what happened to her. Still, she was scared of Frank and what he could do to her career. Years later—after she'd completed a successful postdoc at Harvard Medical

School and gone on to a long and successful career as a full, tenured professor at a prestigious university—that fear hadn't left her.

———————

Several months later, Nancy was sitting in her office when Frank Solomon dropped by. This was unusual. Frank used to come by regularly when he was trying to get tenure; she had suspected that he was campaigning for votes, and sure enough, he had barely been around since.

Frank was tall, with broad shoulders, and usually friendly. Now he was wilting.

He told her the most awful thing had happened to him: Maury Fox, the chairman of Biology, had told him to tell Mary-Lou she would not be teaching Methods and Logic.

"Why can't she teach it anymore?" Nancy asked, her suspicion rising. It was unusually late in the year to remove a professor; the course catalog had already been printed with Mary-Lou's name as an instructor.

Mary-Lou, Frank said, was not a good teacher. She was not smart enough for the students; it was not fair to them. Maury, Frank told Nancy, didn't have the courage to tell her himself, so Frank and the director of graduate studies were going to meet with the graduate students to talk about how to tell Mary-Lou without hurting her feelings.

Nancy could feel her rage rising. Mary-Lou not good enough? She was one of the most important scientists of her generation, smarter, more experienced, more creative than the man standing in front of Nancy now. It was unthinkable that they would meet with students to discuss Mary-Lou. How could anyone think to go behind a colleague's back this way? Graduate students were like sheep; once Frank gave them a whiff of doubt about Mary-Lou, her reputation would be ruined. She'd never draw another graduate student to her lab.

Nancy went to see Maury, thinking he'd see how outrageous it was and do something to stop it.

Maury rolled his eyes. "Oh, you know how Frank is." Maury talked about Frank as if he were a neurotic but charming son. "He's mad because I asked him to tell her rather than telling her myself. He's just being Frank."

This is not about Frank's feelings, Nancy thought. They were degrading Mary-Lou. But there was nothing Nancy could do. She could see no one was going to stop Frank.

———————————

In the end, Maury told Mary-Lou himself she was being removed from the course. Mary-Lou was devastated; she had loved teaching it, the readings and the debates with the students, and she thought the students had enjoyed it, too. She pressed Maury to tell her why he was removing her, but he did not seem to have a reason. She had never seen him look so embarrassed, and he tried to make it better by telling her she didn't have to teach at all if she did not want to. Mary-Lou was embarrassed, too, especially with students, and tried to make it sound as if it were her choice. That fall, when she bumped into a graduate student on the street who asked why she wasn't teaching, Mary-Lou smiled and said, "I know where the bodies are buried."

Nancy heard nothing more about it and didn't think to check to see who was teaching the class. She still barely knew Mary-Lou; they worked in separate buildings. But ever since she'd read *Rosalind Franklin & DNA*, Nancy had been reflecting on how her male colleagues talked about women in the department. One man, a biologist at another university, had remarked how strange it was that Lisa Steiner had ended up in biology. "Why would someone not good at math have gone into science?" he asked. Nancy had told him that Lisa had in fact been a math prodigy, one of the top math students in New York City and a finalist for the prestigious Westinghouse Scholarship, awarded to high school students across the country who show the most promise in science and math. Lisa had graduated with highest honors from Swarthmore and received her master's degree in math from Harvard before earning her medical degree at Yale. Not good at math?

Nancy's colleagues at MIT talked to and about Annamaria Torriani-Gorini as though she were a child, even after she became a full professor. One had told Nancy that Annamaria and Lisa had only been given tenure because of affirmative action.

More recently Nancy had run into Frank getting lunch at the F&T, and he'd complained about Ruth Lehmann, a newly arrived developmental biologist and, to Nancy's mind, the most successful woman hired onto the junior faculty. Ruth was already well-known in her field, awarded big grants and prizes.

Frank had called Ruth a "real piece of work," saying she was greedy for "stuffing her lab with people and money"; he said it was making other junior faculty members jealous. Nancy thought of all the time spent in faculty meetings strategizing about which potential hires would bring in the most grant money and attract the best students. Ruth Lehmann's "greed" would have been prized as ambition in a young man.

Nancy had been thinking of leaving MIT or changing fields ever since George Khoury's visit to her lab a few years earlier. The incident with Frank made her realize she had to make a change.

It was all so much salesmanship, this business of professional science. She had known since her days in Jim's lab that she didn't like the competition. She doubted she could scrap like the men and faulted herself for not being aggressive enough. Now she wondered if any woman could get away with the self-promotion successful science demanded. The rare women she'd heard of who resisted convention or tried to assert themselves—Barbara McClintock, Rosalind Franklin, now apparently Ruth Lehmann—were labeled "difficult." "Greedy" for expecting to be recognized and rewarded in the same way as men.

The challenge of cancer, too, no longer captured Nancy's interest the way it once had. The discovery of human oncogenes had made clear a path for the field: now the way forward was to figure out drugs that inhibited their expression. Nancy's friend Ed Scolnick had become head of research for Merck, the pharmaceutical giant, and put Nancy on the company's scientific advisory board. She was unexpectedly impressed at the quality of the science being done there, and by the way a company could mobilize resources to get behind a drug that looked particularly promising—chemists, biologists, manufacturing teams, all working together. If anyone was going to create drugs to cure cancer, she thought it would be one of these companies rather than university scientists jockeying for position in their departments.

The men she had started with at the cancer center had taken on new roles. Phil had replaced Salva as its director and cofounded Biogen. Bob Weinberg had moved to the Whitehead Institute, which David Baltimore was leading.

Nancy was teaching virology and the women's studies class that Ruth Perry had recruited her to develop, and both were going well. But she wanted to get back to the raw excitement of science, the "real science" she had described in the *Radcliffe Quarterly*. She was interested in AIDS research, and it was a natural move from virology; in 1983, scientists working on retroviruses and cancer had discovered that AIDS was caused by a human retrovirus, HIV. The competition in AIDS was notoriously vicious and, Nancy suspected, harder for women. Two male scientists, one at the Institut Pasteur in Paris and one at the National Cancer Institute, were locked in a legal battle over a patent on HIV, and in all the wrangling over credit the name of the female scientist who'd done the essential work to find the virus in 1983 was rarely mentioned.

So Nancy thought back to the questions that had preoccupied her when she was young, the possibilities she had glimpsed in that first hour-long Bio 2 lecture of Jim's back in the spring of 1963. She had wanted to understand the genetics of cancer and the genetics of human behavior, and she had become obsessed with understanding the repressor.

Maybe now she could study human genetics. Molecular biology had advanced so far in understanding the gene that by the mid-1980s scientists were aiming at what had once seemed a fanciful goal: to map and sequence every gene in the human genome, essentially a blueprint for the human being. In 1988, prominent biologists on a National Research Council panel—among them Wally Gilbert, still at Harvard, and Jim Watson, leading Cold Spring Harbor Laboratory—had endorsed what would soon become the Human Genome Project. But the technology required was expensive. That field, too, was dominated by a small number of aggressive men who all ran big labs. She expected it was as cutthroat as cancer. She could do genetics in mice, but she didn't like working with mice, and they also required a big, expensive lab. You needed three generations to do the Mendelian ratios that proved that specific genetic traits were inherited, which meant a lot of mice and a lot of people to work with them. Nancy didn't know any woman who'd been able to raise that kind of money.

She hinted at her evolution as her twenty-fifth college reunion approached. For every class hitting a five-year mark, Harvard published the "Red Book," inviting alums to note their reflections and accomplishments—the degrees earned, children raised, presidents advised.

"I've been extraordinarily fortunate to spend much of my adult life as a scientist," Nancy wrote. "My good luck was to have been around during a truly great period in science," one that began with Watson and Crick's discovery of the double helix and moved quickly through the cracking of the genetic code. "I got into a lab just in time to participate in the experiments that first demonstrated how gene expression is regulated. Later I went into cancer research where oncogenes were soon to be discovered. Even after twenty-five years in the field, seeing molecular biology explain life still catches me by surprise, takes my breath away, induces euphoria. I'm a junkie for new facts about the incredible cleverness of even the dullest cell. I think the excitement in biology will continue for some decades, with much new knowledge of human genes, treatments for many diseases, and genetic engineering of plants and animals.

"Already a major breakthrough has been made in understanding how genes direct a fertilized egg to give rise to an animal with its legs and antennae in the proper places.

"To ensure fun in middle age I'm considering a shift in research interest to neurobiology, the next frontier," she wrote. "The brain may only prove to be truly accessible to our grandchildren but the stunning successes of the last few decades have made biologists so cocky that we actually dare to sit around and wonder if it might not be possible in our lifetime to understand consciousness in terms of molecules and cells."

She gave credit to her Harvard professors, "particularly J. D. Watson of DNA fame," for encouraging her to be a scientist. She had expected that the gratitude and affection she had for them "would blossom in middle age into a generalized gift giving warmth for the college." It had not. "The reason is probably summed up well enough in a newspaper clipping (from 1986) taped to my refrigerator door. It reports the startling fact that a woman got tenure at Harvard. After 350 years this is still news!"

(In his entry, Brooke Hopkins wrote that he was an associate professor of English at the University of Utah, where life had been good to him: he had colleagues who were friends and intellectual companions; he'd met a philosophy professor and together they had bought a house, thrown lots of parties, and raised her two children from a previous marriage. They had married on New Year's Day 1986, bought a second home under the stars near Capitol Reef National Park, in Red Rock country, and traveled widely, most recently to India. "Reflections on the 'meaning' of the past 25 years would be fruitless," Brooke wrote. "My memories of Harvard (especially the undergraduate years) are fond ones. But, in the end, the best thing about Harvard is leaving it behind.")

Nancy had become close with Susumu Tonegawa, the immunologist who had joined her on the third floor in 1981. She was helping Jim write the fourth edition of his textbook, *Molecular Biology of the Gene*, and Susumu taught her enough immunology to author a new chapter on it. Susumu, too, was thinking of a change in research—he had won the Nobel Prize in 1987 for his discoveries about the genetics of antibodies—and like Nancy he was interested in neuroscience. They talked often about the nature of consciousness and whether molecular biology might someday be harnessed to explain it.

In late June 1988 Susumu and Nancy took a two-week neuroscience course at Cold Spring Harbor. At a cocktail party on the terrace outside at the Banbury Center, an old estate overlooking Long Island Sound, Nancy began talking with one of the instructors, Friedrich Bonhoeffer, who was on the faculty at the Max Planck Institute for Developmental Biology in Tübingen, Germany. Friedrich was a compelling figure. He came from a prominent German family that had fought in the Resistance during World War II—four of his uncles had been executed by the Nazis. He had a movie star's head of hair. Nancy found his science fascinating; he had mapped the cells on the back of the retina that project images to the brain, a process known as the retinotectal projection.

Nancy told him she wanted to do neurobiology, and that she was looking for a new model organism. Friedrich told Nancy about a developmental biologist at Max Planck, Christiane Nüsslein-Volhard, known as Janni, who was going to start studying zebrafish. Nancy knew Janni only by reputation as one of the top developmental biologists in the world. With a colleague she had been the first to discover the genes necessary for the embryonic development of an invertebrate, *Drosophila*—a remarkable advance in understanding how a fertilized egg could know how to grow into a fully formed organism, with wings and legs and a head all in the proper position. This had been the "major breakthrough" Nancy wrote about in her class notes to Harvard.

If any woman could be considered a genius, Nancy thought, it was Janni. Surely her work was good enough to deserve a Nobel Prize.

To find the genes responsible for development in *Drosophila*, Janni and a colleague at Princeton, Eric Wieschaus, had done a large-scale genetic screen, mutating every gene in the fly genome to determine which ones were significant to development. When they started, no one had had any idea whether the fly had thousands of genes or hundreds; their work determined that only 120 were essential to the early patterning of the fly embryo. Still, the discovery could not reveal how humans and other vertebrate animals develop; they had organs and systems that flies did not. No one had ever done a large-scale genetic screen in vertebrates; Janni was going to do it with zebrafish, tropical pet-shop fish—the size of minnows and less than fifty cents apiece.

Few scientists were working with zebrafish. But as model organisms, they offered a number of advantages. They were cheaper and more efficient than mice, the most widely used vertebrate models, and shared 90 percent of the same genetic material as humans. While mice produced tiny litters, and usually only three in their lifetimes, fish bred repeatedly, about every ten days, and could lay hundreds of eggs at a time—thousands in a month. Mice embryos developed inside the mother, which made it hard to manipulate or study them as they grew. Fish embryos grew externally, and because zebrafish were transparent, you could watch them develop, with many of the same features as humans: a spinal cord, intestines, a pancreas, kidneys, an esophagus, even tiny teeth. It took just two days for an embryo to become a fish larva, a couple of months to become a mating adult. You could get a lot of fish quickly, producing the generations you needed to identify the mutations that play an important role in early development. Rather than doing experiments on dozens of mice, you could do them on hundreds, even thousands, of fish.

As Friedrich talked, the change Nancy had been seeking became clear. Maybe she could leave virology behind and shift into neurobiology, using zebrafish as her model organism. Maybe Janni would teach her about the fish.

Nancy thought of the zebrafish as an experiment: she wondered if she could recover her passion for science, the excitement she'd felt when she started in phage and tumor viruses, if she changed fields. And she wanted to answer another question: Would things be different in a field led by a woman?

Chapter 15

Fun in Middle Age

Talking on the terrace at Cold Spring Harbor, Friedrich Bonhoeffer told Nancy that he was organizing a weeklong course on neurobiology in the fall and invited her to Tübingen to take it.

Several weeks later Janni gave a seminar at MIT and stopped in to introduce herself. Nancy liked her immediately. She had a warm, direct gaze, and an elegant, lyrical voice that reflected years of singing lieder. She was obsessed with pattern development the way Nancy had been with cancer—why a hand became a hand and a leaf a leaf. And as Jim Watson had said so often of Nancy, she was interested in only the biggest, most important questions.

Nancy was intrigued. She thought maybe she could learn about neurobiology in Friedrich's lab and how to do a large-scale genetic screen in multicellular animals in Janni's and ultimately combine the work to develop her own large-scale genetic screen to find genes responsible for behavior.

Moving out of virology would mean abandoning years of building a track record, trying to establish herself in a new field all over again. Susumu knew Nancy well enough to suspect that she might not have the courage to follow through. So he paid for her plane ticket to Tübingen and told her he'd go with her to check out Janni's lab; he was flying to Europe for a conference anyway, so would stay two days and then go join his wife there.

They flew out the week before Thanksgiving in 1988. Friedrich and

Janni met them on arrival in Stuttgart and swept them off to dinner. Tübingen was a five-hundred-year-old university town stepped above the Neckar River, a fairy-tale village with colorful half-timber houses and a fifteenth-century church, all nestled underneath a rectangular castle with turrets on either end. Students filled the cafés and streets. Set in the green hills just beyond the town was the Max Planck Institute for Developmental Biology. Like Cold Spring Harbor, it was a research lab only. It spread out like a similar kind of biological paradise, a farm with greenhouses, gardens, a rabbit hutch and chicken yards, and labs in simple barracks-style buildings.

In Janni's fish lab, though, Nancy was underwhelmed. The lab itself was embryonic; most of it was still dedicated to flies. Janni was planning a glassy modern fish house that would be built where the chicken yard was, but she was still trying to figure out how to feed the fish and get them to mate regularly. On the windowsills were dishes of green gunk where she was trying to grow fish food the way school-children grow sprouts from avocados. Nancy wondered if maybe her own ambitions for fish had been too grand.

Then Janni slid a lab plate of fresh fish eggs under a microscope: "Look at this."

The first cell was a clear, colorless lump sitting on top of the yolk of the egg. But quickly it began to cleave into new cells, first like hot cross buns, then like little balls bubbling up in a bingo machine. They kept coming—tens, hundreds, and within three hours a thousand cells in a dome on top of the yolks. It was mesmerizing, like watching the stars at night; the longer Nancy watched, the more she could see. The eyes of the fish forming, the brain, the spinal cord. At twenty-four hours a pumping heart, then a brain shaped into three parts, and by the next day a tail waving. The full-grown fish were beautiful, too: sleek and silver and gold, with different patterns—blue stripes or spotted like leopards. They astonished her.

Nancy could see that it was far too early to think about using the fish to screen for the genes involved in behavior; that could be years or even decades away. She'd first have to identify the behaviors she'd want to study—was there a normal way for the fish to sleep, eat, or respond to a stimulus? It was too soon. But she began to imagine bringing the techniques of molecular biology to the fish, identifying the genes that underlay this multitude of patterns. Zebrafish were too unknown a model to be used in behavioral genetics, but they were ideal for understanding the genes required for early development in vertebrates. Maybe someday the fish could help explain why some embryos fail to develop, why humans end up with birth defects or cancer. Nancy had seen how quickly science could evolve to break open a field that had once been unapproachable—it had happened with the repressor and with cancer. Maybe the same would happen with the fish.

And working with fish was fun. She wanted science to be fun again.

Janni was one of only a few people in the world studying zebrafish. The founder of zebrafish genetics had been George Streisinger, a Hungarian Jew who had fled Budapest as a child and grown up in New York City. He had graduated from Bronx Science, where he'd joined the herpetology club, going on weekend trips to the Pine Barrens in southern New Jersey to collect toads and salamanders. After earning a degree from the agriculture school at Cornell (he chose it because there was no tuition at the time), he took the phage course at Cold Spring Harbor, then studied with Salva Luria and Max Delbrück. When the early adherents of the Phage Church were looking for new challenges—Seymour Benzer had moved on to *Drosophila* soon after Mary-Lou left his lab in Indiana—Streisinger began studying zebrafish. He wanted to identify the genes involved in vision, and the transparent fish were perfect because you could watch the eye develop.

He took a position at the University of Oregon in Eugene to help a new Institute for Molecular Biology. He immersed himself in the area's progressive politics, doing experiments that helped convince the federal government to stop spraying dioxin in Oregon's national forests, marching in Saturday peace walks against the Vietnam War. He also judged goats at county fairs. And in 1981 he had been the first to clone a vertebrate—the zebrafish—a development announced on the cover of *Nature* with a picture of the fish clones surrounded by sparkling bubbles, swimming against a cerulean sea.

Three years later, he died unexpectedly of a heart attack while diving to earn his scuba license. But the people he had trained were still working at the University of Oregon. After her week in Germany Nancy arranged to go there.

Eugene was another university town, this one an hour from the rocky coast of the Pacific and surrounded by tall Douglas fir trees. With its wide streets and small local shops, the city echoed an old American frontier town, and to Nancy it was like turning back the clock. The scientists who'd trained with Streisinger still saw him as the guru of zebrafish, unusually generous and collegial, and Nancy felt a kinship with him even though she'd never met him. She liked that he had come out of phage; she still thought of her days in the Biolabs as her happiest in science. Moving from phage to zebrafish, Streisinger had had to start on a shoestring budget. He'd studied the fish in labs set up in World War II Quonset huts and spent years working out the best conditions to breed and maintain them—he'd sprinkled water against the huts to prevent them from getting too hot in the summer and blasted electric heaters to keep them warm in winter. Visiting those labs now, Nancy was reminded of what it was like to be back on Bungtown Road, kicking around in old clothes and talking about basic science.

Again, she was hooked.

When she returned to MIT from Eugene, she told her colleagues she was going to start a new lab, in developmental biology. She planned to take a semester in Janni's lab to learn how to do a large-scale genetic screen in zebrafish. Her colleagues discouraged her from leaving cancer. Phil had never heard of Janni. "Only second-rate people work in fish," he told Nancy. He warned her it was hard to switch fields in mid-career, especially to a field that was so undeveloped: "You'll never get grants, you'll be out of science in three years."

Nancy took Phil's caution seriously. They were far from their days sharing office space on the fifth floor and now saw each other only rarely. But they were old friends, and his was still one of the opinions she valued most. He himself had taken a risk to start Biogen.

She knew he was right about grants—trying to do a screen on a vertebrate was risky financially and scientifically. While Nancy had taken risks earlier in her career, moving from phage into cancer and from DNA tumor viruses to retroviruses after her postdoc, now she was forty-five. She'd spent most of her career in virology. If she switched fields, she'd have to build a whole new lab, which required a lot of money. With no track record, no publications, not even any preliminary results, she'd have trouble convincing the grant makers to give her any.

Yet she felt compelled, perhaps beyond reason, the same way she had when she approached Jim Watson as a shy nineteen-year-old with no experience in biology to ask if she could work in his lab. Like DNA, fish had quickly developed into an obsession. This was the thing about science. The politics could frustrate her, even enrage her; it was hard work to raise money and run a lab. But she couldn't live without it. She and her one-track mind, as Jim had called it. She had seen something in the fish, an escape, maybe, the joy of the past, or a future. It looked beautiful.

She began her sabbatical in September 1989 with some anxiety. She told herself she could go home after a few weeks if she didn't like Tübingen, but by the end of the month her experiment had produced its initial result: she was having fun in science again. Replying to a letter from the editor of the textbook she had written with Jim, Nancy brimmed with excitement about zebrafish, reporting that she had injected tracer dye into the eggs and watched them turn into fluorescent green fish—"living breathing fish."

"I can't tell whether my extreme delight with this is the incredible pleasure of working in the lab after all these years, the extreme joy of escaping responsibility and the innumerable irritations of my job or a genuine interest in zebrafish," she wrote.

Watching Janni, Nancy was seeing a woman do science at the top of her field. MIT had produced the discoveries of reverse transcriptase, split genes, and the human oncogene, but the science she heard at Janni's weekly lab meetings was as exciting as anything Nancy had seen at MIT in recent years. The atmosphere at MIT had changed. Her colleagues always seemed to be traveling, cultivating their new companies. Janni was almost always in the lab, recruiting new postdocs and students to work on fish, tending to the plans for the new fish house.

Janni lived in an apartment in the millhouse of a fourteenth-century monastery in Bebenhausen, a tiny village not far from the labs. In her backyard she planted flower gardens and stocked a fishpond built by the monks with small native fish—on hot days she swam in it. She ran her lab as an extension of her home, like a family; she loved to cook and baked cakes for the students and held frequent parties.

Nancy was staying in a guesthouse across the street from the lab at Max Planck, and only rarely did anyone else stay there. In the mornings she would linger over her coffee and *Laügenbrotchen* in the cafeteria downstairs, overlooking the institute's gardens, and in the distance the Swabian Alps. Janni tasked her with figuring out how to get the fish to lay eggs reliably each morning. Working in a basement lab, Nancy discovered that the fish laid more eggs if you lined their small shoebox-size tanks with artificial grass. Janni didn't believe her, so Nancy collected the data to show how many more eggs resulted from the grass carpets. She also learned to swat away the fruit flies that had escaped from the lab upstairs.

The two women became fast friends. Janni, too, was single and had no children. Both had mothers who had been painters and taught their children to love music and art. Janni cooked dinner for Nancy many nights, and sometimes they watched movies at Janni's apartment or sat with a glass of wine watching the fish in the aquarium Janni kept in her lab for fun. The flies in the lab loved red wine and landed in their glasses. Janni didn't seem to notice, so Nancy tried to look nonchalant as she picked them out of hers. That November, the Berlin Wall fell, and Nancy traveled there to take in the jubilant crowds celebrating open borders and a new post–Cold War world.

Janni was eager for company; she was the only woman among roughly a hundred faculty members at Max Planck. But in developmental biology she was hardly rare; half the people Nancy considered at the top of the field were women, and it was not unusual at conferences for women to occupy every seat on the dais. Nancy had never seen that at virology or cancer meetings. As a newer field, developmental biology had offered more opportunity to women, and as Janni and the other pioneers worked their way into prominence, younger women sensed the field was friendlier to them.

Janni had developed a fish screen that would damage the genes in order to induce mutations. By watching the fish develop abnormally, she could identify the genes that were responsible for normal development. The goal was to identify a starter kit for vertebrate development. It was what she had done in *Drosophila*. Other biologists had used her work to clone those genes, which allowed them to understand what proteins the genes coded for, and the function of those proteins in development.

But that had taken years—even now, nearly a decade later, not all the fly genes had been cloned and sequenced. Nancy didn't have years. But she also thought she could find a more efficient way to isolate the genes in the fish, with a technique called insertional mutagenesis. She would figure out how to insert DNA directly into the embryos' own DNA. The inserted DNA would disrupt the genetic activity in the embryos and act like a tag on the fish genes. She could then pull out the DNA adjacent to the tag, which would include the gene she was interested in. If a mutation caused a fish to grow a misshapen jaw, for instance, the tag would be on the gene essential to grow the jaw.

Still, she would have to invent the technology to do this; no one had done insertional mutagenesis on a large scale in vertebrates.

Janni sent Nancy back to Cambridge in early 1990 with twenty-two healthy fish to help start her lab, a Noah's ark of mutant fish couples: albino, spotty, long fin, wild-type. Nancy cradled them in a small tank on her lap on the plane home. Like many people, she'd tried to keep goldfish as a child, and like many children she'd found them floating dead on their sides within three days. She couldn't afford to kill these fish. She needed them to produce tens of thousands of families, and for those families to grow a third generation, which she would examine to determine if a mutation was inherited. If the first or second generation got sick or died, she'd have to start all over again. It was too easy for them to get sick—she'd seen it happen in Janni's lab—and it took months to rid the fish stocks of disease.

Arthur Merrill met her in Cambridge. They were no longer romantically involved, but they had remained good friends, and she still took him to occasional work events, not least because he was so interested in science. Now he helped save her from disaster on her very first day in the lab: she noticed her new fish swimming extremely fast, so Arthur called a fish expert he knew at Harvard—Arthur had audited the professor's course and given money to his research. The fish expert rushed to MIT and after one look told Nancy the water in the tanks was too warm.

Arthur gave her $30,000 in matching grants from Citicorp. Phil helped get her a grant to create a new room where she could control the lighting so that the fish had absolute dark at night and then light in the morning—their signal to mate. Phil also gave her money from the cancer center budget to pay a technician for a year and helped her secure a one-year $50,000 grant from MIT. Nancy figured that if she could pull a bit of money from a grant she still had in virology and do a lot of the lab work herself rather than hiring more postdocs, she could survive for a bit without additional money. But she knew she was cutting it close. If she didn't get a result in the next year or two, she wouldn't get grants. It would be over for the fish project, and for her. She'd be fifty, too old to switch fields again.

Other colleagues were still skeptical. But something unexpected happened. As word got out that Nancy was moving into zebrafish, graduate students and postdocs began applying to her lab, unusually talented young scientists. The first was Shuo Lin, who had been born in China and came to the United States to do his PhD. Nancy could tell from the papers and photos of data he spread across her desk that he was a superb experimentalist. She wondered if he understood how challenging it was to develop technology to use in a new organism. "I don't know if I dare take on a postdoc and risk your career," she told him.

"I came here all the way from China with nothing," Shuo responded. "You don't know risk."

Soon there were others, and a humming lab, with rapidly repro-ducing stocks of fish. Everyone understood that this was not a big, rich operation, not yet. They were working in tight quarters, especially as the fish tanks took over more space. They were using old equipment, suited more to virology than developmental biology. But the students and technicians in the lab—"my kids," Nancy called them—shared her excitement about starting something new, and her willingness to work long hours.

Shuo set about trying to find the cells in the embryo that became the germ line—the cells that become the sperm and the eggs and pass on genetic traits from one generation to the next. First he transplanted cells that were responsible for the black pigmentation of the wild-type fish into the embryos of albino fish at the stage when they divided into about one thousand cells. This created chimeric fish with stripes that ran vertically instead of horizontally over their bodies, with different stripes on each side, and confirmed that it was possible to transplant cells from one kind of fish to another. They named the first chimera Greta, after Garbo, the mysterious movie star. Then Shuo mated the chimeras with albino fish and found some fully pigmented wild-type fish among their offspring. That meant he'd found the germ line; it was in the embryo at the thousand-cell stage.

Greta earned Nancy her first grant in fish, in 1991. She had applied to the National Science Foundation, and two women came up to Cambridge from Washington to inspect her lab, to make sure she had the tanks and equipment to indicate she was serious about the new work. It had been intensive work—Nancy had taken just one day of vacation that year—and she still hadn't developed the technology to insert the tags into the genes.

Then, at a meeting for developmental biologists in California, she ran into a German scientist who told her that mouse retroviruses mixed with a virus known as VSV could infect fish cells in culture. No one had been able to grow such a virus at the concentration she would need to cause mutations in the fish, but the German man told Nancy that a lab at the University of California in San Diego had created stocks of it at high concentration. Was it conceivable the virus could act as a vector, carrying the tag to the genes?

Nancy had spent most of her career working with mouse retroviruses. Her lab was already outfitted to work with them safely. It was almost as if she hadn't switched fields at all. She'd have to figure out how to insert the viruses into the fish. There was no way of knowing if the viral vectors would work in fish the way they had in mice. Still, she called back to MIT and told her kids to drop what they were doing and get some of the virus from the lab in San Diego. It was a long shot, a promising lead, and possibly just the break she needed.

Chapter 16

Three Hundred Square Feet

Nancy had returned to Cambridge from Germany at the dawn of a new decade. To judge from the newsstands, women were staggering into the 1990s, exhausted and lonely—victims, apparently, of their own professional success. They had focused so hard on their careers that their personal lives had shriveled. *Newsweek* had set the tone with a cover story in 1986 declaring that "bright young women" who had "single-mindedly" pursued careers were more likely to be murdered than married. "Many women who seem to have it all—good looks and good jobs, advanced degrees and high salaries—will never have mates," the magazine warned, quoting a recent study. If they were still single at thirty they had only a 20 percent chance of marrying. By forty, they were "more likely to be killed by a terrorist: They have a minuscule 2.6 percent probability of tying the knot."

In 1989 the *Harvard Business Review* attempted to analyze why women were still underrepresented in the highest ranks of management. Women were more expensive to keep because they took maternity leave, but more prohibitively, the article argued, the men who still ran the world harbored old stereotypes about women. These men still saw parenting as "fundamentally female" and career as "fundamentally male." ("Why would any woman choose to be a full-time chief financial officer rather than a full-time mother?" asked the chief executive of Pfizer.) The men believed that women or men who took parental leave lacked a commitment to their careers; meanwhile, "women who perform as aggressively and competitively as men are abrasive and unfeminine."

The author, Felice Schwartz, had founded Catalyst, a nonprofit that worked with companies to tap into the talents of women who wanted to combine work and family—helping develop policies on maternity leave, and finding women to serve on corporate boards. She argued that organizations had been built on the assumption that everyone had a wife at home, and they needed redesigning to keep up with the new demographics of the workplace. Women should be allowed flexibility to take time off, do some work from home, go part-time for a while before reentering the competition for the top jobs. Schwartz urged male managers to recognize how hard it was for their female peers when they were still so often in the minority, or even the only woman—excluded from after-work socializing, assumed to be secretaries, unsure whether they could take business trips with men or speak out when they felt discriminated against. "The male perception of talented, ambitious women is at best ambivalent," she wrote, "a mixture of admiration, resentment, confusion, competitiveness, attraction, skepticism, anxiety, pride, and animosity."

The article backfired. Other prominent women derided her proposals for flexibility in careers as the "mommy track." Or "a mommy trap," as Betty Friedan told the *Los Angeles Times*, calling Schwartz's arguments "retrofeminism." For all Schwartz's attempts to explain the nuance, her article ended up reinforcing the notion that what women really wanted was not to work but to be at home with their children. What stuck was that children—not stereotypes—were limiting women from full participation in the workplace. It had been a quarter century since Alice Rossi "immodestly" proposed more day care and more participation from fathers in raising children. If women wanted a solution, they were going to have to figure it out themselves.

In December of that year the cover of *Time* featured an illustration of a woman built like a linebacker, carrying a briefcase in one hand and in the other an oversize, grimacing baby. "Women Face the 90s," the headline announced. "In the 1980s they tried to have it all. Now they've just plain had it. Is there a future for feminism?" Like the mommy track article, this one stirred controversy; the author revealed that the magazine was going to headline the piece "Is Feminism Dead?" but didn't only because a recent cover had asked the same of government. (And, twenty years earlier, of God.) Still, the article read like an obituary for the women's movement.

The movement had been too white, too middle-class, too focused on the wrong goals. Women in their twenties took its hard-earned gains for granted. They saw feminists as strident, loud women who refused to wear makeup or shave their legs. "I'm feminine, not a feminist," one twenty-seven-year-old told the magazine. Women in their thirties and forties might consider themselves feminists but resented the previous generation for pushing the Equal Rights Amendment and lesbian rights rather than fighting for maternity leave, subsidized childcare, or flexible work. Women who worked full-time earned just sixty-six cents to every dollar a man made. The feminists of the sixties and seventies hadn't warned working mothers how hard life would be to juggle; stay-at-home mothers, meanwhile, felt that feminists demeaned them.

Deep in the article were statistics showing that an overwhelming number of American women thought the women's movement had made life better and was still improving it, even if relatively few identified themselves as feminists. Self-described feminists told the magazine that the movement hadn't died, it had only changed shape; rather than one women's movement, there were many smaller groups working for change, groups that included men and took a less confrontational tone. "It is not top-down, it is bottom-up," Hillary Clinton, identified as a partner in a Little Rock law firm and the wife of Arkansas governor Bill Clinton, told the magazine. "Who wants to walk around with clenched fists all the time?"

Fists would clench soon enough. In October 1991, Clarence Thomas, a Black, conservative jurist, was the nominee to fill the seat on the US Supreme Court that had been left vacant by the retirement of Thurgood Marshall, the pioneering Black civil rights lawyer. Democratic aides on Capitol Hill found a former colleague of Thomas's, a Yale-educated law professor named Anita Hill, who reluctantly accused him of sexually harassing her. In televised hearings that were watched by nearly 90 percent of Americans, Hill testified that Thomas had repeatedly pressed her to go out with him even after she'd declined. He regularly made graphic sexual references around the office, talking about pubic hairs on his Coke can and regaling the office about the pornography he watched—group sex, bestiality, rape. The members of the Senate Judiciary Committee, all men and all white, accused Hill of making up the accusations, being unstable or delusional. Thomas insisted she was lying. When other women offered to corroborate her account, the committee declined to summon them, rushing to approve Thomas's nomination in the wee hours of the morning.

Polls taken after the hearings showed that Americans believed Thomas more than they did Anita Hill, by almost a two-to-one margin. Still, the senators' treatment of Hill released a torrent of anger among women who had themselves been sexually harassed or felt unheard in the workplace—not taken seriously, erased. The faces on the Judiciary Committee had underscored how women were still excluded from the highest levels of political power. The next year, a record number turned out to run for Congress, prompting reporters to brand 1992 "the Year of the Woman."

At MIT, there was a new Biology chairman, Richard Hynes, who in his first year pledged to increase the number of women and underrepresented minorities on the faculty. "We have recently made offers to four junior faculty members, three of whom are women and the fourth Afro-American," he wrote in his report that year.

Biology as a discipline had risen in stature at MIT. In 1990 the MIT Corporation announced that it had chosen Phil Sharp as the university's next president. Phil accepted the job, then reversed himself—as he contemplated giving up teaching and dissolving his lab, he said, "I came reluctantly to the realization that I could not fill that void in my life with anything else." Still, choosing Phil signaled a reordering: most of the previous presidents had been engineers. On campus, a new Biology building was under construction, a "special priority" of the university's $700 million fundraising campaign.

In 1991, the faculty voted to add an introductory course in biology as a core requirement for all undergraduates. It was rare to add any academic requirement, and the faculty had debated the proposal for two years, with some objecting that there were too many requirements already. But proponents had argued that the developments of the second half of the twentieth century—the double helix onward—had made biology as fundamental as physics or chemistry. The Human Genome Project was now underway—Jim Watson, who had evolved into what Salva Luria described as "a rambunctious statesman of science," was leading it. A growing number of biotechnology companies were filling in the blank spaces around Kendall Square. As Nancy had recognized when she began teaching the joint women's studies course on reproductive technologies, every student would need a basic understanding of genetics and molecular biology.

Like Nancy's course, the new requirement would include discussions on the ethical implications of modern biology. The department promised that it would be taught by Biology's "best lecturers." Richard Hynes asked Nancy if she would help design and teach it. She was flattered to be included in such an important department responsibility and was at first surprised. But she noted that she'd created the course in women's studies combining introductory molecular biology and ethics, and that the class had earned unusually high ratings. Her experience made her the obvious choice to create the new course. "I know," Richard told her, "that's why I thought of you."

She was also anticipating another big improvement: a renovation of the cancer center was finally going to create more space and shared equipment rooms on the third floor, where Nancy's lab was. Salva had been right when he warned her before she moved downstairs that it would not have the equipment she needed. The move had been worth it to get away from the fifth floor, even if she and her lab members had to run to different floors to find the equipment they needed. Susumu had given Nancy some of his castoffs, which helped. But her fish work required specialized equipment: rooms and water with precise temperature control, the timed lights to control mating. With the grant that the fish named Greta had helped her secure from the National Science Foundation, she had bought new fish tanks, and she needed another light-controlled room to keep them. Her lab was also becoming more crowded; shortly after winning the grant, Nancy had attracted another postdoc, who had arrived with her own funding from the National Institutes of Health.

The renovations would provide three new rooms on the third floor, to be shared by Susumu, Nancy, and Hidde Ploegh, a Dutch immunologist and a friend of Susumu's who was to arrive on the faculty the following year. Susumu was now running a mouse lab to study behavioral genetics, and Phil, the head of the cancer center, had promised him that half of one of the new rooms would be his exclusively. Susumu told Nancy she could take the other half. They agreed to put computers and bioimage analyzers into another room. That left a third equipment room that Nancy and Hidde would share.

It would have been more than enough space for her needs. But how to divide the new rooms turned into a fight that consumed much of the time and energy Nancy would have otherwise spent on her fish research.

In July 1991 Richard Hynes and Phil switched jobs; Phil became chairman of Biology and Richard the head of the cancer center. Soon after, Nancy and Susumu heard that Richard had promised half of the third room to the biopolymers group. Susumu became furious. He called the building manager and instructed him to bring up old equipment from a basement storage room to fill the space. Huge machines soon appeared, dusty and obsolete hulks, some with wires dangling, others that had been cannibalized for parts and would never work again.

The building manager couldn't understand why Susumu wanted the equipment there. "You can't possibly work in the room if you try to put three rows of equipment in there," the manager said. Susumu told him to leave it. As long as the space was occupied, Richard couldn't give it to anyone else. Together, Susumu and Nancy drew floor plans on graphing paper and cut up small pictures of centrifuges and freezers, moving them around to come up with a plan that they could show Richard to indicate that they needed every inch of the newly created spaces—no room for the biopolymers lab! They laughed as they did it; sitting at the table with their scissors, Nancy thought how ridiculous it was: two adults, accomplished scientists, one of them with a Nobel Prize, playing with little bits of paper to prevent their director from stealing their space. But space was a prized commodity in universities. It was necessary for research. It was also a signifier of status—Nancy had once heard it said that a professor would rather give up his mistress than give up his space. Susumu, she thought, was giving her a master class in how to accumulate it.

Other colleagues found Susumu mercurial and difficult at times. But Nancy had always gotten along with him. Now Susumu kept saying he was doing all of this for her and Hidde, that he did not need the space himself. She began to think of him as the father of the third floor, fighting Richard on her behalf. She had new appreciation for how well Salva and Baltimore had outfitted the cancer center; as long as she'd been doing virology, she'd had what she needed. She'd never had to ask for anything before, and now that she did, she was happy to have Susumu show her how it was done.

A few weeks later, the building manager came by Nancy's office and asked if he could move one of her centrifuges out of the room she shared with Susumu. Nancy asked why—she thought there was plenty of space for it. She went to look for herself and realized immediately what had happened: Susumu had taken over the entire area for himself. She would have to appeal to Richard on her own.

Nancy asked Richard for a second room to keep her fish tanks, a place to put shelves for the glassware that was used in her lab, and the promised common equipment room on the third floor. Richard told her she could put her tanks in a room on the first floor, but that it was temporary. She'd have to wait for Hidde's arrival for the common equipment room on her own floor. A year later, Hidde had arrived, and despite several conversations with Richard, Nancy had neither a permanent room for her fish tanks nor the equipment room on the third floor.

The fish project was attracting interest and acclaim; Nancy had just returned from giving a talk about it in Japan and had been asked to apply for a job as chair of the Biochemistry department at Dartmouth. She now had two postdocs, two graduate students, and a technician and was getting at least one new postdoc application each week, some weeks three. But she and the people already in her lab were spending an increasing amount of time fixing or hunting down equipment. They'd had to wrap paraffin around the hoses of the aging gel drier to stop it from leaking. The beta counter had stopped printing, so they were writing down the readings by hand. The centrifuge was old and took twenty-four hours to do what most labs could do in eight. For microscopes, they'd been relying on the kindness of friends in other labs. One of Nancy's postdocs had given up on an experiment because the scope he needed was in another building and several times he'd reached it only to discover that he was too late for the time-sensitive process he was trying to observe. Nancy was convinced he could have made the experiment work if there had been a microscope in her own lab.

Nancy had already purchased three microscopes on her own, but she needed one with higher power. It would cost about $30,000, more than a third of the amount of her grant from the National Science Foundation, and she didn't want to buy it until she had figured out the method for insertional mutagenesis—how to tag the genes—so she would know which kind of microscope worked best with it. She surveyed the cancer center and found three microscopes that would work for now, including a fluorescence scope that was slightly outdated but had the benefit of being largely unused by others. In May of 1992, she asked Richard if she could borrow it for her lab. Richard said no and told her if she needed a microscope, she should write a grant proposal to buy her own.

Richard was about Nancy's age, tall, bearded, and British. Nancy admired him as a scientist and had considered him a friend, but now the combination of his height and his crisp accent made her think he was condescending to her. It was almost unheard of for anyone to write a grant just to buy a microscope. The cancer center had always had a central grant from the National Institutes of Health to buy equipment—Salva and David Baltimore had set this up. Microscopes usually came from this core grant. That's how the one she wanted to borrow had been bought.

Her conversations with Richard about space over the last several months had frayed both their tempers. The discussion about the microscope quickly devolved into more of a playground dispute. Nancy explained why she wanted to wait to buy a new microscope and told Richard that the one she wanted to borrow had been used only five and a half hours that year. Nearly half of that time it had been members of her lab using it.

"That's not true," he said.

"Yes, it is."

"No, it isn't."

"Yes, it is."

She told him she had photocopied the logbook indicating how many hours people used the microscope.

He told her people didn't always sign in to the log.

"You can read the light source," Nancy snapped. "It was in perfect agreement with the log."

Meanwhile she felt Susumu encroaching further. His lab members had been using one of the common rooms to package and receive deliveries, but now they had filled it with more equipment, sending the packing and deliveries into the hallway outside Nancy's office and lab. Getting to work had become an obstacle course.

In July, after Nancy tripped on a delivery box in the hallway and hurt her back, she went to Mary Rowe in the ombudsman's office to complain. Mary had been thinking about sexual harassment in the workplace for twenty years—ever since she'd described the accumulation of slights and disadvantages built up over time as "Saturn's Rings." Listening now, she told Nancy that it sounded as if she was experiencing sexual harassment. Nancy didn't agree; she felt harassed, but no one was making the pornographic jokes or come-ons that Anita Hill had described in the Clarence Thomas hearings the previous year. Mary explained that harassment didn't necessarily involve sex; MIT's policy, like others, defined it as a pattern of behavior based on gender, conduct "which has the intent or effect of unreasonably interfering with an individual's or group's educational or work performance at MIT."

Mary told Nancy about "John Doe / Jane Doe" studies showing that women's contributions were not valued as highly as men's, especially in workplaces dominated by men. Women and men alike, she said, rated the same work more highly if they thought it was done by a man. Nancy took detailed notes—drawing two lines for emphasis under "undervalued." But for the moment she was focused on getting what she needed in her lab. She was already spending half of her time fighting battles over space and was increasingly miserable about the time it was stealing from her fish research.

On Mary's advice she wrote Richard a six-page letter reasoning out her requests and suggesting possible solutions to finding new space. She asked again for a second fish room and offered that she could raise more grant money for renovations if Richard could find the square footage. She wanted a freezer so she wouldn't have to use one on the fifth floor to store her cells. She had recently paid $6,000 to repair a piece of equipment in her lab only to be told that the department was covering the cost of such repairs; she asked to be reimbursed so she could buy a recirculating filter for her fish tanks. She and her lab workers were having to scoop the fish out of the tanks every day by hand to refresh the water, which was time-intensive as well as messy. She had been putting together her own grant proposals at home at night because the central office that was supposed to help had recently told Nancy it was too busy; Nancy had missed a scientific meeting to stay home and finish a recent submission. And at home she had only an old dot matrix printer that spat out blurry images, which she thought made her proposals look unprofessional. She wondered if MIT would pay for a laser printer to replace it.

Mary had also instructed Nancy to ask Richard for a breakdown of resources in the cancer center: how much space each faculty member had; how much money they were expected to raise from grants; and how much MIT contributed to their labs.

Richard requested a meeting in his office, and Nancy asked Mary to join. Richard again told Nancy that if she needed new equipment she should raise more grant money to buy it herself. He presented bar graphs showing that based on the number of people she had working in her lab, Nancy had as much space as eight other senior faculty members in the department. She had more than anyone except Phil Sharp if you compared space based on how much each faculty member earned in total direct compensation. These weren't the numbers that Nancy and Mary had asked for; if some people had fewer resources to begin with, of course they could not hire as many lab members or do the same volume of work. It wasn't clear how much anyone was getting from the core grant to the cancer center.

Nancy left the meeting feeling worn down. "I would not have thought to articulate it," she wrote Mary in a note afterward, "but you are quite right that it is probably the tremendous stress of demanding their share that causes women to retreat too soon and thus to almost invariably end up with less."

Mary helped provide a brief respite, petitioning the dean of the School of Science to get Nancy $31,000 from a foundation to buy a new fluorescence scope. In the meantime, Richard agreed to let her borrow the one she'd wanted, though she'd have to wait another month until another faculty member returned from vacation to sign off. Within four days of receiving the microscope, Nancy's post-doc was able to complete the experiment he'd been trying to do six months earlier.

But in November, Nancy had a new crisis. The technician who had worked in her lab for seventeen years quit to go work for a professor down the hall. The administrator for the cancer center had told the technician that Nancy did not have enough money, and the technician had worried she'd lose her job. Nancy was alarmed—she knew money was tight, but not this tight—and asked to meet with the administrator to discuss her grant funding. She brought along her sister, Ann, who had trained as an accountant and worked as a stockbroker.

As at most universities, lab heads had to pay part of their own salaries by raising grant money, and MIT paid the rest. The administrator began the meeting by saying that Richard had noticed when he started as director of the cancer center that Nancy's salary was low relative to others in the department, and she was paying an unusually high proportion of her own salary. MIT paid 35 percent, but Richard was determined to raise it to 45 percent as soon as the budget allowed. Ann asked what portion of their salaries other professors paid. The administrator told her it was different in each case.

That afternoon, Phil dropped by Nancy's office to confirm that she would teach the biology requirement again the next year. Nancy asked him what portion he thought MIT paid toward other professors' salaries. Phil said it was somewhere between 60 and 70 percent.

"Mine is thirty-five percent," Nancy told him. Phil looked perplexed. He said that sounded impossible.

When Nancy asked Richard, he told her she was misinformed and that she must have written down the wrong numbers. Nancy told him that her sister had written the exact same numbers. She asked again what other professors were paying toward their salaries. She wasn't asking just to see where she stood; if she knew how much others received in grants, she could count how many people they had hired and maybe understand better how many people she could afford for her own lab. Richard refused to say, except that she was close or slightly above the average. He was irritated. He was the one who had noticed she was earning less than others. Couldn't she see he was trying to help her?

On the phone that night, Ann guessed that Richard was lying. Nancy couldn't believe he was. But she wasn't entirely sure she trusted him.

"It is difficult to obtain correct information," Nancy wrote Mary Rowe in December. "I didn't pay attention to these matters for 20 years because I assumed that the director and administrators were only interested in helping me and others—all equally—to do our science. I no longer believe this. I'm sorry to say that I no longer even believe that I will necessarily be told the truth."

Nancy returned in January after the holiday break determined to solve her space problem for herself. She started by asking the staff from the kitchen that cleaned the glassware where she could store the small tanks and beakers and dishes that were increasingly crowding the counters in her lab. They laughed and pointed at a small room that was supposed to be for common storage. Susumu had filled it with his own glassware cabinet. "Nancy," one woman asked, smiling, "why do the others get everything and you get so little?" Nancy smiled back, but inside she cringed. So this was how people saw her.

That afternoon she decided to move some of the scientific journals in her office onto the shelves in a room that was supposed to be a shared conference room, as Richard had suggested in one of their meetings that fall. Susumu appeared in her office and demanded to know why she had placed them there. Nancy lost her temper: Richard had given her permission, she told him, and besides, she'd been waiting a long time to sort out the use of the common spaces on the floor. Her lab had been disrupted for years and she needed to move forward. She had less space than either Susumu or Hidde—they each had a private conference room and she had none—so it seemed reasonable that she might use the empty bookshelves in the conference room they were supposed to share.

Susumu grew angrier and said he would talk to her only if he had a tape recorder. Nancy wrote a note to Richard asking him to confirm that he had given her permission. She gave a copy to Susumu, who ripped it up.

She told Richard that she thought she had less space than any other tenured faculty member. Her science was suffering. "He said this was not true," she wrote that night in an account she had begun keeping. "That he was certain his office was not bigger than mine but that he would measure it and let me know!!!!!!!!!! I think a quick tour of Richard's office (and particularly of available shelf space) and of mine will reveal how close to dishonest this statement of his is. It is ridiculous to the point of being crazy."

Nancy called the director of the MIT safety office, who sent over a custodian to clear the debris from the corridors outside her lab. She sent the director a thank-you: "It is an extraordinary feeling—as if we finally join the real world!" But her triumph was short-lived. Susumu hand-delivered a memo on cancer center stationery, saying his lab would no longer allow the "habitual use" or borrowing of its equipment.

Nancy showed the note to Richard and again said she thought she had less lab space than any other senior faculty member. Richard said he doubted that, but again promised to measure his own office and let her know.

The next week Richard responded with a quick note apologizing for not getting back to her sooner but noting that the hallways had seemed clearer—they were, but only because Nancy had called the safety director. Richard added a note about office sizes: "In fact yours and Susumu's are identical in size; other senior faculty offices are either smaller or larger than yours but not by very much (e.g. mine is 5 sq. feet larger than yours.)"

Nancy wrote to Mary: "Is Hynes crazy? A total liar? Or just a run of the mill male chauvinist?" Nancy had been dreading "the harassment issue," as she wrote, "but I think perhaps I am starting to look forward to it. Imagine how it would feel to be a junior faculty person dealing with this man, or Susumu for that matter."

But Nancy also had good news: she'd received a new grant from the National Science Foundation to develop a method to inject the virus she'd gotten from San Diego into the fish embryos. Shuo Lin, her postdoc from China, had also won more money from two fellowships. "Miracles never cease!" she wrote in a note to Jim. "We're alive for at least another year or two."

If she was being harassed, she wasn't sure it was based on gender. But Richard had planted the seed of an idea. A six-year-old, she thought, could see that other faculty members—male faculty members—had more space than she did. If Richard wouldn't give her the numbers she'd asked for, she'd get them herself.

So she waited until the cancer center was quiet at night, the last graduate students having left to forage for dinner, then used her master key to slip into a lab on the second floor. She left the lights off so no one would see her. She got down on her hands and knees, pulled out a measuring tape, and began to stretch it across the room. She measured every room on the second floor, then the third floor, and over the next weeks every lab space and office and equipment room in the building, maneuvering around centrifuges and desks and liquid-nitrogen tanks to reach the limits of the walls, marking spaces on a floor plan with different-colored highlighters: orange and blue for space belonging to senior faculty, pink and green for junior faculty, yellow for the common equipment spaces. Only once did someone nearly discover what she was doing. She had just let herself into room 537—a common equipment room on the fifth floor—when she heard someone in the hallway outside. She froze, wondered how she'd explain being alone in a darkened room late at night. Should she crouch down to avoid being seen? From the hallway there was only a small window in the doorway. She decided it was best to remain standing, to try to look at ease. *They'll think I'm crazy*, she thought. She was beginning to feel crazy, but if she was, Richard and Susumu were making her crazy. She heard the footsteps pass, then the stairwell door open. The footsteps disappeared. She waited a minute to make sure there were no more. Then she got down on her hands and knees and stretched out the measuring tape again.

Her instincts were right. Richard was not. Even a junior faculty member—a man—had almost double the space she did. Richard had four times as much. Everyone else in the cancer center had more space than Nancy.

Her late-night measuring made her think of what Mary Rowe had said the previous year. Maybe women's work really was undervalued. Her students and technicians had long before noticed that Nancy's lab had fewer resources than others; they joked about their embarrassment, like children forced to wear ratty hand-me-downs. (On a list of equipment needs for the lab, one suggested a psychotherapist.) The conditions in her lab were broadcasting her lack of value.

She started seeing the undervaluing in other ways. At a lunch meeting in the cancer center to discuss job applicants that February, faculty members dismissed a résumé from a woman because she was too old, then agreed to consider the résumé of a man who was two years older, and to Nancy's mind no better. She wrote to Richard, "I confess I had come to believe that it was the year of the woman everywhere else, but certainly not at MIT!"

Her students, she told him, had asked her why there were so few women in the department; were there no serious female candidates? "I sense increasing demoralization among young women scientists,"

Nancy wrote Richard. "And I confess that I can not tell them their perceptions are wrong."

In March, Nancy recalled that Ruth Perry, who had recruited her to teach the women's studies course, had contemplated suing MIT when she did not get tenure. (MIT had agreed to review her case, and she won.) Ruth gave Nancy the name of a prominent civil rights lawyer in Boston.

Nancy met with the lawyer on a Friday morning in mid-March. She took a copy of MIT's policy on harassment, though she still wasn't sure that's what she was experiencing. "Of course it's harassment," the lawyer told her. But he thought she might also have a case of sex *discrimination*—a word that had not occurred to Nancy. To Nancy, discrimination meant you couldn't get a job somewhere. The lawyer told her cases were hard to prove, and she would have to be prepared to leave MIT. But, he said, he had been impressed in Ruth Perry's case with the good intentions of the central administration at MIT. Go to the provost, he told Nancy. The provost was the chief academic officer at MIT; the deans of the five schools reported directly to him. Nancy thought so rarely about university politics that she couldn't even think of his name; the lawyer had to tell her. Don't say you've seen a lawyer, he said, but tell him you've had other job offers, and if he doesn't seem inclined to give you what you want, mention that you think it's discrimination.

Nancy emerged from the appointment into snow flurries that announced the start of what would be called the Storm of the Century, a blizzard that shut down the eastern half of the United States, knocking out power, collapsing roofs, dumping fifty-six inches of snow even in Tennessee. Around Boston, a foot of snow shut down Logan Airport and the T, and the Athena computer servers at MIT. The city was snowbound and quiet for the weekend, and Nancy spent the time at home calling her sister and Jim and other old friends to catch up and talk about what the lawyer had told her. His words had reassured her. She hadn't wanted to sue. Early the next week she wrote Mary, sounding more relaxed than she'd been all year. "The snowy weekend turned out to be a wonderful thing I think," Nancy wrote. "People seem to have great confidence that somewhere in the MIT system there is a person with both the authority as well as the goodwill and ability to solve this problem—possibly in a few hours. I'm going ahead to try to find this individual because fundamentally I really agree with them."

Richard asked to see her a few weeks later, and Nancy brought Mary along. Richard wanted to talk about job candidates. Nancy said she thought the recent search had been discriminatory. When Richard rejected the notion, Mary suggested the cancer center could do its own "John Doe / Jane Doe" study, compare the same résumé with a man's name and a woman's name on it. Richard bristled again and said he saw no evidence of gender discrimination in the department.

Nancy wanted to know—again, still—when she might be able to get more space. She hadn't yet bought the fluorescence microscope with the money Mary had secured from the dean because there was no place to put it in the lab. Nancy would soon have fifteen people working in her lab; it was already so crowded that she worried she'd have to turn down two undergrads who wanted to work for her that summer and turn away postdocs she'd already promised jobs. Richard suggested she could store her glassware in the kitchen across the street in the Biology building. Nancy noted that Richard's own glassware was in his lab. Why should hers be across the street?

She asked if they could walk one flight up to the third floor to look at the space.

"That's a good idea," Mary said.

Richard refused. "I'm too upset!"

Nancy looked at Mary. Mary wasn't going to move without Richard. And in that moment, Nancy realized that she was going to be on her own.

Her meeting with the provost, Mark Wrighton, was scheduled for the next day. She had written him a two-page letter outlining her requests, with a proposed floor plan attached. She was seeking about 300 square feet: she wanted exclusive use of a room that was 113 square feet, and space in another larger room to house large equipment. As the lawyer suggested, she said she had been recruited for other positions, but did not want to leave MIT.

She arrived in Wrighton's office toting a heavy canvas bag loaded with applications so she could show him the caliber of the postdocs who were applying to work in her lab. Wrighton appeared delighted to hear about her work and made Nancy feel he had all the time in the world to talk to her—but he could not understand how the issue had reached his office. No one, he told her, had ever asked him for so little. He handled budgets for the entire university, not disputes about storage cabinets. Why, he asked several times, did her department not support her work? Perhaps Richard, being the head of the cancer center, wasn't interested in developmental biology? Finally, Nancy tried out the word the lawyer had given her: "I think it may be discrimination."

She thought she saw Wrighton sit a bit straighter. That would be serious, he said. He told her she should go see the dean of the School of Science, Bob Birgeneau. Wrighton would call him in advance and was sure the dean could get Nancy what she needed.

Birgeneau was a physicist. A year older than Nancy, he was tall, Canadian, and clean-cut, efficient in manner. He had heard from the provost and told Nancy he would talk to Richard. Nancy guessed Birgeneau had already talked to Richard and that Richard had told him she was difficult. Birgeneau seemed unimpressed by her problem. This time she waited until she was walking out the door to use the word the lawyer had given her: "I think it may be discrimination." Birgeneau followed her into the hallway and told her he'd send an assistant to look at her space.

The assistant came two days later and immediately saw how crowded Nancy's lab was. She reassured Nancy that the dean would take care of things. Within weeks Nancy had the 113-square-foot room she had requested and a new cabinet to store her glassware. It was enough space that she took on another undergraduate to work in her lab—with a recommendation from her, he went on to win a Marshall Scholarship, one of academia's most highly selective. She said yes to more postdocs, provided they could find grant funding, and they did, arriving with money from some of the most prestigious fellowships in the country. That summer, she was one of twelve Americans to win a grant from the Human Frontier Science Program, funded by more than a dozen governments for the study of the brain and biological development.

"It is remarkable what 100 square feet can do for one's ability to function!" she wrote to Wrighton in June, days before her fiftieth birthday. That fall Shuo Lin successfully injected three-hour-old embryos with the retrovirus Nancy had found in San Diego, and the virus had entered the germ cells and inserted a DNA copy of its genes into theirs. It worked, almost incredibly; they had taken the first critical step in doing insertional mutagenesis. If those insertions could cause mutations, they would be able to start the work of isolating genes in the fish.

In October, Susumu had put a lock on the door to his bioimage analyzer, with a note declaring it off-limits. In the entrance to his lab was a sign: LEAVE ALL INCOMING SHIPMENTS IN OUTER BACK HALL. LAB AREAS ARE SIMPLY NOT DESIGNED TO ACCOMMODATE SHIPPING, RECEIVING AND SCIENCE ALL AT THE SAME TIME. The outer back hall was Nancy's front door. On the third Monday in October, Nancy arrived at her office to find the hallway again crowded with heavy boxes, trash, and animal cages. She wrote again to Richard. He apologized, explaining in a note that Susumu's technician had had an emergency with a freezer. That Monday had been an institute holiday, he noted, and the department had been distracted by exciting news from Sweden: Phil Sharp had won the 1993 Nobel Prize.

Richard added that he knew she was looking for more space for fish tanks. "Can you jot down a note about what exactly you would like to relocate with a rough idea of square footage and any special requirements? I do not know what if anything can be arranged, but it would help us to find a definitive answer if you could give us a precise request."

The space wars were not over; it was more a truce than peace. But Nancy was about to be distracted by a new fight, one too big to be solved with her tape measure.

Chapter 17

MIT Inc.

Dean Birgeneau had gotten Nancy the space she needed. But he didn't think the word she'd used—"discrimination"—described the problem. As far as he could tell, hers was a routine faculty gripe, a small one as those went. University leaders liked to consider themselves enlightened; Birgeneau was no exception. He thought of himself as someone who had worked throughout his career against discrimination, who tried to help the disadvantaged the way he himself had been helped.

Tall and trim with salt-and-pepper hair and a high-wattage smile, Birgeneau, now fifty-one, presented as a well-to-do suburban father of four, which he had become. But he had started poor, born in the Beaches, a working-class neighborhood in the far east end of Toronto, to a mother who was potato-famine Irish and a father who was Métis—half-indigenous, half–French Canadian. No one on either side of the family had ever completed high school, and Birgeneau and his siblings assumed they wouldn't, either, though Bob, the third of four children, clearly had an unusual aptitude for math. As a five-year-old he often tagged along with his older brother on his rounds as a delivery boy for a local drugstore. The druggist would lift young Bobby onto the store's counter and give him math challenges to perform for customers, showing off how fast he could calculate the cost of their purchases.

When he was eleven, his family was evicted and ended up in a tenement on the other end of the city. His father ran out not long after, and his older brother and sister dropped out of school to work at age fifteen. Birgeneau assumed he'd do the same. Instead he was saved by a succession of four Catholic priests. The first gave him a scholarship to attend St. Michael's College School, an elite high school affiliated with the University of Toronto. The second pulled him out of a summer job at the Yardley soap factory to tell him that he had an obligation to take his God-given talents to university—the priest had to explain what university was, and how to apply. The third noted Birgeneau's talent in Greek and Latin and tutored him every Saturday to study for the Ontario provincial exams, where a second-place finish won him a classics scholarship to the University of Toronto. Then, the summer before he was to begin, Birgeneau's mother was hospitalized with a nervous breakdown, and he decided he would have to abandon his plan and get a job to support his mother and younger sister. The principal at St. Michael's—the fourth priest—heard this and told him that the school would support his family for the next four years until he graduated from the university. When Birgeneau told his mother, she recovered within a week and was soon well enough to return to work.

His path from university was smoother—a doctoral program at Yale, a job at Bell Labs, and then MIT—but he had never forgotten the sting of an incident in high school, when a wealthy classmate came to the Birgeneaus' apartment to return some class notes and, having had to step over homeless people in the hallways of the building, never spoke to Birgeneau again. At Yale, he had been alarmed to discover that New Haven was profoundly segregated and went to a community center in a poor and largely Black neighborhood to volunteer as a youth group leader. It led to an interest in the civil rights movement and, in the summer of 1965, a job teaching physics at a small, historically Black, Baptist college in South Carolina.

As head of Physics at MIT, he'd helped set up dedicated graduate fellowships for Black students—among the most severely underrepresented in the sciences. Two years later, in 1989, the director of the National Science Foundation had called to tell him that the department had just graduated half the nation's new Black PhDs in physics—only a handful, but progress nonetheless. Now, in 1993, Birgeneau was trying to help the admissions office recruit more young women to MIT. While women made up more than half the college students in the United States, at MIT they were just 28 percent of the undergraduates. He had written a heartfelt letter describing how his daughter Patty had thrived at MIT—that it was, as the 1970s brochure promised, "A Place for Women." The admissions office sent it to thousands of young women who had scored in the top ranks on college admissions tests. The gesture seemed to have helped: Birgeneau got appreciative letters back—"I wish I had a father like you," one wrote—and MIT had a record number of applications from women. It was now expecting an incoming class that was 40 percent female, the highest ever.

All around, signs suggested it was a new world. The Year of the Woman in 1992 had ushered in a record number of women to Congress, including the first Black woman elected to the Senate. It also helped assure a narrow victory for a young Democratic president, Bill Clinton; a majority of women voted Democratic that year, marking what would be a generational shift. Clinton was the first president with a wife who had a professional career, and he had appointed the most diverse cabinet in history, which in 1993 had five women, three of them the first to serve in their positions, including the first female attorney general. He had recently named the nation's second female Supreme Court justice, Ruth Bader Ginsburg, who as a lawyer in the 1970s had litigated some of the most significant cases for women's rights. And for secretary of the air force Clinton had nominated MIT's own Sheila Widnall, who became the first woman to lead any of the military services.

Universities, like other workplaces, were instituting policies against sexual harassment, which before Anita Hill had been a concept discussed mostly in courtrooms. The new rules tried to rein in behaviors that had long made women feel singled out and second-class at work: the come-ons, the put-downs, the gossip about their sex lives. But many people still had trouble understanding that sexual harassment didn't necessarily involve having sex—and even when it did, they weren't sure it could or should be regulated. As campuses debated new rules prohibiting sexual relationships between students and professors—until then regarded with softness or a sweep under the rug—newspaper editorials accused universities of trying to "criminal-ize Cupid." (A *New York Times* story noted that when Harvard passed its policy, John Kenneth Galbraith, the towering economist, diplomat, and public intellectual, had "begged guidance on how to atone" for his nearly five-decade marriage to a former student. The dean of the faculty, the *Times* reported, "replied that there was a statute of lim-itations for unions made at a time when 'amour—instructional and noninstructional—was in fashion.'")

MIT passed its policy in 1993 and that fall distributed twenty thou-sand paper copies around campus. A group of students burned theirs in a bonfire, arguing that the policy violated free speech—"pure censor-ship," one student deemed it, complaining that it "defines harassment as hurting someone's feelings." But their objections were relatively iso-lated; the administration had started working on the policy in 1989, following complaints from a group of thirteen female graduate stu-dents in the School of Science, and the new policy reflected attempts to shape a campus that was more welcoming to students who were not white, male, or relatively wealthy.

MIT also had a new president, Chuck Vest, who was leading the university on a lonely mission to protect needy students: in 1991 the federal government had sued MIT and the eight Ivy League schools to stop what had been an annual meeting to compare notes on how much financial aid they would give needy students. The government accused the universities of inhibiting price competition and wanted them to award more money to wealthier students, based on merit. The Ivies had quickly settled with a consent decree. Vest had decided to fight, despite the risk that if MIT lost, families who had been denied aid could sue for triple the amount in damages.

In his annual report to the campus in October 1993, Vest under-scored that MIT was a place of equal opportunity: "We respect vary-ing views, value the role of our institutions as critics of society, and believe in an elitism that is based solely on talent and accomplishment, rather than wealth or social status." When he had started teaching in engineering in the 1960s, he noted, the field was "a man's world." Now,

Black students accounted for 6 percent of all MIT undergraduates, and women, 30 percent. While the campus sometimes cracked along racial, ethnic, and gender lines, set off by accusations of "political correctness," Vest urged common ground, quoting Alfred North Whitehead's 1925 lectures, *Science and the Modern World*: "Other nations of different habits are not enemies: they are godsends."

"This is true whether we speak of societies, professions or single institutions," Vest wrote. "The electrical engineer and the mechanical engineer are able to build systems together that neither can build alone. Men and women together create a balanced discourse and world view. Black and white . . . brown and yellow . . . red and tan . . . create a campus and a nation far more meaningful and creative than any alone."

Nancy had used the word *discrimination* because the lawyer had told her to. Even she wasn't sure it described what she was facing. She believed the lawyer when he told her that she could fix her problems with a little help from people at MIT whose intentions were good. But events in late 1993 would force her to think differently about what discrimination looked like. She would lead Birgeneau—and the university—along the same path.

Nancy had taken it as a sign of confidence when Richard Hynes asked her to help design the new Biology course that MIT would require all undergraduates to take. MIT rarely changed its core requirements, and the department had fought hard to add this one. The course would have to cover everything from the basics of molecular biology to the newer advances driving the revolutions in biotechnology and human genetics, as well as the ethical questions they created. It demanded a huge investment of resources—faculty time and new teaching assistants. It had to be done quickly: after two years of debate, the MIT faculty approved the new requirement in April 1991, and the first class, a trial run, started that September.

Nancy felt more than qualified. Her career—from phage to virology to developmental biology—spanned the breadth of the course, and she had already designed a class teaching the fundamentals of molecular biology to non-Biology majors with the reproductive technology course. Students had loved that class, rewarding Nancy with high ratings: a 6.6 out of a possible 7, by far the highest in the Biology department, where usual ratings were in the 5s and 4s were not uncommon.

Still, the reproductive technology class had only about twenty undergraduates. The required course, known as 7.012, would have four hundred, and mostly men—67 percent of undergraduates were male. Nancy had not forgotten the admonition from Gene Brown, the former Biology chairman and the department's standard-bearer for good teaching, that students would not accept scientific information if the person at the front of the lecture hall was a woman. She knew it remained true, and that students could be cruel in their evaluations. Of the twenty-five professors teaching required courses in the School of Science, she was the only woman. The new course would have to be spectacular.

Richard asked if she'd prefer to teach with David Housman, a geneticist who worked with her in the cancer center, or Eric Lander. She told him the department knew best, and Richard told her she'd be teaching with Eric.

Nancy barely knew him, but Eric Lander was regarded as a phenom, even by MIT standards. He was brash, bullish, and entrepreneurial, good at seizing opportunities and working connections. Raised in Brooklyn by a single mother—he was eleven when his father, a lawyer, died of multiple sclerosis—he appeared to have moved through life seizing one brass ring after another. At Stuyvesant High School, one of New York City's elite public exam schools, he had been the star of the math team, valedictorian, and won the Westinghouse Science Talent Search with a paper on quasiperfect numbers. He went on to Princeton, where he was valedictorian again—like Nancy, he graduated young, at twenty—and won a Rhodes Scholarship.

He had done his PhD in pure math at Oxford but rejected a career in math as too monastic. A Princeton professor guided him to a job at Harvard Business School teaching managerial economics—a class

Eric himself had never taken. He found the subject shallow. After his younger brother, a neurobiologist, showed him mathematical mappings of the brain, Eric became interested in studying biology. He met Nancy's friend David Botstein, who had been looking for someone who could use math and computer science to map the genes involved in human diseases. Botstein introduced him to David Baltimore, who gave Eric a fellowship and a lab at the Whitehead Institute in 1986. Together Botstein and Baltimore nominated Eric for a MacArthur "Genius Grant," which he'd won the next year, at age thirty. When Baltimore told MIT that Wally Gilbert was trying to hire Eric at Harvard, MIT responded by hiring him with tenure. He'd skipped the typical track; he'd published relatively little, and Baltimore had to do some convincing to get the Biology department to agree to put him on the faculty. Eric had put off some university leaders with his arrogance— he was not above picking up the phone and yelling at them to get what he wanted—but even those who'd crossed him agreed he was a dynamic lecturer, among the best teachers at MIT.

Whatever the risk in hiring him, it had paid off handsomely for MIT. In 1990, the year he was hired onto the faculty, Eric received one of the first grants from the Human Genome Project. The grant allowed the Whitehead Institute to create a new genome center, led by Eric, to explore the mouse genome as a model for mapping the human genome. Two years later, he secured another grant, for $24 million.

The genome project had been controversial from its inception, ridiculed as more ambition than utility, derided as "Big Science" for the money and egos it involved. Eric was fast becoming one of its most prominent faces. He was among the crop of men starting new biotechnology companies based on the project's work. That made him a symbol, too, of the new Cambridge. Signs of the city's countercultural and manufacturing past were fading; still nicknamed the People's Republic, it was expanding in distinctly capitalist ways. In Kendall Square, biotech companies were budding on the land once cleared for NASA, and the air no longer smelled like rubber but like money. The spaces of the old Boston Woven Hose plant—now defunct—had been transformed into a loft-style office complex that included a brewpub and a restaurant where Julia Child was a regular. Venture capitalists visited in search of the latest, most promising protein or growth factor inhibitor.

When Eric first called Nancy to discuss the course, he sounded amused, almost confused about having to teach introductory biology— another course he himself had never taken: "I can't teach introductory biology, I don't know any."

"Don't worry, you're thinking of old bio courses." Nancy described the kind of course she had in mind, based on the one she had taught earlier.

That was different, Eric said. "That actually sounds like fun."

Nancy took the lead on designing it. She read all the student evaluations for a previous introductory biology course to figure out why it had received low scores and convinced Phil to let her hire the one teaching assistant who had won high praise. She worked with the assistant to draw up problem sets and tests. She met with Eric in his office to write the syllabus, then picked and ordered the textbook, and toured the available lecture halls to find the right one. She settled on 10-250, one of the largest classrooms on campus, near the president's office and where the university gathered to announce its Nobel Prize winners.

She was concerned only when Eric insisted on teaching the first half of the semester. He had left his final preparation until the end of the summer, which made it hard for Nancy to plan her lectures. She worried about how students would respond to a middle-aged woman following a thirty-four-year-old man; already, she knew, students tended to think of the first lecturer as the boss.

The first year of the course in 1991 got strong reviews from students. Dean Birgeneau and the Biology department were thrilled with its success. Gene Brown, who had always taken close interest in the quality of teaching in the department, sat through every lecture and gave Nancy only one critique: you can't be self-deprecating, he told her, the students will see it as weakness. Eric had scored a 6.5 and won a teaching award. Nancy scored 5.1 and told Phil she would continue teaching and would throw herself into making it better; she wanted the course to score at least a 6. She set about improving, taking notes from Eric and others about how to lecture to a large group—how to modulate her voice and use the blackboard to signal key moments in her presentation. Her scores rose, to 5.3, then 5.6. In 1993, the first year the course was required of all incoming freshmen, the course scored a 6—she'd hit her goal.

She was feeling upbeat the week before winter break in December 1993, having finished her final lecture and filed the students' grades. She was confident that once she had more practice with the blackboards, her scores would continue to rise. So she was surprised when she ran into another professor, Tom RajBhandary, in the hallway, who told her he'd heard that she wasn't teaching the intro course the next year; that another professor, Harvey Lodish, would be taking her place. Nancy confidently assured Tom he was wrong.

The next day she was sitting in an office in the Biology building reading letters for upcoming tenure cases when the assistant chairman ducked his head in and asked her to see Phil when she finished. She could think only that Phil wanted to congratulate her for a job well done. Just recently he had sent her a bar graph breaking down her scores from student evaluations. Across the top he'd written, "Great job!" Nancy was excited to see him; they were old friends but rarely saw each other now that he was chairman of Biology. He was fresh from Stockholm, where he had been presented with his Nobel Prize.

"What do you want to teach next year?" Phil asked.

Nancy clenched. Maybe Tom was right. "We already agreed on what I was teaching," she said. "The bio class."

Phil told her he wanted to bring in Harvey Lodish to teach her half.

"What?" Nancy had to stop herself from yelling. "You promised me I could continue indefinitely." She had never talked to Phil—or anyone else—about the challenges of being a woman teaching a large undergraduate course; she didn't think he'd understand, and she didn't want to draw attention to it. But she'd worked hard to develop the course, and to teach it. It had been a hit. Now he wanted to remove her? There had to be more to it. There was something odd in Phil's manner, she thought, as if he couldn't look her in the eye.

Phil asked if she would consider developing another version of 7.012 for the spring semester.

"I have already developed three highly innovative courses, including one of major importance to the department," she reminded him. "It took enormous effort, and I can't do it again."

"What do you want to teach?" Phil asked again. Nancy could tell this was a done deal. No wonder Tom already knew about it.

"If you take away this course, I will never teach at MIT again." It was a stark threat, but she wanted Phil to understand how much the course meant to her. She wanted to stop him from shoving her aside.

"Never is a long time." Phil told her he would get back to her.

The holiday break passed, and Nancy ran into Phil in the hallway in January, where he again asked what she wanted to teach. She was trying to remain involved in the course, somehow, so she suggested she could work with Harvey and Eric, perhaps sit in on the lectures to suggest ways to improve it. She told him she'd call Eric.

She and Eric were not close friends, but they had become friendly teaching together. She had told him how much she'd learned from watching him. He'd invited her to his home for the bris following the birth of his son. Now on the phone, she told him she was upset and asked what he wanted to do.

Eric took an uncharacteristically long pause before he spoke. Finally, he told Nancy that he was excited about teaching with Harvey. They were good friends: Harvey had been at MIT since 1968 and was now also at the Whitehead, recruited by David Baltimore as a founding member. Like Eric he was an entrepreneur, having been a founder of Genzyme. Eric said he thought it would rejuvenate him to rethink the class with someone else; he was feeling a little stale after three years.

Nancy expressed surprise. Harvey had told her he wasn't interested in teaching the intro course.

Another pause. "That's all changed," Eric said. A company had approached Harvey about writing a textbook for introductory biology. "He got all excited and talked to me, and I got all excited, and now we want to write a book together."

So there *was* more to it, Nancy realized. She told Eric how much time she had put into the course—seven years, if you considered that she had built the intro bio course from the course she'd developed for women's studies. "I can't afford to walk away from all that work. I don't have the time or energy to start another course all over again."

"It's not that I didn't like teaching with you. It's just that I'm excited about teaching with Harvey."

Nancy suggested maybe they could all teach the course together.

"Yeah," Eric said. "Maybe something like that."

Eleven days later she had a letter from Phil with her assignment for the next fall: she would sit in on the lectures, as a kind of consultant, not as a teacher; her name would not be in the course catalog or on the syllabus.

She was devastated. That weekend, she sat at the desk in her condo on Chauncy Street, trying and failing to focus on writing a grant proposal. She looked out at her neighbors' garden, its glorious fruit trees now bare in the winter, and beyond at the rooftops of the city where twenty-three years ago she had been a newly minted PhD. She couldn't kick the feeling that she had failed, that she wasn't good enough, that that was why Phil had removed her from the course. But the more she thought about it, the more that did not make sense. She'd pulled this course off in record time; it had succeeded more than the dean and Phil could have hoped for. She had bought it when Richard Hynes tried to explain that other faculty members had more space or higher salaries because they had won the Nobel or other prizes or been rewarded to try to keep them from moving to other universities. That didn't apply here. The scores for the course indicated that she had created a hit. Objectively speaking, it was crazy to think she was not good enough.

By Sunday she had written several versions of a letter to Phil, finally settling on one that she hoped struck a balance between righteous and rude, accusing him of an "astounding breach" of her agreement to teach the class. "The so-called teaching assignment you describe for me is not a teaching assignment as far as I can see. My role as an outsider would be another colossal waste of time with no value to the course, the students, or the department."

She was on her way to see Phil the next week when she passed

Gene Brown. She showed him her letter from Phil. Gene was now sixty-eight. "If I received such a letter at my age, I would walk down the hall and hand in my resignation," he told Nancy. He admitted he'd been surprised sitting in on the class what a good teacher she was. Now he was puzzled at Phil. "I wonder how bright people can be so unbright in these ways," he mused.

Phil was unusually chatty, so much that Nancy cut him off to say she was there to register her objection. It wasn't just that he was removing her from the course. She didn't think the department should be choosing professors based on who wanted to sell a textbook. Phil seemed surprised to hear that Harvey and Eric were writing a book based on the course. Once again, he said he would talk to some people and get back to her.

Harvey had left Nancy a message to call him at home. On the phone, she asked him about the book.

"Well, let me come clean, it's gone a lot further than that." He and Eric were starting a company and had already hired a president, a woman who had been a book editor at the company that published Jim's classic textbook, *Molecular Biology of the Gene*, as well as one Harvey had written on cell biology. The idea was to do not just a textbook, but videos, CD-ROMs, and other teaching devices for introductory biology. "We don't expect to make money for the first year or so, but by the fifth year we expect to make millions," Harvey told Nancy. "Do you know how many first-year biology students there are in this country?"

Still, he was bothered by her distress: "Where does this leave you?" He wanted her to know that he hadn't pushed her out; it was Phil who suggested bringing Harvey in to teach with Eric.

Harvey and Nancy said goodbye without any resolution. But Nancy was relieved, even a bit proud of herself. *Imagine*, she thought, *the course I designed is worth millions!* It really had succeeded. She called her friend Blair, a classmate at Harvard who had cofounded Charrette, a chain of architectural and art supply stores. She asked him if he thought there was money in the kind of venture Harvey described.

"Yes, I think there is," he replied dryly. "And I think there's a ten-million-dollar lawsuit in it for you. Either you're a saint or I'm missing something. Why aren't you outraged? They stole your course!" Blair wanted Nancy to talk to a lawyer friend of his. Soon other lawyers—also friends of friends—were calling. Nancy phoned Jim to ask whether he thought the course could make money. He doubted it could make much, and he, too, was angry on her behalf: "It's enough to make me a feminist." He told her she should ask for 20 percent of the company or $100,000. And she should write down everything as it happened and be prepared to take a record to President Vest. "Since I have always followed Jim's advice," she wrote that night in a computer file she titled "History of 7.012," "I will keep a diary on the affair from now on."

Nancy told Phil the news about Eric and Harvey starting a business, and it seemed to surprise him. MIT had an office of licensing technology, but he was not sure it had the rules around licensing courses. Again, he said he would look into it and get back to her. Nancy flew to Germany, and when she returned two weeks later, she had not heard from Phil, but the president of Eric and Harvey's new company had left a message saying she was coming to Boston and wanted to take Nancy to lunch. Harvey had also left a message to say he thought they should write the textbook together.

Why hadn't he thought of it before? he wondered aloud on the phone the next morning. Nancy would be perfect since she was already a coauthor on the most recent edition of Jim's textbook. Harvey and Eric had had dinner over the weekend and had gone over the notes from the course. "It's a great course," Harvey told her. He had spoken to Phil, who had raised concerns about how MIT might look allowing faculty members to profit off a course. Harvey said he'd offered to give MIT shares of the company and put Phil on the board.

In mid-February, Nancy met the president of the company in a booth at the Legal Sea Foods in Kendall Square, across the street from the cancer center. The president said she had approached Eric and Harvey with the idea to start the company; the introductory biology textbook would be its centerpiece, and they hoped to get it out in three years. "Please don't tell anyone," she said, "because people might steal this idea."

Was she joking? Nancy wondered.

Nancy wasn't sure who was telling her the truth, who wanted her out of the course, or what she should do about it. She was growing increasingly angry about the venture she was now referring to in her diary as "Eric & Harvey, Inc." In late February, she made an appointment to see Dean Birgeneau, who had helped her get the additional 113 square feet the previous year. Birgeneau didn't think it sounded right for one person to market a course that had been jointly developed. He suggested that Nancy needed to be clearer about what she wanted. She should tell them she wanted to write the book with them and get Phil to sign on to the idea. Eric and Harvey would be more inclined to work with her if they knew the department was backing her.

She told Phil, who looked irritated that she had gone to the dean. She suggested he could put a positive spin on the idea of cooperation. Again, he said he'd get back to her. She didn't hear from him. But a week later, she ran into Eric on the steps of the Whitehead. He greeted her cheerily: "Do you want to write a book with us?"

"I think I do."

Eric replied that it would be fun. He had seen her book. It was brilliant.

Nancy thought something about his manner was too ingratiating. Her friend Blair insisted she see a lawyer. So at the end of March Nancy met with Daniel Steiner, who had been general counsel to Harvard and was now a lawyer at Ropes & Gray, one of Boston's white-shoe firms. Universities had been changed by big money, he told her, and no one was unaffected. No one could ignore the potential to make money on courses or could afford to alienate someone who was hauling in grants the size that Eric was. Steiner thought her problem could be solved without a lawsuit, at least for now. He told her to start by going back to Birgeneau and coached her on what to say. Her removal was, as she wrote in her notes, "UNPRECEDENTED" and "UNWARRANTED," and her work had been appropriated by two other faculty members seeking to profit.

She arrived at the dean's office prepared, only to discover he was out with stomach flu. On her way back to her office, she noticed an article in MIT's internal newspaper announcing that Amgen, one of the world's largest biotechnology companies, had agreed to give the university up to $3 million a year for ten years in exchange for licensing rights on MIT research. The article quoted Richard Hynes, as head of the cancer center, saying that faculty interested in the grants would apply to a committee led by Birgeneau.

Nancy's lab was pushing on the next step of insertional mutagenesis: to see whether the retrovirus that had found its way to the germ cells would make mutations in the genes as the fish developed. She needed a grant to pay for an additional postdoc, so she stopped by Richard's office to ask him how she might apply for the Amgen money. The article had said he was on the committee to consider proposals. Richard told her he didn't know what the procedure would be to apply.

Nancy had been thinking of asking her colleague David Housman to collaborate, so she asked him if he knew anything about the Amgen grant.

"Richard has been after me to apply for that," David told her.

Nancy was now fed up. She wasn't just out of the loop, she was being kept out of the loop. Meanwhile, she continued to have her work validated outside of MIT. In April, she went to a small, two-day scientific meeting in Switzerland. The group loaded onto a bus that began to curve around narrow mountain roads. Nancy was afraid of heights, so she struck up a conversation with her seatmate to avoid looking out the window. He worked for Glaxo, the pharmaceutical giant, and was fascinated by her work with fish. On the spot, he told her he wanted to give her $100,000 to buy new fish tanks.

When she returned to MIT the next week, Phil pulled her aside as she was entering a lecture and asked if she would develop a new version of the biology requirement to teach in the spring. Nancy erupted: "I created a good course for you, and the thanks I got was to have my work taken and given to others. If I create another good course, you'll take that and give it to some other men to make money on.

"I'm seeing the dean tomorrow and I've already seen a lawyer. We're all busy people. I think the best thing is just to turn this over to lawyers."

In Birgeneau's office the next day, she told him she had met with the lawyer and had been inclined to try to work the problem out within MIT. Now she'd changed her mind. She just wanted to send lawyers after Harvey and Eric and be excused from teaching for three years to focus on her lab.

"I don't think you'll be happy," Birgeneau said. "You're a true academic."

Nancy asked him about the Amgen grant, and if he could help her get a proposal in front of the company. Birgeneau agreed; it was not a big request. But he was puzzling about the course: "There's still something about this that doesn't make sense."

"Yes. It's called discrimination. I feel powerless and out of the loop, and you told me I'm not aggressive enough, but to behave like this makes me sick." No wonder there were so few women in science. "There never will be because normal women wouldn't live like this."

Birgeneau said he agreed with her about powerlessness. In her diary, Nancy wondered if he even understood what she meant—how could he?

Her meeting with the dean again prompted a phone call from Phil, who asked to come by her office the next day. Maybe she and Eric and Harvey could all teach together, he said. He'd arrange for them to have lunch and said he thought it would all work out.

But three weeks later she had not heard anything more.

She had typed up her proposal for Amgen and hand-delivered it to Birgeneau's office; she asked his assistant to stamp it RECEIVED, which reassured her he would really see it. In early May she wrote Birgeneau again, suggesting they take the question of the course to President Vest. "I personally do not see how people can be asked to work for the Institute if their administrators can take their professional work away and give it to others, possibly for financial profit," she wrote. "A logical next step might be to have our students sell the answers to problem sets. Why shouldn't those who are good at it make money on it? Or better yet, if a student can steal the answers from someone else, they could sell those. Where does it end?"

Her letter to the dean again prompted word from Phil. He said he wanted to make Nancy happy. If it meant putting her back in the course with Eric, he would. But Harvey, he said, did not think three people should teach the course. So Nancy called Harvey, hoping to change his mind. The conversation, as she later wrote in her diary, was "a disaster." Harvey told her the course he and Eric had designed was completely different from the one she had taught, though when he read her the outline, it sounded to Nancy like 80 percent the same. She asked about writing the book together, and he said it was too soon to know. They were looking for someone to write about immunology. Great, Nancy said, noting she had written the chapter on immunology for the most recent edition of *Molecular Biology of the Gene*. Harvey demurred. What they really needed, he said, was someone to do neurobiology.

Nancy felt blown off, so she told him she was still considering Phil's offer to let her teach the course in the fall. That would get his attention.

"What do you mean?" Harvey asked. "Phil asked me to teach it. Period."

"It's possible I will be teaching it in the fall."

Now Harvey erupted. Phil had told him he was pressuring Harvey to step aside only because the university was afraid that Nancy would sue for discrimination.

That got Nancy's attention. "Did Phil say that?"

"Oh, come on, Nancy, you know what this is all about. I can tell you Eric will be very unhappy."

Eric was indeed unhappy. He called Nancy the next day and told her she could have the whole course—he was taking the year off. He was not going to be treated like this. Phil had never asked him if he was willing to teach with her. "I didn't like teaching with you." Phil had told him that Nancy said Harvey had stolen her course. That was ridiculous, Eric said. "My part was one hundred percent me. What did you ever contribute? Your role was just to come whimpering to me for help."

All the fighting was costing Nancy time away from her lab—her calendar some days showed three or four hours devoted to meetings or phone calls looking for ways to stop her removal from the course. The fish experiments were going well: a scout from Amgen had called to say the company was interested in her proposal and he wanted to come visit her lab. She was discussing a partnership with Massachusetts General Hospital and Harvard Medical School to collaborate on a large-scale screen to identify the genes responsible for basic development. Glaxo had come through with the $100,000. But Nancy had not had time to order the fish tanks the money was supposed to buy.

"All I know is that I can not spend any more time on this problem," she wrote the day after her disastrous phone calls with Harvey and Eric. "It is ruining my ability to run my wonderful lab."

She made one last push in late May. A colleague gave her the name of an intellectual property lawyer in MIT's technology licensing office. On the phone, the lawyer told Nancy that courses developed by two faculty members constituted joint intellectual property. The faculty member who wanted to commercialize the property would be expected to buy out the other, compensating her for the value of her work. It was also a matter of collegiality, the lawyer said, governed by the section on fraud and misconduct in the institute's academic handbook. The handbook said that faculty had an obligation to protect the intellectual property developed within the university. Therefore it would be inappropriate for one faculty member—or an administrator—to take away a course without mutual consent; professors would refuse to coteach if they thought their colleagues could take their work without asking.

"I assume these characters are no longer at MIT," the lawyer said.

"Oh, no, they are my colleagues."

The lawyer put Nancy on hold to talk to the director of the licensing office. Then the lawyer got back on. "Nancy, this is not the time to be having tea with the dean. It's time to bring charges against these people for academic misconduct and fraud. No wonder lawyers are calling to take your case."

The next day was the hooding ceremony to award graduate students their doctorates, and as Phil and Nancy walked back to the cancer center together, she told him what the lawyers had said. He could not disagree. But he needed his course filled. He told her he would set up a meeting for the next week with Harvey and Eric to work out a solution.

They met the Thursday after Memorial Day. Birgeneau had left campus for two months. Nancy knew that without him, Phil would not stop Eric and Harvey if they wanted to teach the course, which they still did. They agreed that Nancy would sit in as an adviser. Nancy got small concessions: Phil would write a memo saying that the course could not be commercialized because it had been developed "collegially," and Nancy insisted that her name be on the syllabus. Eric seemed friendly, gracious, Nancy thought. They walked back together from Phil's office toward the cancer center and the Whitehead.

In her diary that night, Nancy wrote, "I am heart broken."

Over the next few days, her despair turned to indignation. She kept hearing what Eric had said: "What did you ever contribute?"

She felt like a fool, duped. She thought of all the years she had blamed herself for whatever problems she'd had in science, thought she could fix them on her own. Now she saw it didn't work like that. She could try to innovate, try to be the best, try to please everyone; it would not matter. A woman's work would never be valued as highly as a man's. It had taken her twenty years to see it—she'd understood it about other women before she'd realized it was true for her, too. But now that she did, it was as obvious as the clearest scientific result, like seeing the repressor bind to DNA. She had taken all of this personally, but she now realized it didn't have much to do with her. These men barely saw her; she was a nonentity.

She drafted a letter to President Vest. "I am a professor in the Biology department," she began, because she didn't expect he knew who she was. She itemized the fights of the last four years—the space, her salary, the course—and said that she believed "strongly" that they were "the result of discrimination."

She showed the letter to her friend Arthur Merrill. He urged her not to send it; the president did not know her and without context had no way to evaluate her complaint; he might assume the problem was a personality conflict, or Nancy herself. Nancy thought Arthur had a point, so decided to run the language past someone who knew the department, someone whose opinion she trusted. She asked Mary-Lou Pardue if they could have lunch. Nancy did not know her well but respected Mary-Lou's experience and knew others did, too.

They met at Rebecca's, an upscale soup-and-salad place that had replaced the F&T Diner—now closed, another artifact of Old Cambridge—as Nancy's regular spot. Sitting at a small table near the back, the clamor of dishes from the kitchen around them, Nancy briefly explained what had happened with the course and asked Mary-Lou to read her letter to the president.

Nancy watched for what felt like forever as Mary-Lou read. She could see Mary-Lou's eyes moving along the page, but her face registered no emotion. Nancy began to worry she'd made a mistake even showing Mary-Lou the letter.

Mary-Lou was now one of the more senior members of the department and had sat on many of its committees, and others outside MIT evaluating scientific prizes and awards. Mary-Lou was thinking about all the faculty meetings she'd sat through to evaluate talks by job candidates, how if a woman was giving the talk and hesitated when asked a question, the men would pounce on it as weakness, but if a man did the same thing, they'd excuse it, saying he hadn't come into his own yet—he reminded them of themselves at his age. She thought of how much more women had to achieve to even have their names mentioned for a prize. She thought about Nancy's tenure case, and the letter reversing the names.

"To these personal experiences, I have to add the disturbing fact that although about 40–50% of our graduate students are women, the number of women faculty in my department as a percent of total is essentially unchanged over the 21 years I have been at MIT," Nancy had written the president. "It remains about 15%.

"I have seriously considered legal action against MIT as a way of drawing attention to and ultimately correcting this situation and am still considering whether to ask the Provost to initiate an inquiry into my chairman's actions," the letter continued. "I dislike the very negative aspects that inevitably would accompany such actions. I would much prefer to find a different more collegial way to achieve the same positive outcome."

Finally Mary-Lou looked up. "I'd like to sign this letter. And I think we should go see the president."

Mary-Lou had long thought that there was a problem for women in the department, she told Nancy, senior women in particular. Mary-Lou didn't share the story about David Baltimore's interference in Nancy's tenure case, and Nancy didn't tell Mary-Lou about Frank Solomon coming around to say he planned to tell the graduate students that Mary-Lou was not up to teaching Methods and Logic. Nancy was too excited to think about that, and she was thrilled and relieved to discover she was not the only woman who felt this way.

Then she had a thought: "Do you think there might be others?"

Part **3**

Chapter 18

Sixteen Tenured Women

By the time she had called Mary-Lou to arrange the lunch, Nancy felt she'd run out of hope—hope of getting her course back, of feeling like a valued colleague in her department. Now, sitting at the small table at Rebecca's, Nancy suddenly felt power. That a woman of Mary-Lou's caliber and renown saw the problems that Nancy thought she alone could see—MIT would have to confront this, have to respond. Mary-Lou was a member of the National Academy of Sciences, universally respected and liked in Biology. Phil or Richard Hynes might dismiss Nancy as "difficult," but no one could deny Mary-Lou.

Nancy still didn't think she'd get the course back. But this felt like the possibility of something bigger, better—a change, though she had no idea what that change might look like.

She and Mary-Lou had parted without knowing exactly what they would do—or what they could accomplish—beyond talking to some other women. Right now, Nancy couldn't name many women on the faculty at MIT, even after being there for more than twenty years. So that night, she went home and called Ruth Perry, who had recruited her seven years earlier to teach the women's studies course. Perry had been saying since the space wars that Nancy was up against discrimination, but she knew Nancy had not been convinced. Now Perry got right to work and within a week had gathered a small group of women after dinner at Nancy's condo on Chauncy Street. They came from across the university: Mary-Lou; Karen Polenske, an economist; Lorna Gibson, an engineer; and Judith Jarvis Thomson, from Philosophy.

Arrayed on the sofas in Nancy's airy living room, the women told of being abruptly removed from teaching assignments and shut out of decisions and powerful jobs in their departments—even in the humanities, where for years at least half the students had been women. Women there made up less than 25 percent of the faculty. Lorna recalled a male colleague in Engineering who had been hired around the same time as her telling her that he'd bought a house; when Lorna told him she was still saving for one, he'd explained that his mentor had secured him a loan from MIT. That was when she realized the men were tapped into a whole set of privileges she didn't know existed. Other women suspected they were paid less than men in comparable positions; they couldn't know because salaries were secret. Judith said she thought MIT simply had a lower salary schedule; a former student, she said, a man, had just been hired as an assistant professor at Stanford and was making more money than she did as a full professor. Nancy thought to argue but stopped herself. Judith was fifteen years older and had been hired in 1964; Nancy understood that she came from a cohort of women who considered themselves lucky to be hired onto university faculties at all. Sometimes it was easier not to think about how you were treated.

The women suggested they all sign a letter to the president and ask to meet with him. Judith suggested they start with Mary Rowe in the ombudsman's office. Others said no; the space wars had shown Nancy the limits of Mary's power. The evening left Nancy and Mary-Lou wondering if it was too ambitious to try to understand the issues for women across the faculties of MIT—much less figure out solutions. MIT had five schools—Architecture, Engineering, Humanities, Management, and Science—each with its own culture that might explain differences in salaries and status. In any experiment it was best to limit the variables; they thought they'd better stick to Science, which is what they knew. Mary-Lou had heard that the provost's office had an initiative to increase the number of women on the faculty. Nancy knew little about how the university worked—eighteen months earlier she hadn't even known the provost's name—so she called Vera Kistiakowsky, who had been the first woman hired onto the physics faculty and had just retired. Vera came from science royalty—her father had led the team developing the explosives for the Manhattan Project, been a professor of chemistry at Harvard and science adviser to President Eisenhower, and received the National Medal of Science. At MIT, Vera was long treated more like the help, working as a research associate for years despite having the same qualifications as the men she worked for. She had done the first report on the status of women for the American Physical Society, in 1972.

On the phone, Vera told Nancy there were two committees to deal with women at MIT: the one in the provost's office that Mary-Lou had mentioned—it was more focused on hiring—and a new committee on affirmative action that Dean Birgeneau had started in the School of Science. The committee had representatives from each department, but Vera couldn't recall who was on it from Biology, so she looked up the name while Nancy hung on the phone: it was Frank Solomon. Nancy gasped. In all her conversations with Dean Birgeneau over the previous year, how could he not have told her about his committee? And how could the Biology representative be Frank, when Nancy thought he'd been so demeaning to the women in the department? In her diary she wrote, "This is ultimate joke on women (biologists) at MIT."

She and Mary-Lou decided they'd start in the School of Science by talking to Lisa Steiner, who had been the first woman hired in Biology, in 1967. Lisa told them she'd found out a few years earlier that her salary was about 40 percent less than what men of comparable age in the department were earning. The discovery had sent her into a long depression. More recently she'd realized she was the only immunologist not included on a group grant that had funded the work of the others in the department over the last twenty years. Nancy asked why she hadn't asked the others about it; one had been Lisa's postdoctoral adviser and remained a friend.

"Every time I tried, I started to cry," Lisa said. The practices in the Biology department struck her as "undemocratic," she said, but she couldn't say for sure it was discrimination. She suggested they make a list of all the women in the School of Science and survey them, ask them if they thought there was discrimination against women on the faculty. Maybe they'd want to meet or sign a letter to the president. Lisa took down the course catalog from her shelf and flipped to the listing of professors by department, and together the three women began poring over the names to identify the women.

There seemed so few. "Do you think the women are listed separately?" Nancy asked. "Maybe in the back of the book?"

There was no other list. They counted: 15 women had tenure, and 197 men. Their department, Biology, had the largest number: the three of them plus two younger women, Ruth Lehmann and Terry Orr-Weaver, who'd earned tenure that year. Brain and Cognitive Sciences had the second most, with four. Chemistry had two, and EAPS—they had to look it up to see that acronym stood for Earth, Atmospheric and Planetary Sciences—had three. One woman was tenured in Physics. None in Math. Adding the seven women without tenure—the junior faculty—the total number of women was twenty-two. Less than 8 percent of the entire faculty in the School of Science.

At least it made their job easier. They decided that the women without tenure would not want to speak up; it would be unfair to even ask them. Sitting around the small table in Lisa's office, they divided the names into two lists. Mary-Lou and Lisa set out to talk to one list of names, and Nancy to the other.

Nancy set out toward the heart of campus, the neoclassical limestone buildings along the Charles River, to find Sylvia Ceyer in Chemistry. This was the grander, quieter part of MIT, so different from the cancer center with the now-constant construction in Kendall Square. As Nancy walked across the shaded diagonal paths of Eastman Court, she started to feel the same uncertainty, almost embarrassment, that she had felt waiting for Mary-Lou to finish reading the letter over the table at lunch ten days earlier.

They had not met, but Nancy read about Sylvia regularly in the MIT newspaper, for winning this award or that honor. Sylvia had been the first woman to work her way up to tenure from within the Chemistry department, known as a difficult proving ground even by MIT standards. Nancy assumed she must be content, and perfect. Sylvia was young and blond, her hair neatly curled at the end, with porcelain skin and big blue eyes like a doll. In her office now, she sat perfectly poised, hands folded, framed by the full-height windows looking out onto the trees in the quadrangle below. A large drafting table was next to the chair where Nancy sat; glancing at the drawings on it, she guessed that Sylvia was designing a large machine of some kind, but her work was a mystery.

Sylvia regarded Nancy as a foreigner, too; she wondered from Nancy's voice if she was British. Sylvia listened carefully as Nancy spoke, saying nothing and showing no emotion. Sylvia was thinking that Nancy was eloquently describing some of the same problems she herself had been thinking about lately. Like Nancy, Sylvia had devoted her life to science, starting at an even younger age. She had been ten, making her way through a series of *What Is* books in a suburban Chicago library, when she fixed on the picture of an atom in *What Is Chemistry?*—she pronounced the *ch* as in *chew*. It seemed so logical at an age when much of the world didn't make sense, so she took the book home, copied it out with her peacock-blue fountain pen, and told everyone she was going to be a *chem*ist. She was a physical chemist, studying the reactions between molecules and surfaces in ultrahigh vacuums, and there were almost no other women in the field. The men on the senior faculty had welcomed her to MIT like a daughter early on. But since she'd gotten tenure, she told Nancy, she felt like more of a competitor. It was a constant battle for resources and recognition.

For her experiments Sylvia relied on machines big enough to stand in—the drawings on the drafting table were her designs for one of them—but the department had not given her space big enough to fit one. She'd discovered that a man hired after her had twice the space she did, and another more than that. More recently Sylvia had sat on an institute-wide committee on the budget, chaired by a male biologist, who had never acknowledged her presence. Then he had lunch with her department chair to solicit his ideas on the committee's work and afterward sent a follow-up note remarking that it was a shame there was no one from Chemistry on the committee. Sylvia's department chairman replied with a letter noting that a chemist—Sylvia—*was* on the committee and that "her input will be valuable." The chairman had shown Sylvia the exchange of memos; to her, it was proof in black-and-white that she was invisible. The department chair could see it. Sylvia used a word Nancy hadn't thought of: Sylvia felt "marginalized."

"Do you have something written I can sign?" Sylvia asked Nancy.

Nancy thought she might spring out of her chair. Yet it didn't take long to discover more women who felt the same way—more power. In an office one building over, Nancy had barely gotten the words out before JoAnne Stubbe—a biochemist who had been hired with tenure and elected a member of the National Academy in 1992—laughed knowingly. "I've seen it all," she told Nancy. "There's nothing you can tell me that I haven't seen or experienced." Back at the Biology building, Lisa and Mary-Lou had returned with good news from EAPS—another distant land, headquartered in the Green Building, a poured-concrete I. M. Pei tower that loomed over the center of campus and the city as the tallest building in Cambridge. They had met an impressive young geophysicist, Leigh Royden, who had said she was happy in her career but wanted to be part of any discussions about women on the faculty. She thought her colleague in the neighboring office, Marcia McNutt, would want to join, too. Leigh had given them a number at the Woods Hole Oceanographic Institute, where Marcia was working for the summer.

Within a day, ten of the fifteen women on the faculty of the School of Science had agreed to meet to discuss their common issues. Nancy and Mary-Lou decided to increase their sample size, at least slightly, so added to their list two women in Engineering who had joint faculty appointments to Science: Millie Dresselhaus and Penny Chisholm.

Penny called Nancy right back. She was a biological oceanographer in the Department of Civil and Environmental Engineering who had been given the appointment in Biology recently after she suggested to a colleague in the department that the new bio requirement could include a unit on ecology. She'd been surprised to hear that Nancy was removed from the course, Penny told her now; she had sat in on Nancy's lectures and couldn't imagine being as good a teacher as she was.

When Penny had arrived at MIT from the Scripps Institution of Oceanography in San Diego in 1976, she was the only woman in her department and looked more like a student than a professor, neither of which eased her entry into Engineering. Even now she looked much younger than her forty-seven years, with her sandy-blond hair and an unencumbered smile. Early in her career at MIT she led the team that discovered a new microorganism she named *Prochlorococcus*—or primitive green berry—the smallest, most abundant photosynthetic cell in the ocean, responsible for a sizable fraction of the oxygen supplied to the earth's atmosphere. Still, she told Nancy, she'd recently had to fight for promotion to full professor. A new department chair told her she hadn't shown him examples of "leadership," so Penny asked to go above his head and compiled her own case, which she took to the dean, and her case sailed through. Four years ago, she'd turned out to be so underpaid that she got a raise midyear.

Penny had grown up around boys and considered herself a tomboy, but for years she'd felt overwhelmed in faculty meetings in her department. She'd watched as the men tore each other apart over whom to hire and where to spend money, whose research was bold and whose was a waste of time, only to walk out into the hallway to talk about the Red Sox or head out together for a beer. For a long time, she had blamed herself for not being aggressive enough to join in. She'd been on the tenure track with a Puerto Rican man—they were awarded it the same year—and he used to joke about how they were the "diversity" in the department. But like many of the men on the faculty in Engineering he had come up through the ranks as a student at MIT; he had the playbook. She thought maybe she didn't fit in because she was a biologist in a school of Engineering. But after tenure, she told Nancy, her "feminist needle" shot up, and she suddenly got it: as women, she said, "we aren't even on their radar screen."

On July 7, a month after Nancy and Mary-Lou's lunch, twelve of the seventeen women in the School of Science gathered in Mary-Lou's office in the biology building, which they chose because it was out of the way. Mary-Lou had set out plates of carrots and grapes. When she couldn't find enough chairs to drag in, the women settled on the floor or perched on the long windowsill, their backs up against the blinds. Leigh Royden's six-month-old son crawled on a blanket laid out on the carpet.

Most of the women did not know each other, but they found their stories were startlingly familiar. A neuroscientist told how an Institute professor—the highest rank at MIT—had forced her to vacate a lab where she had five people working by informing her he was changing the locks in three weeks and would resign if anyone tried to stop him. Her department head declined to intervene, so she moved rather than fight. Two women—one a member of the National Academy—had gone to their department heads with job offers from other institutions, the kind of thing men did all the time to get more money and bigger titles. In the women's cases, MIT had declined to make counteroffers.

The women realized as they talked that no woman had ever been head of any department, or head of any center. There hadn't even been a woman as an assistant head of any department.

The women had spent their careers trying not to think about being women, hoping they would be seen as scientists. But as the first or only in so many settings, they felt they had to live up to a higher standard. As Penny said, a woman couldn't fail, because everyone expected her to. She obsessed over the smallest details of writing grants, afraid that being turned down would be another reason that no one would take her seriously. When she was awarded grants—which happened so often that one of her colleagues said she had a "golden ass"—she saw it less as success and more as the absence of failure. When she received a prize, she knew people assumed it was because she was a woman; she joked that she wanted to be the second woman to win, or better yet the tenth. Sylvia told how male undergraduates had challenged the math she put on the blackboard. Giving seminars to faculty or at conferences, she found herself interrupted regularly; the custom was to wait to ask questions at the end, but she noticed that people felt free to interrupt the women. Trying to be perfect was like dancing on the head of a pin, she said. Exhausting.

The women ranged in age from thirty-eight to sixty-four. Terry Orr-Weaver was the youngest; she had been a postdoc at the Carnegie Institution with Allan Spradling, who years before had been Mary-Lou's "stealth postdoc" and considered Mary-Lou a mentor. Terry listened to the stories the others told the way Nancy had listened to Barbara McClintock at Cold Spring Harbor years earlier; it was important to know what it had been like for women before her, but it was history. She believed the older women, she told them, but she also thought the problems had largely been solved for her generation. Terry had been the first woman hired as a professor at the Whitehead Institute—there had been ten men before her—and she felt welcomed and supported there. Only one thing troubled her. She had asked Mary Rowe about taking maternity leave, and Mary had urged her to do so. No one else had ever taken it and gotten tenure, Mary said; Terry could be the test case. Terry declined, recognizing it would be a stigma. She'd worn baggy clothes to hide her pregnancy.

When she'd won tenure, there'd been a dinner to celebrate, and Birgeneau was there with his wife. Birgeneau's wife had greeted her, "You're the one who was brave enough to have children before tenure!" The dean had smiled and confirmed it: he wouldn't have mentioned whether a man had children, so he hadn't told the members of the Science Council that Terry did, until after the vote. Then he'd dropped it on them: "You just voted to give this woman tenure and she has two children." He'd done the same thing with the Academic Council. Terry recalled her mouth falling open, and that moment as the most disillusioning of her time at MIT. She'd later told the story to Gene Brown, the former dean of Science, and he'd agreed that Birgeneau had only been trying to help her; it was extraordinary that she'd been given tenure with children. Meanwhile, Terry suspected that men in the department were spending at least as many hours tending to their biotech companies as any women were to their children.

Molly Potter, a psychology professor in the Department of Brain and Cognitive Sciences, was among the older women. Now in her sixties, she had been hired in 1968—to the Department of Urban Planning, because that was where she could find a job. She had been skeptical when Nancy first approached her about the idea of gathering the women. She had been a veteran of the efforts to make MIT more welcoming to women in the 1970s, a regular at Millie's lunches; she thought they'd done the work. But listening to the other women in Mary-Lou's office, she could see they had a point. MIT had made enormous strides in increasing the number of women among undergraduates, but it wasn't supporting the women on its faculty.

Nobody, though, wanted to just complain. They were scientists, so perhaps obviously they wondered if they could use data to quantify the problem: Did the women's sense of disadvantage show up in salaries and grants and other resources? Nancy had seen how powerful her tape measure could be. She suggested putting a woman in every department, an administrator who could monitor the allocation of resources and make sure women were being treated fairly. The other women shot down that idea; department heads would never agree to create the position, they'd stymie the woman—who would want to take that job? ("I would do it," Nancy said.)

They decided instead that they would propose a committee, established by the president, to examine the data on space, salaries, resources, and teaching assignments and make sure that women were being treated fairly compared to men. The committee would meet with each woman on the faculty once a year to determine any problems, then recommend ways the dean could solve them. Nancy worried a committee would be toothless, too inconsequential to apply pressure on the administration; they needed someone in a position to demand the data. But others argued that MIT prided itself on collaborative governance; faculty committees could get things done, not be merely ceremonial. Nancy went along; she was worried about alienating the group by seeming too radical. As Penny kept reminding her, "We just have to get on their radar screen."

Molly recommended that they write Dean Birgeneau instead of the president; they risked alienating him if they started by going over his head. She had been chair of the faculty in the 1980s, so the other women trusted her judgment. Over the course of July, a subset of the women met again several times to write a proposal to the dean—they met next in Penny's office in the Civil Engineering building, built like a bunker and even farther from the center of campus than the Biology building.

Nancy could still barely believe how quickly the women had come together. Mary-Lou had made Nancy feel empowered; the group was that times ten. She didn't want it to fall apart, so she ran every draft of the letter to the dean past every woman, insisting that everyone sign off on every word changed. She marked drafts CONFIDENTIAL, and as she assured Marcia McNutt in a note that July, "We've been shredding the documents until they are in final form."

Nancy later thought of it as a summer of catharsis, occasionally of rage. She was looking at a directory of the labs in Biology when she noticed that Mary-Lou did not have any graduate students in her lab. It was odd, given her stature. One of her former postdocs had won a Nobel Prize five years earlier, which would ordinarily have made her more attractive to students. Nancy recalled Frank Solomon's visit to her office and, over lunch one day, gingerly asked Mary-Lou why she had no graduate students. Had she decided not to accept more?

No, Mary-Lou replied. "They choose not to come to my lab. I think the type of research we are doing now is not mainstream enough for them."

Nancy asked whether Mary-Lou had requested to stop teaching the graduate seminar in Methods and Logic. Mary-Lou said no, Maury Fox had made the decision. Nancy told her that Frank had mentioned it to her at the time. Mary-Lou told Nancy about trying to help the graduate student who'd had the difficulty with Frank, and Nancy asked if Mary-Lou thought that had anything to do with her being removed from the course. No, Mary Lou said, she had helped the graduate student after Maury removed her.

A few days later Nancy returned from traveling to find messages from Mary-Lou on her phones at home and at work, a note taped to her office door, and another slipped underneath. Mary-Lou had checked the dates, she said, and Nancy was right. Mary-Lou had been removed from the course after she tried to help the graduate student. She was angry and humiliated all over again.

But the women were also excited, optimistic. In the space of a month, they found they had become a group of friends, sharing lunches and dinners and near-daily phone calls. To some it was like finding water in a desert. Penny's friends in her early days at MIT had been among her students. She now thought it had been inappropriate, but she hadn't been able to make friends among men in Engineering. Nancy hadn't had a group of girlfriends since Spence; she realized now how she'd missed it. These were women like her, with minds full of questions and passion for their science. JoAnne liked to say that whenever she felt discouraged, she just read a new paper and felt inspired all over again. Penny had recently read *The Girls in the Balcony*, a book about a group of female journalists who had sued the *New York Times* in 1972 after finding themselves locked out of prestigious assignments and paid less than men. Penny gave it to Nancy, who pronounced it "exhausting."

"Parallels are truly startling," she wrote in a note thanking Penny for the book. "Is MIT better than or same as NYTimes of 20 years ago? I think I'll give to Molly Potter, maybe also Lisa and JoAnne. Or is it too depressing?"

The women made an appointment to see Bob Birgeneau on August 11 and, by the last day of July, had sent Nancy's secretary to his office to drop off a letter and a two-page proposal for a women's committee that would examine how resources and teaching assignments were distributed. Their tone was polite, conciliatory, collaborative.

"There is a widespread perception among women faculty that there is consistent, though largely unconscious, gender discrimination within the Institute," they wrote.

"We believe that unequal treatment of women who come to MIT makes it more difficult for them to succeed, causes them to be accorded less recognition when they do, and contributes so substantially to a poor quality of life that these women can actually become negative role models for younger women," their proposal said. "We believe that discrimination becomes less likely when women are viewed as powerful, rather than weak, as valued, rather than tolerated by the Institute. The heart of the problem is that equal talent and accomplishment are viewed as unequal when seen through the eyes of prejudice. If the Institute more visibly demonstrates that it views women as valuable, a more realistic view of their ability and accomplishments by their administrators, colleagues, and staff will ultimately follow."

They mentioned nothing about maternity leave or childcare in the letter to the dean because they worried that, if they did, the administration would say that the reason there were so few women in science is that women chose to have children. The women themselves had assumed that for a long time; most of them had chosen not to have children because they saw it as the price of doing science. They wanted MIT to realize that children weren't the only problem. They asked that all the women in the School of Science rotate onto the committee so they could learn how departments function, and to help younger women understand. "There are so few women in some departments that it is easy for them to become isolated and excluded from the types of information that provide professional advantages." Ultimately, the women could learn how to serve as department heads. And if it was successful, the model could be used to examine the other schools at MIT.

The women attached a confidential memo detailing "some of the disturbing experiences" in Biology, where their movement had begun: the inequities in salary and space; Frank Solomon's denigration of Mary-Lou; and Nancy's removal from the Biology requirement, an episode that had been, she wrote, "more traumatic than an attempted rape some years ago by one of my colleagues."

By the time they delivered the proposal to the dean, fifteen of the seventeen women in the School of Science had signed it. Millie Dresselhaus had not. Millie and Sheila Widnall, some of the earliest women on the MIT faculty, were still the most celebrated on campus, famous and beloved for paving the way for others. But for some of the women, their example had become more of a burden. No one doubted they were extraordinarily accomplished and inspiring. But it was as if two female stars were enough. Men held them up as proof that women at MIT could be all they wanted to be if they just worked hard enough—"Millie did all this and had four children, what are you complaining about?" A colleague had told Penny that when they discussed her promotion to full professor, others compared her to Sheila and Millie, even though they worked in entirely different fields; the only thing they had in common was that they were women.

Sheila was on leave, as secretary of the air force. Millie was hard to reach. She traveled often, having become so prominent nationally. In 1990 President Bush had awarded her the National Medal of Science. She had been president of the American Physical Society and in that role started a visiting committee that worked with high schools and universities to improve the climate for women at the entry levels of physics.

When Nancy finally found her, Millie hesitated about signing the letter to the dean. She was surprised by the low numbers of women on the faculty in the other disciplines, more than two decades after MIT had called on her to help increase them. But she still believed in the approach she had embraced then: to encourage more young women into the pipeline. She told Nancy she knew that she'd earned less than some men, including her husband. "But I had enough money," she told Nancy. She still worried that hiring more women might mean lowering standards.

Then Millie looked at the list of women who had signed the letter to the dean: "JoAnne Stubbe. She's supposed to be good."

With that, Millie became the sixteenth and last woman to sign. Nancy hadn't heard back from June Matthews, the tenured woman in Physics, but Lisa Steiner said that was okay, maybe even good; if one woman had refused, the dean would see that no one had signed under pressure.

In the week before the meeting Nancy became more and more anxious. This would be high-stakes. She was sure the dean must have told the president of the university, and the president had probably alerted MIT's lawyers. The institute could not afford to have the women in the

School of Science—women working at the top of their fields—say it discriminated against them.

They had decided that six women would represent the group in the meeting with the dean and, two days beforehand, did a dress rehearsal in a conference room near Nancy's office.

Penny picked up Nancy in the cancer center an hour before their appointment with the dean on August 11. They walked across the street to collect Lisa and Mary-Lou in Biology, then to the main campus to pick up JoAnne and Sylvia. Nancy thought Sylvia looked pale and thinner than usual, almost translucent. Lisa was wearing a skirt, which signaled the significance of the occasion. As they walked as a band across the shady expanse of Eastman Court, the women imagined that everyone must be watching them. It was ridiculous, Penny thought: here she was, a full professor, feeling as uncertain as a freshman on her first day of class.

They pulled open the heavy doors to Building 6, one of the white monuments of the elegant original campus on the Charles, and began walking down the long, cool corridor. On a summer day, without the usual crush of students, they could hear their steps echo against the marble floors and the tall, painted cinder-block walls. No one said anything.

The dean's assistant showed them into his conference room. Nancy had always been curious to see it; this was where the Science Council argued over tenure decisions. It was a stately room, with high ceilings and wood paneling. Nancy's eyes went to the long, polished-wood table that dominated the room. She thought of the opening scene of *The Girls in the Balcony*, which described the newly formed Women's Caucus of the *New York Times* meeting the publisher and other men of the newspaper's masthead across a twenty-five-foot table, an obdurate, gleaming mahogany symbol of the 121-year-old institution the women were challenging. To the journalists in the book, it had seemed overpowering, "to go on as long as the eye could see." This table was smaller, Nancy thought, but no less daunting.

Someone had set out soft drinks, coffee, and cookies on a credenza next to the table. Above it was a large photograph, and Nancy could see that the other women's eyes had fixed on that. It was a picture of Birgeneau and the five heads of departments in Science. They were all men, as department heads had always been, and all grinning. One was wearing a tuxedo. They were holding their forefingers aloft to say, "We're number one!" Suddenly all Nancy could see of the room was the photograph. She felt sick. This had all been a bad idea. She remembered what Penny had said all summer: "We're not even on their radar screen."

Chapter 19

X and Y

Penny was right; they were not on the radar screen.

When Bob Birgeneau walked into his conference room for his three o'clock on August 11, 1994, he didn't even know what the meeting was about. He hadn't read the letter or the proposal the women had so carefully written, shredded, and rewritten over the last month. He was just back from Brookhaven National Lab, on Long Island, where he spent the better part of every summer running experiments on neutron scattering in the High Flux Beam Reactor. He had spent his early career avoiding administrative jobs, and while he liked his role as dean, he preferred being in the lab, especially at Brookhaven, where he did his own research without postdocs or graduate students to manage. He had returned recharged, as he always did. To the six women who sat waiting for him, he showed a picture of confidence and ease, a late-summer tan, and a broad smile.

If he had to, Birgeneau would have guessed they were there to talk about Nancy being removed from the Biology course, a dispute he knew well from the spring. Instead, Nancy explained how they had come together over the summer, that they wanted to work with the university, and explained their idea for the women's committee. She had typed out notes, knowing she'd have trouble keeping her nerves in check. In bold she'd typed, **Progress at universities comes when committed faculty meet up with a committed administration. Opportu-**

nity exists now at MIT to do something important about this very important problem.

The women went around the conference table, starting with Sylvia, then JoAnne. They described the arc of their careers: their optimism coming to MIT, only to end up feeling isolated, ignored, frustrated over resources. Lisa talked about salaries, how some women realized they were underpaid only after they got sudden raises. The women had known when they chose careers in science that they would have to make sacrifices in their personal lives, but they had not expected to be underpaid. None of the women in the room had children, Nancy told him, "They aren't even married."

"My personal life doesn't exist," Sylvia said. "I can't even buy a house."

At this the dean jumped in—Nancy thought he might lunge across the table. "Why didn't you come and see me about that?" Male faculty members had been getting loans to buy homes for years; none of the women in the room had realized they could ask. A whole world existed for men that the women were only now glimpsing.

Birgeneau had experienced a few eureka moments in his thirty-year career, times when he was struggling to make sense of a set of facts that didn't seem to fit together, and suddenly like a thunder-clap everything moved into place to reveal a fundamental truth, like a shift in the weather. He still vividly remembered the time in 1978 when he'd been working on a problem about the phases of smectic crystals—an unsolved question first raised by French physicists a century earlier. He was driving south along the Connecticut Turnpike when the answer struck him, with such force that he pulled off the highway to find a pay phone and call his collaborator: "I've got it!"

Listening to the women now, one after the other, Birgeneau felt the same sudden clarity, a feeling so strong he later described it as a religious experience.

As dean of science, it was his job to know all the faculty members and the challenges they faced, so the women and even some of their stories were not unfamiliar to him. Had any of them come to him individually, as Nancy had in the spring, he would have explained their complaints as the idiosyncrasies of a department, a situation, or relationship, a budget dispute or internal politics. Now he had six women in front of him, and a letter with sixteen signatures. Seeing the women all together and hearing the uniform unhappiness in their stories, he suddenly realized, *We've got a big problem.* This wasn't just about lab space or a course, it was a pattern. A problem in the system. These women were not difficult. He was struck by how much they'd managed to accomplish despite the environment they'd been working in. He hadn't realized how few of them had children. Few men had made that personal sacrifice, he thought—he himself was the father of four.

Birgeneau asked the women if it was all right for him to speak now. He told them that when Nancy had come to him the previous year, he didn't know her, so couldn't evaluate what she was saying. He didn't think it was discrimination, but he'd asked his daughters and his wife, a social worker, and they had started him thinking. And now he understood what they were saying, what Nancy had said: women were—here he borrowed the word Sylvia had used earlier in the meeting—"marginalized."

Nancy asked if he thought their committee would work. Birgeneau was doubtful. What they were feeling was disrespect, and that was hard to quantify. He told the women he thought the issue might be tangled in the competitive, male-dominated culture of MIT, and no committee could fix that. But he told them they could try. He suggested they keep it small—three people—but agreed when they asked for four or five. He told them to meet with his assistant to draw up a charge outlining the committee's role—that was standard practice for establishing any new committee, like setting a hypothesis. His assistant would be back from vacation in two weeks.

"Our meeting with the dean went extremely well," Nancy wrote the other women. "In fact it's hard to see how it could have gone better. He was receptive, concerned, and prepared."

Her jubilation was short-lived. Two weeks later, Birgeneau appeared in the doorway of Nancy's office in the cancer center. It was the other end of campus from his own. He laughed awkwardly. "I'm lost."

She invited him in.

"There is a snag."

"I'm being fired?" Nancy was still exulting from the meeting, and half joking, though it occurred to her that maybe she shouldn't be.

Birgeneau had told the Science Council—which included the department heads and Richard Hynes as head of the cancer center—about the proposal for the women's committee, and some department heads were annoyed. They thought there were too many committees already, which Birgeneau thought was a concern he could work around. They also didn't like to be second-guessed. He was surprised at how vehemently they had resisted the idea.

"So what?" Nancy said.

"They'll resign."

"Good."

Birgeneau laughed. "You've probably noticed that deans don't have much power." It was true, the power in the School of Science had always been with the department chairmen because they controlled teaching assignments and resources such as space and internal grants. Phil—Nancy suspected he was especially irritated by the proposal—was particularly influential. He had a Nobel Prize and but for his own reversal would have been the president of the university. Birgeneau told Nancy he had chosen strong chairmen on purpose. "Weak ones are boring. I like strong people, that's why I like you." But he had to rule by consensus.

The chairmen had reacted the same way Birgeneau himself had when Nancy told him her problem was "discrimination." They had so few women in each department that they couldn't see any pattern. They could explain all the reasons this woman or that woman was unhappy; as far as they could see, her difficulties were tied up with individual circumstances that had nothing to do with her being a woman. And MIT was just like all other elite universities in having so few women on its science faculty. All but three of the top ten math departments in the entire country had not a single woman on the faculty. As Phil pointed out about the Biology department, Harvard was worse.

A couple of the chairmen wanted to sit in when the committee's charge was drawn up. Birgeneau told Nancy he would let them; it would help get them on board.

Nancy asked if the women should go to the president instead—maybe with Birgeneau. Birgeneau said no. "In universities things don't work from the top down. Your movement is working because it's grass-roots. You have to get the chairs on your side."

"Will it work?"

"I think so."

"Can you promise?"

"Promise?" Birgeneau laughed again.

———————

Birgeneau had already gone to see President Vest. Chuck, as the president was known, was a tall and rangy West Virginian, soft-spoken and self-effacing. His father had been a celebrated professor at West Virginia University in Morgantown, and his own classmates recalled Chuck Vest as the smartest kid in every class, but his colleagues appreciated that he never needed to be the smartest man in the room. He was warm and unpretentious and still thought of himself as a small-town boy with small-town values. He'd arrived as president four years earlier from the University of Michigan, where he'd been ever since finishing college in his hometown, and risen rapidly from professor through a succession of high-ranking posts. He had arrived to a cool reception at MIT; the faculty preferred presidents who had risen through its own ranks, and Vest had the additional stigma of being the second choice, having taken the job after Phil Sharp reversed himself. Vest confronted any skepticism head-on, joking that he'd gotten two letters from MIT in his life: one rejecting his application for assistant professor, the other hiring him to be president. His kidding aside, many faculty members sniped that MIT had hired a president who couldn't get tenure there.

Vest had proven himself a prolific money raiser among private donors and in Washington, where he saw it as his responsibility to explain the importance of research universities for American innovation in the post–Cold War era. He had opened MIT's first office in the nation's capital. And he had recently succeeded in fending off the federal government's attempt to force universities to give more financial aid based on merit rather than need, the fight the Ivies had declined to take on.

Standing up for needy students had made Chuck a hero to many faculty members, including Birgeneau, who had been among the early doubters. Birgeneau had begun seeking Chuck's advice often. Now, Birgeneau told Vest that he thought the women had a good idea to look into salaries and other resources, but the department heads were pushing back.

Vest liked to seek a lot of opinions before he made decisions, which could sometimes vex his lieutenants. But in this case, he didn't hesitate. He told Birgeneau to go ahead, that he'd back him against the department heads if it came to that. If there were inequities, MIT needed to fix them. Birgeneau quoted Vest's exact words to Leigh Royden, one of the three women in Earth, Atmospheric and Planetary Sciences. Leigh relayed the words to Nancy, who wrote them on a sticky note that she attached to her computer monitor: "The president said, 'Do it.'"

Vest was not surprised by the low number of women on the science faculty, or any faculty. No one could be. The number of female professors in the United States had doubled since Title IX was signed in 1972, but they still made up only 33 percent of the total. In Vest's own field, engineering, the number of female students was still dismally low, and the number of Black scientists remained minuscule. In 1973, when Shirley Ann Jackson became the first Black woman to earn a doctorate at MIT, she was the second Black woman in the United States to earn a PhD in physics; twenty years later, fewer than twenty Black women in the history of the country ever had. Avowals of good intentions on affirmative action were now routine on many campuses. At MIT, every department had to certify that its job searches had attempted to include candidates who were women or from minority groups. Their annual reports included some version of what had become a standard disclaimer: "In our hiring of staff we make every effort to locate qualified minorities or women, in full compliance with MIT's affirmative action policies."

While Vest was willing to allow the women's committee, he didn't expect it to find that women on the faculty were treated differently. Like most people, he thought the problem, and certainly the solution, was not in universities but started earlier, in middle and high schools. To end up on an elite faculty of math or science or engineering, a woman almost invariably started young.

The patterns for girls were well-known: they started out doing as well or better than boys in math and science, but by middle school began to fall behind, and by high school scored lower on standardized tests of math and science. Girls' scores had improved over time, suggesting that proficiency was malleable, not the result of innate ability, but a gap with boys persisted. Research suggested the gap had something to do with attitudes: middle school girls liked science just as much as boys, but by senior year of high school they were more likely to say they didn't like it. Girls were less likely to say they had a computer or a microscope at home, more likely to say they didn't take math in senior year because a teacher had discouraged them, or that they themselves didn't think they were good at it.

Not everyone agreed that girls were the ones who needed worrying about. They had the edge in many other ways: boys were more likely to end up in remedial math or special education and to repeat a grade, and they were less likely to go to college. After years of pushing to open opportunity for girls, some educators were starting to argue that schools were failing boys. (In 1993, the Ms. Foundation had held the first Take Your Daughter to Work Day, in an attempt to encourage girls' career ambitions. Almost immediately, parents and schools and workplaces asked why boys should not be included, too, and within a decade the event would officially be renamed Take Your Child to Work Day.)

Whatever the reason that women lagged, the men in the School of Science thought it was unfair to blame them. They were not misogynists. This all seemed like an overreaction.

Tom Jordan, the department head in Earth, Atmospheric and Planetary Sciences and Leigh's boss, called Marcia McNutt to demand, "What the hell is this?"; he'd heard the women wanted their own ver-

sion of the Science Council, with the power to oversee tenure and promotions. He liked what he called the entrepreneurial culture of MIT,
that people made deals to get what they wanted. He planned to argue
against the committee. Frank Solomon had turned up in Mary-Lou's
office, upset, saying that people were telling lies about him. He insisted
he had nothing to do with Mary-Lou being removed from the graduate
seminar; he knew nothing about it. Mary-Lou called Nancy to make
sure she had recorded the account of her interaction with Frank years
earlier in her diary. (Nancy assured her she had.)

By mid-September, Birgeneau had met with the heads on the Science Council and failed to convince them to agree to the committee.
He told Nancy that he wanted the women to present the idea—one
woman from each department. It didn't matter which woman, but
Millie should be there, given the respect she commanded. So late that
month, Nancy and Millie and four other women gathered in the dean's
conference room again, this time for a lunch with Birgeneau and the
department heads, to make their pitch. This was a once-in-a-lifetime
opportunity to fix a problem that had existed for generations, Nancy
told the heads. Who better to fix a problem than MIT, the world's
experts in science?

The meeting went badly. The department heads sat mostly stone-
faced. Only one—Bob Silbey, the chairman in Chemistry—supported
the idea even in principle. The others said that allowing a committee to
examine how departments distributed resources would be outside the
normal chain of command. The committee would just nurture grievances. Birgeneau concluded the lunch by saying that the committee
would be "advisory only"—it would have no administration authority
to fix any problems it found.

"I told you it would never fly," Millie said to Nancy afterward.

Birgeneau resorted to an ur-academic solution: he would form a
subcommittee from the people in the room to write the charge to the
women's committee. In all the bureaucratic nonsense he had seen in
his career, this was a first: a committee to form a committee. But he
thought it would ease the department heads' concerns. He chose two
men for the so-called charge committee: Silbey, the Chemistry chair-

man, and Tom Jordan, from EAPS, who remained firmly opposed. Including Tom had been Leigh Royden's idea. As much as he came out strong on his opinions, she knew he could reverse course if he was presented with good arguments and advocate just as forcefully for his new position. Birgeneau agreed that if other department heads saw Tom change his mind, they would, too.

The other women, though, worried Birgeneau was not really on their side. Was this a committee to kill the committee? On Penny's advice, Nancy sent him a copy of *The Girls in the Balcony*. "The situation has a startlingly familiar ring to it," Nancy wrote in a note to him. The book ended with the *New York Times* settling the suit, with a consent decree that put more women in high-level positions but did not raise salaries. Nancy suggested Birgeneau share the book with the department heads: "We are counting on you for a far better outcome than was obtained at the *Times*!" she wrote. "However, I must say that some (*not all*) of us found the meeting with the Department Heads concerning. It seemed to me not just a matter of understanding, but a question of whether this item was on their agenda?"

In smaller conversations, the women themselves were trying to change the department heads' minds. The five tenured women in Biology met with Phil, who opened by noting that the department had more women than any other, so it must be doing something right. The women pushed back. Ruth Lehmann asked him to explain how group grants were awarded. Phil said it was out of his control; the faculty who applied for them decided the faculty members to include. "That's not true," Nancy said. "If there is a grant and the only excluded person is a woman, that is discrimination, and you can object." How could he recognize it? It was like smoking and lung cancer, Nancy said. "Not the one case, the pattern."

Phil said that when he gave raises he tried to think about young people starting out. They had to establish themselves in the community, buy houses, they had children. "That is where the women's viewpoint needs to be heard," Nancy said. "Maybe single women need more money."

Richard Hynes, her adversary in the space wars, had heard about the confidential memo attached to the letter to the dean, which detailed mostly problems in Biology, and asked to meet with Nancy in his office. He listed all the things he had done for women as department head—he had left the salaries "perfect," he told her. When he took over as director of the cancer center, he immediately noticed that Nancy was underpaid and gave her a raise. He couldn't correct all the low salaries at once because the department only got so much money, and he had to give raises to the men, too.

Nancy told him that she'd finally gone to see a lawyer about her efforts to get more space, and that the lawyer had said she was being harassed.

Richard looked shocked. "It was stressful for me, too."

Nancy was exasperated, but she couldn't blame him. It had taken her twenty years to understand. Maybe it was too much to expect Richard to do the same.

Nancy was still casting about for solutions, in case Birgeneau or the other men killed the committee. She and Penny went to see Lotte Bailyn at MIT's Sloan School of Management. Nancy knew her from a committee on family and work they had served on together in 1988 and knew that Lotte was an expert on careers and consulted for companies and foundations looking to help employees balance their work and personal lives. Lotte had recently published a book based on that work, called *Breaking the Mold: Women, Men, and Time in the New Corporate World.*

She had come to believe that improving workplaces for women required a redesign of the very idea of work. The problem she had diagnosed in the 1960s had not gone away: work was still considered the realm of men and home the realm of women. Work was built on the assumption that employees had a wife at home to take care of personal matters. With so many women in the workplace, the job of caring for children and home and elderly parents had become the "second shift." It made it hard for women to be ideal workers, since compa-nies' ideal workers were people with no responsibilities outside the company. Subsidies for day care and family leave, where they existed, barely touched the problem.

She thought back sometimes to the conference at the American Academy on "The Woman in America" in 1963, and Erik Erikson's ideas about how men and women approach work differently. She had thought it bizarre. But she had come to agree with his argument that women couldn't win by trying to jam themselves into a male model of work; true equality required what he had called "revolutionary reassessment." Her book had argued that employers needed to stop measuring commitment by the number of hours spent in the office. Instead, they had to define the tasks, let employees figure out how and where to get the work done, trusting their intrinsic motivation to meet expectations. All the talk of "quality time" with family got it backward; women and men would be better off with quality time at work and more quantity time at home. Instead of work-life balance, she pre-ferred "work-personal life integration," though she acknowledged it was a clunkier phrase.

For many years Lotte was the only tenured female professor at Sloan, a lonely position made worse by the low status of her field—organization studies—in a school that emphasized finance and eco-nomics. At meetings to discuss tenure and promotions, the men chatted with each other, not her. A young male colleague—they got tenure the same year, though he had been hired right out of his doc-toral program—had made a poster with headshots of all the senior professors and assigned students to make a computational model of the organization. He left off Lotte, and when she went to him, furious, he explained that he worried that students would have too hard a time doing the model if he introduced sex difference.

Lotte was unsurprised to hear the women talk about discrimination; she knew it existed everywhere, not just at MIT. But listening to their stories, she was surprised at how egregious they were. Like Birgeneau, she was struck by how much the women had accomplished despite the environment they described. But she told them she doubted they'd be able to dent the consciousness of MIT. Her own committee on work and family had recommended basic benefits such as allowing parental leave and days off to care for a sick relative. Yet people—especially women—were still afraid to take the time.

Lotte's book had been based on the successful experiences of companies she consulted for. But she told Nancy and Penny that MIT did not take her work seriously. She had just turned sixty-four and was thinking about retiring.

Nancy hadn't had much interest in the earlier committee she'd served on with Lotte, about family and work. Nancy had no children and figured she'd been asked to serve because there were so few tenured women to choose from, and she'd had to quit when she went on sabbatical in Germany. She still believed, as she had written in the *Radcliffe Quarterly*, that high-level science required seventy-hour weeks. But she was surprised by Lotte's pessimism. Nancy and Penny left discouraged. Nancy wrote that night that Lotte had seemed "underwhelmed" by their idea for the committee.

The work of getting the women's committee past the Science Council landed with Leigh Royden, the youngest of the sixteen women who'd signed the letter to the dean. Birgeneau made her head of the so-called charge committee. Nancy was impressed by Leigh, thought she would be a university president someday. Still, she wondered if Leigh, at thirty-eight, was too young to understand the problems they were trying to fix.

Known to friends by her childhood nickname, Wiki, Leigh had broad shoulders and the confident posture of an athlete. She had been a champion swimmer as a teenager in California, winning state and national titles and a world ranking. She liked to joke that she majored in rowing at Harvard—she had won the US championship in the single scull in 1974 and rowed on the women's team that brought home the silver from the world championships in 1975, a triumph celebrated by *Sports Illustrated* and *Time*, under a headline borrowed from the team's nickname: the Red Rose Crew.

Direct and self-possessed, she was also the daughter of a well-known mathematician and dean at Stanford. Her father had set high expectations; he forbade her to talk about swimming at the dinner table and instead began the meal by posing a question to her and her two siblings, sending them scrambling to the encyclopedia to be the first to answer. All her athletic training had prepared Leigh well for her career; she thought of defending her data like being on the starting blocks. Living with her father had been perfect preparation for dealing with the older men in her field; she learned early to show respect and interest, but also to stand up for herself when she knew she was right.

She'd found supportive male mentors and, in the next office, Marcia McNutt, a geophysicist who had infinite energy and a glamour unusual in science—while most professors at MIT dressed in pleated khakis and rubber-soled shoes, Marcia wore impeccably tailored jackets and arrived, in heels, on a cherry-red motorcycle. She had been widowed with three young daughters—two of them twins—yet still managed, with the help of a loyal housekeeper and precision scheduling, to spend several months a year at sea, often as chief scientist of the mission. Marcia had arrived at MIT two years ahead of Leigh, been tenured a year ahead, and had her children four years ahead, so Leigh figured she'd just keep following Marcia's example. It had worked. Leigh had been tenured at age thirty-three and married Clark Burchfiel, a professor twenty years her senior. They now had two young children. The only discrimination Leigh felt in her department were the whispers from colleagues that she had gotten tenure because of her husband. But Birgeneau had told her that her case was "extraordinarily strong." It had impressed him enough that he had introduced Clark at a party shortly after as "Leigh Royden's husband."

The charge committee, made up of Leigh and Nancy, Bob Silbey, and Tom Jordan, met for the first time in November of 1994. On the first and most fundamental question—should there be a women's committee at all—they all quickly agreed yes. Nancy and Leigh said it was important for the women's committee to be able to collect data, especially on salaries. Silbey worried about privacy, and he also doubted the data would show any discrepancies. Birgeneau had sent an assistant to the meeting, who interjected to say that the women in Biology had all received a 20 percent raise one year to make up for an inequity. That seemed to satisfy Silbey.

The committee met three times for two hours at a stretch, debating whether the women's committee would be advisory or administrative—would it have power to fix inequities, or just to propose fixes that the dean would be free to ignore? The question of who would choose the members of the committee—would it be the women, the dean, people recommended to the dean by the women?—took up an hour and a half. But by late November, the charge committee had agreed on a proposal to send the Science Council, for a nine-member committee on the status of women, with three men and six women, including a representative from each department.

Nancy had argued that the committee should be all women. Birgeneau told her no; it was critical to put well-connected and influential men on the committee, who could help the women get the information they needed and help make their case to the other men. Sylvia agreed; without men, the committee would have no status. JoAnne, too, said they had to compromise: "They"—the men, that was—"are very nervous right now," she told Nancy.

The Science Council approved the charge in January 1995—five months after the women had first met with Birgeneau. The department heads changed only one thing: the charge committee had wanted the women's group to be "the avenue of address for women faculty who are concerned that they are being treated inequitably," and to have the authority "to recommend a possible course of action"; the Science Council did not want it to become a grievance committee. Birgeneau changed the line about membership, after all the debate; he wanted to choose it "after consultation with" women on the faculty rather than based on a list proposed by them.

As the dean wrote to the School of Science faculty, the committee would "collect data to be used in assessing the status and equitable treatment of women faculty in the School of Science." It had little to no administrative power; its function was merely advisory, to "facilitate communications between the women faculty and the dean and department heads" and "act as a resource" for the deans and department heads, and "to the MIT community as a whole to provide advice about issues of concern to women faculty at MIT."

The last question was who should run the committee. Birgeneau had assumed it would be Nancy; she had started this, and the women trusted her. Some of the department heads complained that she was a radical, a troublemaker. Bob Silbey and Tom Jordan disagreed: Nancy, they said, was a consensus seeker. Still, Birgeneau was worried the whole plan could blow up; he wondered if he should choose Leigh, or Molly Potter, as a well-liked former chair of the faculty. But Nancy had shown him that she was open to compromise. That impressed him. He began to survey the other women—first Penny, then Molly, then Leigh—to see if one of them might want to lead the committee. They all told him what he already knew: Nancy was the obvious choice.

Birgeneau offered Nancy the job as chair in a meeting in his office in late January. He told her June Matthews would represent Physics on the committee. June hadn't signed the women's letter because she'd been in Los Alamos over the summer, he said. She didn't believe she had been discriminated against. But Birgeneau knew that she was underpaid, and that another faculty member had given her trouble. He thought she would be a good addition.

Then he turned to his concerns about Nancy. He wanted to warn her about one flaw, one he said he'd had to learn to check in himself.

"I'll write it down," Nancy said.

You react too quickly, Birgeneau told her. Wait twenty-four hours to respond when you are upset. And remember that your strength is in the group: "Stay consultative."

"Did you ever see me get mad?" she asked.

"No. But you have said that you do."

She knew Richard and Phil had told him so, too. "But you know those are very difficult men," Nancy said. "If you say something to Richard seventy-two times and he doesn't listen, on the seventy-third time you get mad."

Nancy reported this back to Sylvia, who was angry that the dean had made Nancy sound irrational. Nancy was consultative, Sylvia said: look at the way she had passed around all the initial drafts of the women's proposal, making sure everyone signed off. Nancy did not mind the advice. She knew Birgeneau had a point. She knew he was right about the group. It wasn't clear yet what if anything the committee could accomplish. But they were on the radar screen.

Chapter 20

All for One or One for All

The Committee on Women Faculty met for the first time at the end of February 1995, in the conference room around the corner from Nancy's office in the cancer center—no more sneaking around to out-of-the-way locations. Nancy realized she had never run a committee before and worried over the logistics; she sent out an agenda, then a revised agenda, and wondered if she should set out wine to encourage attendance. Given that the meeting was scheduled for eleven o'clock in the morning, she offered cookies instead. Seven of the eight other members showed up. Around the table were six women—Nancy, Penny Chisholm, June Matthews, Molly Potter, Leigh Royden, and JoAnne Stubbe—and two men, Bob Silbey and Jerry Friedman. The men began talking first. Leigh shot Nancy a look, then interrupted, "Wait a minute, women are supposed to be running the show." Everyone laughed. They started over, this time with Nancy leading.

They made a list of the data they wanted to collect: on salaries, teaching assignments, grants, and space; the number of women at each stage in scientific careers at MIT, from undergraduates to faculty. They still weren't sure the administration would share data on salaries. Penny thought it was important for the committee to hear stories, too; it was hard to understand marginalization without anecdotes. So they planned to interview all the women on the faculty in Science, tenured and untenured, and the department heads, who were all men.

Having insisted to Nancy that the committee include men, Birgeneau had selected three who commanded respect around the university. Jerry was a Nobel laureate in physics—he and two colleagues had proven the existence of the subatomic particles known as quarks—and had been chairman of his department. Bob Silbey was chairman of Chemistry and known as a shrewd navigator of university politics. There were no women with tenure in Math, so Birgeneau had asked the one untenured woman to recommend someone to represent the department; on her suggestion he chose Daniel Kleitman, who had been at MIT for thirty years and was another former chairman. (Not long after, someone would also recommend Kleitman to two aspiring filmmakers, Cambridge locals Ben Affleck and Matt Damon, who were looking for a mathematician to advise them on their screenplay about a genius named Will Hunting, who worked as a janitor at MIT.) The men understood how the system worked in a way none of the women could because no woman had ever led a department. After the committee's second meeting, the next month, Nancy wrote the dean, "You were right."

Silbey was a physical chemist, like Sylvia Ceyer, and had been one of her mentors when she arrived as a junior faculty member. He appreciated the concerns of the department heads—the "proceduralists" as he described them to Nancy—but he also liked to tweak at authority. Birgeneau, knowing Silbey typically wore nothing more formal than a tattersall shirt, had instructed him to put on a tie for the "We're number one!" picture that hung in the dean's conference room; Silbey arrived in a tuxedo. In late March, he showed up in Nancy's office and announced himself by dropping a large volume the size of a phone book on her desk. It was the book of sponsored research activity—known as the Brown Book if you knew about it, which Nancy did not. It showed the distribution of grants and many other resources to the schools, departments, and laboratories at MIT, money that came from outside and inside the university. It also revealed how much grant money individual professors had to raise to pay their own salaries—the question Nancy had put to Richard Hynes during one of their arguments three years earlier.

In May, when Nancy worried that the administration was stalling on confirming the numbers, Silbey accompanied her back to Birgeneau's conference room and banged his fist on the table: "You have got to get this woman the data she's asking for!" Nancy knew she could not have done that—or gotten away with it.

The data in the book proved invaluable. In Biology and in Brain and Cognitive Sciences, the only two departments where faculty members had to raise grants to pay their salaries, women were funding a greater percentage of their salaries, sometimes twice as much as men. Mary-Lou was paying more than anyone else in Biology, both in dollars and as a percentage, despite being one of the most senior members of the department, and a member of the National Academy. Women's salaries in Biology were 17 percent lower than the men's, even after Lisa and Nancy had received raises to compensate for previous discrepancies.

The data hinted at how marginalization worked: men and women often started out equally, but while women struggled against a thousand cuts, men were accruing small benefits—a modest grant here, a crucial piece of equipment there. No one had necessarily intended to discriminate; it was more what the department heads had defended as the entrepreneurial culture of MIT: you could get what you wanted if you knew to ask for it. The women did not know; they took what they had been given and assumed everyone was following the same rules. Or they were too polite to ask: some men received raises to keep them from accepting job offers outside MIT; some women who'd had outside offers either had not been able to move because of their husbands' jobs or did not use the offers as leverage because they did not want to operate by threat.

Other numbers showed a similar pattern. By their own accounting, men were spending more time on outside professional activities— running companies, giving talks, serving on committees for professional societies—that were compensated. Women were doing the same work, but more often for free.

As the women in Science had suspected when they first talked in the summer of 1994, the number of women on the faculty had not budged in twenty years, since the push for affirmative action in the early 1970s. In Biology, the average age of the women was fifty-eight, and just three women were on the junior faculty. In Brain and Cognitive Sciences, the average age of the women was fifty-six and there were no junior women to rise through the ranks. There were no junior women in Chemistry, either. But it wasn't as if women weren't interested in scientific careers: the number of female students had been growing steadily for a decade. In Biology, Chemistry, and Brain and Cognitive Sciences, there were more women than men among the undergraduates. (The proportions were roughly the same in EAPS; only in Physics and Math did male undergraduates vastly outnumber women.) But the pipeline was leaking: the number of women dropped off between undergraduate and graduate school, and precipitously at the next level, when women would have been expected to move on to postdocs or junior faculty positions. Among Chemistry postdocs, there were three times as many men as women.

The numbers, and the women's stories, surprised Jerry Friedman. "You have to write this down, get it into a report for the dean," he told Nancy. Many committees and task forces at MIT had put their conclusions into reports, but there had been no plan for this committee to do the same. "The dean needs it in writing to be able to act," Jerry kept saying. "The dean is a scientist, he's driven by data. If you can show him the data, it's persuasive."

Nancy made a mental note to write a preliminary report by June and a final report the following January.

And by spring Birgeneau had already begun fixing inequities. He had started his own review of salaries soon after the women appeared in his office the previous summer; now he had given large raises to several of them. The committee had found that women were paying a bigger share of their secretaries' salaries in some departments; he evened out the subsidies MIT paid to make them equal. He hired two new fully tenured women, one from Johns Hopkins, the other from Cornell. One of the women had said she could not move unless her husband found a job; MIT had helped him. Birgeneau was in discussions with a third woman, at Berkeley, to form a new center on biochemistry with JoAnne Stubbe. He had also found money for an expensive piece of equipment that JoAnne needed.

For the women, even small changes were life changing, if only because they were creating a new sense of fairness and belonging. The committee had made them feel they had a say in how the university operated. They no longer felt isolated within their departments, which lifted morale as much as the material changes. Together, the women had started a file of other outstanding female scientists from around the country, to help in job searches. They had also begun regular dinners with the women on the faculty in Engineering. At a dinner in March the women in Engineering sought advice on a new maternity leave policy they were proposing. In May, the School of Engineering adopted it.

Nancy was busy with the committee, which was meeting monthly, and still trying to hunt down information from various departments and offices around MIT. Still, now that she was no longer fighting over the course or resources, she could devote more time and energy to her work in the lab. It was flourishing. The representative from Amgen had visited her lab in December 1994 to see the technology she had developed for insertional mutagenesis. He'd told her he believed that her ambitions to isolate the genes involved in normal embryonic development of the fish could lead to therapies for genetic diseases, and that of all the proposals at MIT, hers was the one the company was most interested in funding. She had asked for $30,000, to help pay a postdoc to study blood formation in the fish. Amgen offered more than that: $250,000 as a start, with the potential to increase to a million dollars a year if the work went well.

Nancy had used the money to hire more people to her lab, which meant more people doing insertions to try to cause the mutations that would help identify the critical genes. Birgeneau found her seven hundred additional square feet to fit the new lab members and a growing stock of fish tanks. By mid-1995, her lab had completed the next step, confirming that they could make the mutations. But the virus Nancy had found in San Diego was not concentrated enough to reliably introduce mutations. It would take too long to tag the genes, so she had a new challenge: to develop a more potent virus.

She and Birgeneau had lunch in July 1995, and afterward Nancy wrote him, "Except for that one absolutely unfortunate problem we discussed about teaching—and its underlying causes—everything else is fine for me at work now—wonderful space, resources, lab, etc etc. Also, the committee *has* made a difference, at least to me, by creating a community of women faculty. They are truly exceptional, and have been an entirely unexpected bonus in my life at MIT."

That "one absolutely unfortunate problem" was the one that had inspired the women's committee in the first place: Nancy's fight over the introductory biology course. By fall, it was again consuming her time.

The fight had resumed in March of that year with a phone call from Harvey Lodish. As they had worked out with Phil the previous summer, Nancy sat in on Eric's and Harvey's lectures in the course in the fall. They had been cordial, but a residue of bad feeling remained.

Now, on the phone, Harvey was annoyed: "Three people have told me that you said I stole material from your course for the book Eric and I are writing. While in fact our book has nothing to do with the course."

"What book?" Nancy said, her voice rising a register.

"Come on. You know we're writing a book."

"I do not." She thought that Phil had put an end to their book with his memo the previous spring, saying that no one could commercialize the introductory biology course. From what she had judged watching their lectures, the course was largely the same as the one she and Eric had taught. If they were now writing a book based on the course, she told Harvey, she wanted to meet with the two of them in the president's office as soon as possible.

Harvey said something about lawyers and court.

Nancy thought of Birgeneau's twenty-four-hour rule and immediately broke it: "Oh, grow up. This is a matter for the faculty. This is a question of collegiality, integrity, fraud, and misconduct. MIT's own lawyers said so."

"The book is unrelated."

"That's insane. You threw me out of my own course so you could have it for your book. Do you think everyone is an idiot?"

"I don't want to speak to you anymore."

Nancy hung up before he could.

She hadn't expected the women's committee would be able to get her course back; she'd thought it would be a conflict of interest to use the committee to even try. But she also thought this was an ethical breach by Eric and Harvey, that Phil's memo had clearly established that no one could write a book on the course without her consent.

She called the lawyers in MIT's licensing office again, who agreed with her. They urged her to go to the provost and take a group of women with her for support. On the phone, Jim Watson told Nancy not to go without a lawyer. Take one MIT administrators can relate to, he said: "The lawyer needs to be fifty-five, white, and male."

Nancy was reluctant to take that step. While she found Eric and Harvey's behavior "despicable," as she wrote to Molly Potter, what upset her most was how the department had pushed her aside after she'd worked so hard, over years, to make the course a success. "It is that that I still firmly believe could never have happened to any man," she wrote Molly. "I believe that had I been a man they would have very likely given me a prize for this contribution. Instead I was fired.

"In some ways I think it would be best to let go. Particularly now when life at MIT is fun for the first time in 20 years. But I wonder if one can stand to stay in the place."

So Nancy started, again, by trying to fix it herself. She met with Birgeneau and Phil instead of the provost and took Leigh Royden instead of a lawyer. She had come to value Leigh's good sense and negotiating skills. Birgeneau greeted them by saying that his assistant would be there to take notes to send to MIT's outside counsel. Nancy was alarmed; she hadn't realized MIT had engaged lawyers. Phil told Nancy she was correct that his memo had been intended to stop anyone from writing a book based on the introductory bio course. However, he couldn't stop books from being written. It was Harvey and Eric's prerogative to write one. And it was Nancy's to sue if she thought it overlapped with the course. But she would not be able to determine whether it did, Phil said, until Harvey and Eric published the book.

They talked past each other—Phil about the copyright question, Nancy about collegiality and academic misconduct. Finally, Leigh asked to speak and said it was in everyone's interest to settle the matter quickly. The fastest way to do that, she said, was to get everyone into the same room. Birgeneau said he'd start by meeting alone with Eric and Harvey.

Five days later, Eric called Nancy at home. It was nine o'clock at night and he was still in the lab, preparing to go to Japan the next day. He told Nancy he had not appreciated being called into the dean's office. Nancy was ruining his reputation, he said. People outside MIT were asking about it. Joan Steitz, at Yale, had confronted one of his colleagues from the Whitehead Institute and demanded to know, "What are you doing to Hopkins?" Eric told Nancy he was calling as a friend and a colleague to say that if she continued talking about misconduct and fraud, he would have to consider libel charges. His wife, he noted, was a lawyer and could file papers. Great, Nancy said. "My brother-in-law is a lawyer. They can have fun together." She stayed calm, and Eric tried to do the same. He came off by turns angry and warm, cajoling and confidential. He told Nancy that Phil had told him in the spring that Nancy had threatened to sue; that's why he and Harvey had to try to include her in the course.

"I didn't threaten anyone with a lawsuit," Nancy said. "I said I thought it was discrimination, however."

Eric said he hated fighting; he preferred collegiality. So they agreed to meet with Birgeneau and Phil to try to work things out when Eric returned from Japan.

It was six weeks before they could even confirm a meeting, and when the meeting happened in late May, it only reminded Nancy of how little they had valued her work. Harvey told her she should not be insulted they didn't ask her to help write the book because they had considered and rejected other eminent faculty members, too. They did not see that she was not just any faculty member, but one who had developed the course. Eric said he would not have agreed to teach the course in the first place if he had known that he could not commercialize it. Phil said he would ask MIT's lawyers for a ruling, and Eric said if it did not go in his favor, he would not teach at all.

"Does one need a profit motive to teach?" Nancy wrote to Bob Silbey after. "Having gotten me thrown out of my course and having destroyed my commitment to it, can Eric now threaten to leave the department without any course at all?

"To you it must be hard to see why for me and many other women this story is so clearly, even primarily, a women's story, but it is. I would rather have you spend your valuable time and skill solving the general case of women faculty's unhappiness than my special problem. But then perhaps they are one in the same after all?"

Once again, Nancy heard nothing more for weeks, and by August, with the matter still unresolved, she hired a lawyer. Not a middle-aged white man, as Jim had suggested, but a thirty-four-year-old woman, an employment-discrimination lawyer in Boston.

President Vest had tapped Leigh Royden as his emissary to Nancy. In September, Leigh reported that the president wanted her to hold off on any lawsuit in return for a promise that the administration would fix the problems in Biology: remove administrators who treated women badly, put more senior women into decision-making roles, have teaching assignments done by committee to make sure they were done fairly and not used to play favorites. The president had looked at Nancy's teaching evaluations, Leigh told her; he saw how good they were, he wanted to keep her and was worried she would quit. He wanted Nancy to wait until the Committee on Women Faculty finished its work. He promised that the university would fix the problems it identified.

When Mary-Lou and Lisa had approached Leigh the previous summer to discuss forming some sort of women's group, Leigh signed on readily because she thought it was important to look into the women's concerns. But she had thought to herself as she listened to the two older women that their problems were problems only in Biology. Leigh considered herself tough—the result of her athletic training and the family dinner table challenges—and believed she'd gotten to MIT on her merits. If she stayed tough enough to play the game, she figured, it wouldn't matter that she was a woman in a male-dominated institution. But now, more than a year into the committee's work, she was seeing that the problems weren't isolated to Biology. And the other women were plenty strong themselves.

It was obvious to Leigh—and to Nancy—that MIT was not going to budge on the course. Phil was too powerful, though not powerful enough that he would take on Eric. Eric was threatening to cause problems if he did not get what he wanted. By late fall, Leigh told Nancy she had little choice but to sue.

Jim, too, told Nancy she should move ahead with a lawsuit. Ask for $5 million, he told her; MIT would settle for a million. Nancy's lawyer began to prepare a complaint to file with the Massachusetts Commission Against Discrimination. It was not against Harvey and Eric, but against Phil, one of Nancy's oldest friends in her professional life, for age and gender discrimination in removing her from the course. The complaint argued that she was being treated differently because she was a woman. Her evaluations were proof that she was not removed for bad performance; Eric and Harvey's class had received a lower score on evaluations than the one she and Eric had taught.

Nancy dreaded suing, as many women would. Many were scared off by the relatively small number of lawsuits against universities in the 1970s, typically filed by women who had been denied tenure. The cases were notoriously difficult, drawn out, and usually unsuccessful. And they were expensive, because tenure decisions were influenced and made by layers of people and committees; disputing their judgments meant deposing everyone involved, and finding experts in the woman's field to argue she had been improperly denied. Decisions about merit were subjective, even in fields such as math or science that were assumed to be governed by hard numbers. Lawyers warned that courts did not look kindly on the cases; liberal judges worried about interfering with academic freedom, while conservative judges were dubious about antidiscrimination charges. Even when the woman won, she found herself ostracized by her colleagues, in the university and in her field.

Nancy had read about the recent case of Jenny Harrison, a mathematician who sued the University of California in 1989 after being denied tenure at Berkeley, which had one woman and nearly seventy men on the math faculty—a unusually high number of women, considering most of the other top math departments in the United States had none. It had been the first time in nearly fifteen years that the department had denied anyone tenure. Harrison made a complaint common among women in academia, that her colleagues had held her to a higher standard, undervaluing her accomplishments and magnifying her weaknesses. She had published as many papers as her male colleagues who got tenure and had two "major results," in the parlance of the field, including one that some of her defenders praised as more elegant than a similar result by two men who had won the Fields Medal, considered the Nobel Prize of mathematics. Another man in the department had been tenured without a single major result.

Harrison had sued only after exhausting the university's channels of appeal, which took three years. Once in court, she watched as her case was also litigated in newspapers and journals, where she was described as a "sore loser" who was crying sexism to cover up for subpar scholarship. Even neutral observers hinted that she was pathetic; a 1993 *Los Angeles Times* story introduced her as "pale-skinned, wearing a long black skirt and a baggy purple sweater." Readers judged that she "simply wasn't good enough," as one wrote in a letter to the newspaper. "I'm tired of the irritating, incessant whining from the women's camps. Grow up ladies! If you don't wish to engage, then withdraw from the field."

The university and Harrison had settled out of court in 1993. After she had a third major result, the university allowed her to reapply to a new, independent tenure review committee, which made her a full professor. But that, too, was a rare outcome, and even in "winning" she had alienated many of her colleagues, who were hurt at being called sexist and felt she had wronged the department. As a female math professor at another University of California campus wrote, "If we start to confuse honesty with sexism, the future of mathematics is bleak."

Nancy had to decide whether to sue by February 23; otherwise she'd run out the clock on the statutes of the state discrimination laws. All fall, she agonized.

The falling-out with her colleagues and the fights with Phil had left her distressed already. Her lawyer warned her that it would only get rougher when the argument spilled out into the newspapers, as it inevitably would.

The dean was continuing to fix many of the grievances the women's committee had revealed. That fall, he increased the size of pensions for women who had been underpaid compared to men of equal status, one by 35 percent. He put all salaries on "hard money," meaning that no one would have to raise grant money to pay themselves. After the women in Biology noted that no woman had ever held a named chair—a mark of high prestige and security in the department—he created a new chair and gave it to Mary-Lou. It was named after Boris Magasanik, who had brushed off Mary-Lou's first application to MIT back in 1971.

In December, Nancy gave a talk in Japan, then flew to Stockholm to watch Janni accept the 1995 Nobel Prize in Physiology or Medicine for her work in the early 1980s identifying the crucial embryonic genes in *Drosophila*. The guests dined under the soaring ceilings of Stockholm City Hall, seated at long banquet tables strewn with gold-wrapped chocolate replicas of the Nobel medal. They looked up to watch the procession of laureates entering down the enormous staircase. Janni was at the front, arm in arm with the king of Sweden. Nancy cried quietly at the triumph: a woman of her generation, recognized for the genius of her work. Nancy's colleagues in Biology could not help but know Janni's name now.

In January, Amgen extended Nancy's grant, giving her $250,000 a year for the next three years, and told her it would consider increasing the amount, to a million dollars a year for ten years. JoAnne Stubbe was given a named chair in the Chemistry department, and she and Nancy celebrated over dinner. Nancy wrote to Birgeneau afterward, reporting that JoAnne had been thinking of leaving for Brandeis a year earlier and now was happy enough that she planned to stay. "Saving the best (who just happens to be a female), at what I imagine to be zero cost to MIT, seems to be an outstanding administrative move," Nancy wrote. "It was also a high point of my year."

Still, her lawyer and Nancy worried the dean was fixing things only under the threat of her lawsuit. The lawyer warned that the more MIT fixed other women's problems, the weaker Nancy's case would look. She wanted to know when Nancy was going to be able to identify the genes in the fish; if the experiment was a success, she would have leverage in a lawsuit or could threaten to leave and embarrass MIT in the newspapers.

The women's committee was now meeting almost weekly to conduct interviews with faculty members and department heads. The work had become stressful. June Matthews, the only tenured woman in Physics, was proving the most skeptical member of the group. She and Nancy became impatient with each other: June thought Nancy was generalizing about MIT in her determination to prove discrimination; Nancy thought June was ignoring the pattern in her determination to prove discrimination didn't exist. June's grandfather had been Harlow Shapley, a Harvard professor considered the dean of American astronomers, who had led the protest that forced the American Academy of Arts and Sciences to finally elect women in 1942. June wanted to believe that for the most part MIT was a meritocracy. She believed there was discrimination in some departments, but thought physics was well-run—it was flush with federal funds, so she'd never worried about writing grants. One obstreperous colleague tried to take a disproportionately large share of department grants, but he had been difficult with everyone, not just June. She had been underpaid—she had been among the women given large raises by Birgeneau the previous spring—but she didn't think it had to do with gender. She did not think it was fair to blame the department for the low numbers of women on the faculty; several women it had tried to recruit had been unwilling or unable to move, often because a husband had a job in a different city.

In early February, Nancy was still undecided about a lawsuit when Phil appeared before the committee, as chairman of Biology, along with Bob Sauer, the assistant chairman. Nancy pressed them to explain the dearth of women on search committees. There had been three in the last year, which had included eleven men and no senior women. None of the searches, she noted, resulted in an offer to a woman. Sauer said there were no senior women because Mary-Lou Pardue was on sabbatical that term. Nancy wasn't sure this was correct, but that still left her and Lisa Steiner. Nancy asked why she had not been included. Phil said he was not at liberty to answer that. Nancy became furious; either she and Lisa were invisible, or Phil now saw her as too toxic to involve in department activities.

Nancy stormed out of the meeting. Bob Silbey and Penny worried that Nancy was becoming unstable. Nancy wrote to Penny that she was still "tossing and turning" about whether to file suit. "I know that, as before, the answer lies in the group," she wrote. "I can be out there alone (which I hate) but if the group supports me I'm safe. Otherwise, dead. I guess my fear is that by the time June Matthews has divided the issue by the square root of the width of an eyelash, and agonized over whether there might be one millionth of truth on the other side and therefore how could we say this is discrimination, the bodies (mine in this case) will already be accumulating in the halls."

Not all the other women knew that Nancy was considering filing suit—she was trying to keep the committee work separate from her own dispute. But in early February, as the deadline to file suit approached, Mary-Lou organized a letter supporting Nancy's case. Six women signed, as tenured professors "concerned that female Professors are not provided with employment opportunities and benefits that are equal to those provided to the male Professors." The letter endorsed Nancy's "integrity and fortitude in presenting the differential treatment of male and female Professors to the administration for the School of Science and seeking the correction and cessation of these practices."

Nancy was heading to Germany to meet with Janni on a proposed collaboration. Nancy was scheduled to fly out February 22, a Thursday, and the day before the deadline to file suit. On Tuesday, another woman from the original sixteen, Paola Rizzoli, a physical oceanographer, added her name to Mary-Lou's letter. That day Nancy stopped by Phil's office to tell him, as a courtesy, that she intended to file suit against him with the state commission on discrimination. Phil told her he'd do anything to change her mind; anything but stop Eric and Harvey from writing their book, which he said he couldn't do, legally. But he said he recognized that the department had a problem with gender equity and offered to put Nancy on an advisory committee he was establishing, and on the graduate committee. He promised to work out a teaching assignment to her satisfaction.

Wednesday morning, Nancy wrote Birgeneau she was "cautiously thrilled with the progress Phil and I made yesterday." That night, Silbey told Sylvia he'd seen Phil and couldn't believe his "change of heart." Nancy woke up Thursday morning thinking she would not sue after all, and wrote her lawyer, "I am inclined to think that my best bet is simply to drop the whole thing now. Take the loss and call it quits."

A couple of hours later, her lawyer had a new letter from MIT's lawyer, which did not recognize any of the promises Phil had made two days earlier. Nancy reversed course again, stopping by her lawyer's office on the way to the airport to authorize the complaint. She wrote a note to Sylvia, who was newly upset about being removed from her department's committee on appointments. In it, Nancy sounded ready for the long fight: "We will keep working on it until it is solved. I know it seems impossible to solve it, but that is the nature of this problem— to seem impossible but not to be entirely so. Remember—50 years just to be allowed to VOTE!!!!!!"

Nancy did not sleep on the plane. In Germany, her friends in Janni's lab urged her not to sue; the cost to her work would be too great. Sitting alone in her room at the guesthouse at the Max Planck Institute, she spent more than an hour on the phone with her lawyer, going over the pros and cons again. The lawyer warned her the case would play out in the pages of the *Boston Globe*, Nancy's teaching evaluations and character held up for judgment in court and by public opinion. Then Nancy called Birgeneau, who was in a hotel room outside Washington. They had come a long way together since she had first arrived in his office in the spring of 1993 wielding the word *discrimination*. Nancy had changed Birgeneau's mind, and he told her that she was changing others' attitudes, too. It took time, he told her, but her group was proving that she could change things by working with the system. The men were starting to understand the extra challenges that women faced. If she sued, he told her, she risked halting that progress. She would end up only more frustrated. He also told her he'd have to remove her from the committee if she sued, to avoid any appearance of conflict.

It would be hard enough suing Phil, her old friend. Filing suit would make an adversary of Birgeneau, who had shown himself to be her ally.

Nancy recognized that even with the women standing behind her, it was a choice between going it alone if she filed suit, and putting her trust in the group if she did not. If she did not sue, she was giving up the last hope of winning the fight over the course, which had started the women's committee to begin with. But even if she won the lawsuit, it would devastate her—emotionally, professionally, possibly financially. Even if the lawsuit fixed the problem of the course, she knew some other, unforeseen problem would arise soon enough—it had happened so many times before. With the women's committee, she had a chance to make a dent in the larger problem for women in science, without the personal damage.

The fighting had exhausted her. She called her lawyer and said she would not sue.

She wrote to Jerry Friedman, who had advised her against suing: "I can't say I'm happy about it. Relieved perhaps, sad, maybe but in any case ready to move on."

She focused on writing a final report of the committee to give Birgeneau, as Jerry had instructed her to do. She was already three months beyond her self-imposed deadline. She sent it off in August. An inch thick, it opened with a *New Yorker* cartoon Bob Silbey had given Nancy, of four well-off, well-fed men in suits sitting around a table, one asking, "Anyone here not a feminist?" It included individual reports from all the departments. At the back was an appendix with the original letter the sixteen women had written to Birgeneau, and a four-page personal statement from Nancy, explaining her complaint over the course—the problem, her attempts to solve the problem, and how it led to the committee's formation.

"All such problems pose a dilemma," she wrote. "Is it better professionally to fight for a correction, or better to absorb the outrage and try to move on?"

She concluded under the headline "On a positive note": "When the women faculty initiative began, I had come to regret that I had spent my professional career in an institution where I had so often watched and experienced the degrading and unjust treatment of women faculty.

"The formation of the tenured women faculty group and the support the group received from Dean Birgeneau dramatically improved my life at MIT. Thanks to them, this spring I experienced a workplace free of the chronic stress of harassment and intimidation. I feel extraordinarily privileged to know the tenured women faculty in the School of Science and I am grateful for Dean Birgeneau's interest in this issue and for his sense of fairness."

She sent the report to Leigh with a note: "More revealing than the report was a diary I kept at the beginning of our group efforts. We've come a long way baby. So have they!" Nancy was on her way to vacation in Olympic National Park with her sister, Ann, but first she was heading back to the University of Oregon in Eugene, where she was collaborating on zebrafish experiments, and to Amgen headquarters in California. She was happily back to science, full-time. She thought that what she called her "women's work" was done.

"The Greater Part of the Balance"

"Social progress, like that in the** sciences, is inevitably the result of many experiments." That was what Jerry Wiesner had written when he took over as president of MIT after the tumult of the late 1960s, shortly before Nancy arrived to work in the cancer center. It was also true, then as now, that for all the careful planning and controls, experiments could be shaped by serendipity and circumstance. And but for a sequence of events and lucky breaks that started in the winter of 1997, the report that Nancy had written the previous summer would have ended up like so many others before it: unknown much beyond its authors, who'd be left to lament that universities did little more than nod at the need for more experiments, more progress.

The report had included data and an analysis of the problems in each department, and while it did not name names, they were obvious from the details. Before the whole women's committee had signed off on it, the members had debated whether what Nancy had written was too frank; Daniel Kleitman, from Math, wondered if it was "more effective to be a little bland." After a long talk, they agreed to let the report stand as the "historical record," and that they could later produce something "more hushed for the general community." "While I fear hushing things is exactly what has produced the problems we have been studying," Nancy wrote to the rest of the committee, "I agree that something that could be shown around would be helpful."

Nancy had returned from her trip to the West Coast in August of

1996 feeling unburdened, like "a new person," as she wrote in her diary, "extremely happy." Having turned in the report earlier that month, she had completed her term as chair of the committee. Six years after her start in fish—begging for grants, changing the water in the tanks by hand—her lab was now flush with cash. Amgen had increased its stake again, committing to give her $1 million a year for three years, with the promise of raising the amount to $3 million a year if the work went well; the money the company had spent on Nancy's lab had exceeded its entire original grant to MIT. One of her graduate students had developed a virus that allowed the lab to make more mutations, faster, which sped up the pace of the work. They began injecting the new virus into the fish embryos and by the end of 1996 had identified the genes responsible for six mutations. Five of those were considered significant to embryonic development; on the shelf in Nancy's office were five champagne bottles the members of her lab had emptied in celebration. The experiments confirmed that the technique Nancy had designed would allow the lab to isolate the much larger number of genes she was hoping to identify. She published eight papers that year.

The women's work was continuing without her. Dean Birgeneau had agreed to renew the committee for another two years and named two new cochairs: Molly Potter, from Brain and Cognitive Sciences, who had been at MIT since the mid-1960s, and Paola Rizzoli from Earth, Atmospheric and Planetary Sciences, who had arrived in 1981. Leigh Royden had started an offshoot based on what the original committee had heard from junior women on the faculty, working with the provost to expand day care and establish policies around maternity leave so that women would feel secure in taking it. Birgeneau said he planned to share the report Nancy had written with the Academic Council, which was the university's highest governing body, and he agreed she could write a summary to be published in the faculty newsletter.

But in mid-October 1996, Molly called Nancy to report trouble. Birgeneau had told her that the department heads in Biology and Chemistry were unhappy with the report. Phil Sharp, as the head of Biology, wanted to write a rebuttal and make his case before the original committee on women—Nancy included. Molly said this meant that MIT might refuse to accept the report or not allow it to be shared with the faculty; the administration would not want to signal that it agreed with conclusions that were in dispute. Frank Solomon had also asked to see a copy of the report and wanted to register an objection; he was upset that Mary-Lou continued to believe he was responsible for her removal from the graduate seminar.

By February 1997 the committee had met with Frank and Phil. They essentially agreed to disagree: Phil acknowledged the feelings of marginalization among the women, especially the older women in the department. "I am sure it is difficult to be a pioneer, to be a member of an underrepresented group and to perform at the level of an MIT faculty member," he wrote in his rebuttal to the report. He agreed that the department and MIT should work on the problems, but not with the implication that administrators in Biology had ever practiced discrimination.

Other men, though, had welcomed the work of the women's committee as a revelation. The report from Brain and Cognitive Sciences had noted that women in the department thought bias was especially hurting efforts to increase their numbers; the percentage of women on the faculty was far below the percentage of those who earned degrees in the field. While the department had recently hired one woman, it had taken three years, which the women felt reflected a lack of urgency. They complained that female applicants were scrutinized more harshly than men. One tenured woman had recently left MIT after she suggested the name of a potential hire in a meeting and was ignored, only to have a male faculty member mention the same name a few moments later and have everyone else in the room say it was a great idea. The head of the department said that no one had ever raised the issue of marginalization with him before; he was grateful the women had. "There's little time to think in these jobs," he told the committee. "It's important you bring it to my attention." (As Leigh noted to Nancy, "He's definitely educable.")

Still, the work had started in secret, and even now, two years later, few people on campus knew about it.

The number expanded, slightly, in March 1997, when Larry Bacow, the chair of the faculty, invited Molly and Paola and Bob Birgeneau to present the report in a confidential session of the Faculty Policy Committee, made up of sixteen professors from across the university's five schools. Paola and Molly presented the findings and the recommendations: the need to monitor salaries, distribute resources fairly, include more women in positions of authority and on search committees. Birgeneau credited the women's committee with changing the way he looked at the problem of discrimination, and helping him find ways to fix problems the committee had identified. The problems were not unique to the School of Science, he said. He thought other schools at MIT could learn from the women's approach.

Nancy was at the table listening, no longer with the women's group but as a member of the Faculty Policy Committee, which Bacow had asked her to join several months earlier. Her successors' presentation thrilled her: Molly so steady, as always; Paola fervent, her words punctuated by her heavy Italian intonations. Nancy had worried when she stepped down that no one would have the time or interest she did to run the committee, which had, after all, started with her struggle over the bio course. "Today all fears were laid to rest," she wrote Birgeneau after. "I thought Molly and Paola were magnificent—eloquent and more radical than me!—as were you."

Lotte Bailyn, too, had been listening carefully. She was now in line to succeed Bacow as faculty chair. She was daunted by the prospect. Bacow, an economist, had been an undergraduate at MIT and returned as a professor after receiving his law degree and doctorate in public policy from Harvard. He seemed to know everyone and exuded effortless charm. Lotte was more reserved and had for two years been taking notes from Bacow in the hopes she could learn to run a meeting as skillfully as he did. But she thought he'd made a mistake with the presentation from the women's committee. He'd scheduled it for only the first half of the faculty meeting and stuck to the schedule, cutting off the discussion well before the faculty had run out of questions. Lotte could see that others were as intrigued as she was.

She had been skeptical when Nancy and Penny arrived in her office at the Sloan School three years earlier, the summer they were trying to push the dean and the department heads to allow the women's committee in the first place. In Lotte's long career studying work and family, she'd seen too many similar studies on women—she had analyzed the data for Alice Huang's study of disparities between male and female microbiologists twenty years earlier—and she was unconvinced another one could accomplish much. Now, listening around the table at the faculty meeting, she was impressed by how the women had tackled the problem: how they had convinced the dean, and the ingenuity of comparing space and grants and teaching assignments. Lotte had some idea of what it took just to survive as a woman at MIT; that these women had done so and achieved at the highest levels in their fields amazed her. She made a mental note that when she took over as chair, she'd find a way to share the report with the other schools at MIT.

Lotte took over as faculty chair in July 1997. She turned sixty-seven that summer and was no longer thinking of retiring. This new work, seeing up close how the university was run, rejuvenated her. She had studied organizations her entire career, but there was special satisfaction in studying one's own. She had long written about the need to integrate work and family, how organizations tended to define the ideal worker as one who had no commitments outside the professional sphere, and how that disadvantaged women. Universities, with their unforgiving tenure clocks and blindness to childcare needs, were particularly ripe for improvement. She thought she might be able to start a conversation about the subtle biases against women, in hiring and tenure and evaluation.

She quickly had reason to wonder if that was too ambitious.

As faculty chair she sat on the Academic Council, the university's top governing body. She arrived early to her first meeting that fall and distributed copies of the first chapter of a new book by Virginia Valian titled *Why So Slow? The Advancement of Women*. Valian had marshaled decades of research to argue that the glass ceiling was propped up by unconscious and largely unspoken assumptions about men and women that took root in childhood—Valian called them *schemas*, rather than *biases*, because *bias* made them sound intentional when they were largely cultural. In these schemas, men were independent and assertive and women were emotional and nurturing. Because the most prestigious professions prized the male traits, a man started out with the benefit of the doubt, while a woman was "at least slightly unsuited to her profession."

That meant men are "consistently overrated, while women are underrated," Valian wrote. "Whatever emphasizes a man's gender gives him a small advantage, a plus mark. Whatever accentuates a woman's gender results in a small loss for her, a minus mark." The pluses and minuses—whether in everyday encounters or formal evaluations— added up to large disparities in pay, promotion, and prestige. Women who complained—about the slight in a meeting, the invitation not extended, the credit not given—were often scolded not to make a mountain out of a molehill, but for these women, mountains were an accumulation of molehills. "A woman who aspires to success *needs* to worry about being ignored," Valian wrote. "Each time it happens she loses prestige and the people around her become less inclined to take her seriously."

Valian recognized that people didn't like to think of themselves as biased. Men and women alike wanted to believe that a meritocracy rules. Confronted by evidence of bias, they could point to professional women they admired, those who had succeeded—or, if they were successful women, to themselves—to suggest that bias didn't exist. "These exceptions should not, however, obscure the rule," she wrote.

Lotte had to conclude that no one on the Academic Council was interested in a discussion about gender schemas; the chapter handouts remained on the conference room table after the meeting cleared.

Soon after, a campus tragedy pushed less urgent matters off the agenda. In late September 1997, Scott Krueger, a clean-cut freshman who was going out for crew, died of alcohol poisoning after an "Animal House Night" party at an off-campus fraternity where he was living as a pledge; fraternity members had given the pledges a bottle of hard liquor each and forced them to drink and left Krueger alone after he became unconscious. The district attorney convened a grand jury to consider charges against the fraternity and MIT for not properly supervising the Greek houses, and Krueger's parents were planning to

sue. As his agonized mother had told the cameras at a news conference, "We sent our son to MIT for five weeks and came down here and
picked him up in a box and took him back in the back of my station
wagon." His death reignited an old debate about housing: by tradition,
freshmen rushed to choose a place to live during the chaos of their first
week on campus, with many settling in fraternities or independent living groups. Students and alumni argued that four years with the same
housemates created lifelong bonds. They complained bitterly whenever the administration suggested requiring freshmen to live in dormitories where they could be more closely supervised and mentored.

The Academic Council approved such a requirement in fall of 1998.
Lotte had less than a year left in her term as faculty chair, and she had
not forgotten about the report on women. In February 1999 she called
a meeting of the Faculty Policy Committee with that as the only agenda
item, hoping to continue the conversation Larry Bacow had started
nearly two years earlier. No one thought it was a good idea to distribute
the original report to the entire faculty at MIT; the women in the School
of Science had spoken to the committee in confidence and might be
embarrassed by the details in the report. But Lotte still thought that
women and men in the other schools could learn from what had happened. She asked Nancy to try her hand at writing another version,
a summary of the committee's findings that would also describe how
the women had gone about identifying the inequities, and what the
dean had done to solve them. A process report that other schools could
use as a model. Lotte's instructions reflected the lessons of her parents' work on the unemployed villagers in Marienthal, of the work she
and Bud had done in 1959 on early-American shipping: the data were
important, but a story moved people. Tell a story, Lotte told Nancy.

Nancy went home and wrote it in three hours: how the women's
work began as a conversation between her and Mary-Lou and then
Lisa in the summer of 1994, each "uncertain as to whether their experiences were unique, their perceptions accurate." How the group of sixteen women signed the initial letter to Birgeneau, despite fearing "they
were putting a lifetime of hard work and good behavior at risk." How
the dean won the approval of the department heads for the committee,

and what its members had found: the low numbers of women on the faculty despite the steady increase in the number of women entering the fields, the differences between men and women in salaries, space, and resources, the women's slide into invisibility after tenure.

"Each generation of young women, including those who are currently senior faculty, began by believing that gender discrimination was 'solved' in the previous generation and would not touch them," Nancy wrote. "Gradually however, their eyes were opened to the realization that the playing field is not level after all, and that they had paid a high price both personally and professionally as a result."

She described the leaky pipeline, how the percentage of women on the faculty had remained flat for two decades and was likely to remain flat if MIT relied, as it had in the past, on the promise that more female students would organically lead to more female faculty. "The inclusion of significant numbers of minority faculty will lag for even longer because of the additional problem of a shortage of minority students in the pipeline."

She talked about the changes: thanks to Birgeneau's new hires, MIT now had more than 10 percent women on its Science faculty for the first time ever. In 1999, there would be a "remarkable" 40 percent increase in the number of women with tenure in Science, even if the overall number was "still tiny"—in 1994 there had been twenty-two women on the faculty, now there were thirty-one. "Even if we continue to hire women at the current increased rate in Science," Nancy wrote, "it will be 40 years before 40 percent of the faculty would be women."

Nancy opened with an abstract, as if she were writing for a scientific journal. She ended with more of a manifesto: How had the inequities come about? How was it that after three decades of affirmative action, these women renowned in their fields had been excluded rather than cherished as the rare commodity they were? "One would have assumed that all tenured women would be treated exceptionally well—pampered, overpaid, indulged," she wrote. "Instead, they proved to be underpaid, to have unequal access to the resources of MIT, to be excluded from any substantive power within the University." She struck down the notion that had once nagged her: that the women were "simply not good enough." Nearly half were members of the National Academy of Sciences or the American Academy of Arts and Sciences, which made them, pound for pound, more accomplished than the men in their departments.

"Indeed, it should be almost obvious that the first women, the first blacks, the pioneers who break through despite enormous barriers must be exceptional," she wrote. "While the term 'affirmative action' is sometimes used to mean letting people in simply because they are women, minorities, that is the opposite of what affirmative action means at MIT and most emphatically, to women faculty at MIT."

Discrimination had been allowed to continue, she wrote, because no one, not even the women themselves, recognized it: "*It did not look like what we thought discrimination looked like.*

"Women faculty who lived the experience came to see the pattern of difference in how their male and female colleagues were treated and gradually they realized that this was discrimination. But when they spoke up, no one heard them, believing that each problem could be explained alternatively by its 'special circumstances.' Only when the women came together and shared their knowledge, only when the data were looked at through this knowledge and across departments, were the patterns irrefutable."

Lotte liked this version, with one improvement. Nancy had written that the dean and the women, as a group, had "made a discovery" about the contours of gender discrimination on the cusp of a new century. You couldn't just say "discrimination," Lotte told her, you had to say what it looked like. She and Nancy and Molly had one last meeting, over lunch at the Faculty Club, and together came up with a phrase. Molly, who studied human perception, started off: "They found that discrimination consists of a pattern of powerful but unrecognized assumptions and attitudes that work systematically against women faculty." Lotte chimed in to finish the sentence: "even in the light of obvious goodwill."

The women arranged for Nancy's summary to run in the faculty newsletter, which at MIT was the equivalent of the scrappy alternative newspaper. It had been started in 1988 after the administration's surprise decision, reported in the *Boston Globe*, to shut down a department that included twenty-four faculty members and eighty-six graduate students. (Vera Kistiakowsky in Physics had solicited checks from faculty members to sponsor the new publication, since, as she argued, neither the official university newspaper nor faculty meetings served as a channel to communicate information the faculty needed to know.)

Lotte showed what Nancy had written to Birgeneau, who offered to write a short commentary to accompany it. Lotte then asked President Vest if he would like to do the same.

President Vest and his top lieutenants had worried about sharing the report, fearing it might invite lawsuits from other women on the faculty, or from those who had retired, seeking back pay and pension adjustments. They worried about the data: the numbers of women were so small that comparisons were not statistically meaningful. Would the old "scientific school" look soft? But Vest, like Birgeneau, had been convinced by the stories, how consistent the women's experience was across the departments. He had met with a few of them around the coffee table in his office after the report was finished and kicked off the conversation by noting the good news, that the junior women felt relatively well supported in their salaries and start-up packages. He had been jolted by the response from Sylvia Ceyer in Chemistry, who had just been elected to the National Academy: "I also felt very positive when I was young."

Vest was in his study in the president's house along Memorial Drive late at night when he received Lotte's email asking him to write a short preamble to the report. He sat down and wrote out what he was thinking, without advice from advisers or anyone else. He had learned to trust his own instincts after Scott Krueger's death. Vest had been deeply affected by the loss. He had two children not much older. He had gone to the funeral in Upstate New York and had wanted to approach Krueger's parents to offer his condolences, but his advisers convinced him not to, warning that even his presence at the funeral might trigger anger, or violence. Ever since, he regretted that he had not done what he thought was right—and he would not repeat that mistake. Vest wrote just three brief paragraphs, but they included lines that would soon be repeated more widely than he or the women could anticipate. He sent his addition to Lotte and Birgeneau, and to Nancy. She opened the email, then sat back, overwhelmed by what she read. It had taken her two decades to recognize how she had been disadvantaged as a woman in science, and in all that time she would never have imagined that the president of the university might understand it, too.

Here, Vest had: "I have always believed that contemporary gender discrimination within universities is part reality and part perception. True, but I now understand that reality is by far the greater part of the balance."

A few weeks later Nancy gave a presentation on zebrafish to the Knight Fellows, a group of journalists studying at MIT for a year. In the question period, a reporter for *Science* asked what it was like being a woman in her field. Funny you should ask, Nancy replied. She had spent much of the last four years looking into the treatment of women on the faculty at MIT and had just finished writing a report that would appear in the faculty newsletter. She stayed after class talking with Boyce Rensberger, the director of the Knight fellowships and a longtime science writer himself. The report sounded like news to him, and he asked Nancy if he could alert people he knew at the *Boston Globe* and the *New York Times*. Nancy demurred; the women had always said they did not want to air their grievances in the press. But when she got back to her office, Nancy reconsidered. She thought about the president's statement, and how extraordinary it was. She called Boyce and told him about Vest's remarks, which she thought might indeed make the report newsworthy. Boyce agreed it was. To him it was a story about heroes: the women who had come forward, and Chuck Vest for acknowledging the report's conclusions. And it was a story of MIT, where logic and numbers reigned supreme. The data had won.

Boyce's tip reached me thirdhand in the newsroom of the *Bos-*

ton Globe, where I had been covering higher education for about six months. Though my father, a physicist, had talked about the dearth of women in his field, and the work that Millie Dresselhaus was doing to try to remedy it, a story about gender discrimination sounded unlikely, out of step with the times. It was March 1999. *Time* and *Newsweek* had kicked off that month with cover stories about the brave new life ahead for Hillary Clinton, in her second term as First Lady. Having been beaten up in the polls for decades—she had changed her last name, her ambitions, and countless times her hairstyle to help her husband win and keep the White House—she now commanded such approval that Democrats were recruiting her to run for the US Senate. It would be her first elected office and inevitably lead to speculation that she would run for president. Elizabeth Dole, a former transportation secretary and wife of the last Republican nominee for president, had just announced her intention to run for president the next year. For the first time a woman, Madeleine Albright, was representing the United States on the world stage as secretary of state.

Discrimination had settled down, or settled in; no one would have said it did not exist, but nor did it seem an animating force. The number of women running for Congress had declined since the Year of the Woman. A *Time* cover nine months earlier had asked, "Is Feminism Dead?" Not quite a decade since the magazine had last asked the question, the answer was a resounding yes. Not because women didn't want feminism, but because younger women thought they no longer needed it. A previous generation had earned them the right to be "sex positive" in their feminism—with more fun, more makeup, less preoccupation with social change or careers. Inheriting the mantle of Mary Tyler Moore on television was Ally McBeal, a fictional Boston lawyer in her thirties who spent her days scheming about her ex-boyfriend in the next office and her nights haunted by visions of a dancing baby, keeping time, presumably, to Ally's ticking biological clock. The runaway bestseller in a new genre of "chick lit" was *Bridget Jones's Diary*, an updating of *Pride and Prejudice*, with the heroine's quest for the elusive Mr. Darcy measured out in daily weigh-ins and units of chardonnay slurped.

But the women of MIT were describing something fresh. In 1995, two psychologists had written what would be later recognized as a seminal paper that described how perceptions are shaped by conditioning, stereotypes, and "traces of past experience," so that even people who explicitly disavow prejudice bring bias to social interactions. That notion, however—of implicit or unconscious bias—was not yet in popular use. Lotte's introduction to the report described it this way: "The key conclusion that one gets from the report is that gender discrimination in the 1990s is subtle but pervasive, and stems largely from unconscious ways of thinking that have been socialized into all of us, men and women alike. This makes the situation better than in previous decades where blatant inequities and sexual assault and intimidation were endured but not spoken of. We can all be thankful for that. But the consequences of these more subtle forms of discrimination are equally real and equally demoralizing."

That MIT was admitting gender discrimination made it a man-bites-dog story.

When I called Nancy, I asked how all of this had started. She told me about the space wars, how she had needed more room for her fish tanks and been told no, only to discover that men with less experience than her had more space than she did.

How did you know the men had more? I asked.

"I measured."

"With a tape measure," she added, as if it were the most obvious answer in the world.

And maybe it was. *Mens et Manus*, "mind and hand." What else to expect from MIT? A tape measure was more subtle than putting a police cruiser on the Great Dome, but it struck me as a hack all the same. It would lead my story.

Lotte, though, had no interest in talking to the *Globe* and wished that Nancy had not. Lotte had never intended for the report to be publicized beyond the campus. She worried that Vest and Birgeneau would feel betrayed. She didn't want the faculty reading about the report in the newspaper, either—this was exactly the kind of surprise that the faculty newsletter had been designed to prevent. The newsletter with Nancy's report hadn't yet been printed.

She asked if the *Globe* would wait on the story, until the faculty newsletter could be distributed. The *Globe* was worried the *Times* would catch on to the story—some faculty member would call if Boyce hadn't already—and didn't want to delay. Someone raised the solution of posting the report on the fledgling website of the faculty newsletter, but Lotte reported that they were having trouble with this. This seemed suspect, given that this was the Massachusetts Institute of *Technology*. The *Globe* insisted on running the story that Sunday.

On campus, students were clearing out for spring break. Lotte spent the week miserable, rushing to pull together the final presentation for the newsletter—adding tables and graphs, including what the women had come to think of as the pancake graph, showing how their numbers on the faculty had stayed flat for twenty years. She worried about the *Globe* headline, that it would underscore discrimination and ignore what MIT had done to fix the women's complaints. The president would hold her responsible; how could she show up at the next week's Academic Council meeting?

She finished writing her comments on the report Friday afternoon, then wrote an email to the faculty, attaching the report and saying that she expected the *Globe* would have a story on Sunday. Around five o'clock she hit SEND and went home to fret for the weekend.

The article ran March 21, above the fold on the front page of the *Boston Sunday Globe*. Lotte cringed at the headline:

> MIT Women Win a Fight Against Bias; In Rare
> Move, School Admits Discrimination

Nancy had only to glance at the paper before she felt the pit in her stomach. She rushed to her lab because she knew at home she'd only sit and worry. She couldn't resist checking her email when she arrived. There was a note from Chuck Vest. Subject line: "Congratulations." It was a tough story, but fair, he'd written, and had given credit to the women, where it was due.

Then came other emails, from women she did not know. "Thank you, thank you, thank you, thank you, thank you," wrote one, a young woman who had recently switched out of a doctoral program in physics at MIT. "Discrimination may often be subtle but it is undeniable. I hope that women coming behind me will find MIT a much more pleasant place to work as a result of your action."

The full impact of the story would not begin to strike anyone until Monday morning. Mary-Lou woke to her clock radio announcing the news that MIT had acknowledged discrimination against the women on its faculty. Was she dreaming? Birgeneau arrived in Building 6 to discover a camera crew from the *CBS Evening News* waiting for him in the hallway. High above campus in the Green Building, Leigh Royden was in her office when a colleague, a Scottish geochemist several years her senior, burst in waving the front page of the *Sunday Globe*. "Above the fold!" he proclaimed. "You are above the fold!" It was a fitting metaphor, Leigh thought, for the journey the women had traveled together.

Nancy's phone was ringing when she arrived in her office at the cancer center. On the other end was a voice from Radio Australia. The phone barely stopped all day, with calls of congratulations, and from more reporters. On Tuesday, a front-page story in the *New York Times* declared the report "an extraordinary admission" by MIT. The newspaper's editorial board hailed the breakthrough: "Women on other campuses and in other professions have called attention to gender bias and the glass ceiling. But the MIT study is unusual because it examines tenured women who have excelled in the male-dominated sciences, and whose collective experiences with bias cannot be explained away by special circumstances.

"The study has significant social value," they wrote, "because it documents with unusual clarity how pervasive and destructive discrimination can be even when there is no blatant harassment or intimidation."

After the *Times* story, the emails came in a flood, from women who recognized their own story in the report—women in Florida and Kentucky, Singapore and China. A chemist at Wayne State in Detroit wrote that in the twenty-four hours since the *Times* article, "I have received five e-mails from five women faculty members at four different universities, urging me to read and discuss the article and the report."

A cancer researcher in Seattle described the conversations among her group of female colleagues: "There are days when we just hate to come to work—days when we know we'll have to face the smug, self-satisfied, self-righteous all-boys network that unconsciously, but systematically puts us down. What an incredible waste of our energy and talents to disenfranchise us. Lots of times you can think about your lab and ignore it, but not all the time.

"Seeing the *Times* this morning reassured me that I am not a paranoid, mentally ill person, that this is really happening."

Nancy's tape measure had given women a new way to quantify a feeling they hadn't quite known how to explain, a measure of all the indignities that added up over time. "The attitude is so prevalent and undermining that even women graduate students do not treat all of their faculty equally," wrote a biochemist in New York. "Subtle things such as the size of lab space, the number of people in a lab, state of the art equipment, who is on committees that make the really important policy decisions, etc. all contribute towards a regard of women faculty as not quite equals." From the University of Alaska, a marine biologist wrote that she and her female colleagues hoped to complete a similar analysis: "It's uphill, but if it is a data-based study and done as you describe, we have a chance. If we carefully pattern it on something the famous MIT did we're above the charge of being petty and silly and irrelevant. It's been clear forever that there's a problem, just no way to address it acceptably."

The White House called to invite Nancy and Birgeneau to sit on the dais the next month for Equal Pay Day. Nancy sat listening, astounded, as President and Mrs. Clinton both gave speeches encouraging other institutions to dig into the data to fix the inequalities. Nancy would be asked to deliver hundreds of talks at universities across the country. Late that year, the *Chronicle of Higher Education* proclaimed, "Nancy Hopkins has done for sex discrimination what Anita Hill did for sexual harassment."

Other universities almost immediately began fixing a gender gap in salaries, a disparity that some had known about but seen no pressing reason to fix; some women referred to them as their "Nancy Hopkins raises." No one wanted to be embarrassed by discriminating against women on the faculty. But the report had not embarrassed MIT; it made it look bold. As Leon Eisenberg, a prominent psychiatrist at Harvard Medical School, wrote to Chuck Vest, "It is one thing to be known for respecting data when it is a matter of 'neutral' scientific research and another when the results deal with internal affairs that are a betrayal of professed beliefs and policies."

Practically overnight, MIT became the pacesetter for promoting gender equality in higher education. The Ford Foundation and the Atlantic Philanthropies gave the university a million-dollar grant to work with others to increase the number of women in science and engineering. President Vest responded by convening the leaders of eight other elite institutions—Berkeley, Caltech, Harvard, Princeton, Stanford, Yale, and the Universities of Michigan and Pennsylvania—who together pledged several measures to make sure women were treated equitably, and that the number of women on their faculties reflected the number of students in the fields. Robert Berdahl, the chancellor of Berkeley, greeted Nancy at one of the meetings by paraphrasing Abraham Lincoln on Harriet Beecher Stowe: "So you're the lady who started the great war!"

The National Science Foundation created a new program to scale up the MIT model across the country. For years it had directed grants to individual women; the new program, known as ADVANCE, gave money to universities to collect data to identify any disparities and make changes to the system to even out other differences—in evaluations, hiring practices, teaching assignments—that had interfered with women's ability to do their science. Over the next twenty years the foundation spent $365 million to establish programs at 217 institutions nationwide.

In the fall of 1999, MIT's provost, Bob Brown, instructed the deans of MIT's four other schools to form committees to analyze the status of women, to collect data and interview women on the faculty. By the time the reports were completed in 2002, the number of women on the faculty was already increasing. The most striking change was in Engineering, the largest school at MIT, where Lorna Gibson had been watching the women in Science and was eager to repeat their effort. The number of women in Engineering nearly doubled within four years, thanks largely to the dean's insistence that the résumé pool for every job search be broadened to include more women. In Science, the number of women had grown from twenty-two to thirty-four. The number of women in top administrative jobs in Science and Engineering rose from one woman in 1999 to ten in 2003. Even the Math department—the null set when the women's work began—had four tenured women on its faculty.

Birgeneau was named the president of the University of Toronto—his alma mater—in December 1999, and the women in Science feared they would lose their biggest ally and, with him, the progress they'd made. Bob Brown, the provost, responded by creating a new council on faculty diversity, which, at the women's insistence, included a new mandate to increase the number of underrepresented minorities as well as women on the faculty. As cochairs Brown named Nancy and Phil Clay, the associate provost and a former head of the Department of Urban Studies and Planning, who was Black. With the new position Brown also gave Nancy a seat on the Academic Council, the highest policy-making board, and on a smaller group within the council that reviewed every hire and promotion, and faculty salaries, across the university. Nancy, again, was astonished by the progress. Brown, a mustached and plainspoken Texan, saw it as a practical question: If science wanted to solve the world's hardest problems, why even risk closing it off to half the world's population? "We're engineers," Brown told Nancy. "We fix things."

Women in science had been trying to fix the problem for years: Johns Hopkins University had convened a committee on women after a similar report in 1990; the year after, a computer scientist at Harvard, Barbara Grosz, had led a panel that concluded the university was having "serious difficulties" attracting women to its science graduate programs and faculty. Another group of women at Harvard had been writing annual reports since 1994; in 1996 they had reported that the number of women on the faculty remained "ridiculously low," that women were less likely to be given tenure from within, and even when they were, they were paid only eighty-eight cents to every dollar male professors made.

The MIT report proved more powerful for the reason the *San Francisco Chronicle* identified in an editorial: "The difference this time . . . is that the respected president of MIT—one of the most prestigious universities in the nation—not only did not ignore the report, he acknowledged existence of the discrimination and took steps to redress it."

Ruth Perry chided Nancy for allowing the men to look like heroes. But to Nancy, they had been. The women could not have accomplished what they did without Birgeneau and Chuck Vest any more than Nancy could have accomplished anything without the fifteen other women who signed the first letter to the dean.

The report also disrupted the old assumptions about why there are so few women in science: that women were not smart enough or were not succeeding because they were distracted by children, that the solution was to encourage more girls in high school and college into science and math courses. The challenge presented by the MIT report was to allow women as full participants, to support them not just as daughters but as colleagues, bosses, leaders.

As Nancy wrote in the report sent to the faculty in March 1999, it was a discovery—"like many discoveries, at first it is startling and unexpected. Once you 'get it,' it seems almost obvious." It did result from a series of experiments. Choosing a career in science had been an experiment. So had been moving to study zebrafish, asking Mary-Lou to lunch, surveying the women, and showing up in Birgeneau's conference room on that August day in 1994. They'd had the good fortune of a dean and a president willing to widen their perspectives, a chair of the faculty who had been thinking about the challenges for women decades before the rest of them. All the experiments had been a test of the hypothesis the postdoc had put to Nancy all those years ago in Jim Watson's lab: "We think you might be the one." She had proven that she was. So had the other women, and so many more whose names were not on the letter to the dean, who came before them and would follow in their footsteps. The exceptions had been many.

Epilogue

In January 2005 Nancy had promised herself that she was finally done with the "women's work." As a favor to a friend, though, she agreed to one more talk, at a small, daylong conference in Cambridge—just fifty or so participants—on the topic of diversifying the science and engineering workforce. The event was at the National Bureau of Economic Research on Massachusetts Avenue, around the corner from the apartment she had shared with Brooke. The lunchtime speaker was Larry Summers, the president of Harvard, a former secretary of the treasury, an MIT graduate, considered one of the most brilliant economists of his generation.

Summers was impatient with politesse and conventional wisdom, and he opened by saying as much, telling his audience that when the organizers invited him, he asked if they wanted a talk on Harvard's policy on diversity, which he "didn't feel like doing," or "some questions asked and some attempts at provocation," which he relished. He did provoke. But the hypotheses he offered about the "very substantial disparities" in the number of women in science had been trotted out for decades, since the middle of the last century.

First, he stated that science demanded a high level of commitment—eighty-hour weeks—that married men had historically been more willing than married women to make, and that women in their twenties were rejecting. Second, he cited a "different availability of aptitude" between men and women, given that more boys than girls score at the very highest levels on standardized math tests, perhaps a predictor of a future in elite science.

Nancy, seated at the large horseshoe-shaped table at the center of the room with the other presenters, had pricked up at "eighty-hour weeks" and was now listening closely to make sure she heard Summers correctly.

"So my best guess, to provoke you, of what's behind all of this," he continued, "is that the largest phenomenon, by far, is the general clash between people's legitimate family desires and employers' current desire for high power and high intensity, that in the special case of science and engineering, there are issues of intrinsic aptitude. . . ."

At "intrinsic aptitude" Nancy turned to the woman next to her: Shirley Malcom, the author of the 1976 report on the "double bind" of being a minority and a woman in science, who now worked at the American Association for the Advancement of Science.

"I can't take this. Should we leave?"

"I can't," Shirley said. "I'm the next speaker."

Nancy deliberately closed her laptop and slipped it into her bag, took a last sip of coffee, then pushed back her chair and rose. She stood for a moment so no one would think she was just ducking out to the restroom or for a previously scheduled engagement. Then she left the room and walked out into the cold and rainy afternoon, past Summers's black Town Car idling at the curb—license plate 1636, for the year of Harvard's founding.

Back in her office, she had an email from a *Globe* reporter checking in on a story they had discussed the previous week and asking, by the way, how the conference had been. "I just walked out on Larry Summers," Nancy wrote back. "I'm still shaking."

That began another round of newspaper stories about women in science, and the revival of an old debate: Were women in short supply at the highest levels of science because they were not good at it and did not like it? Or because they had repeatedly been told that they were not good at it and did not like it?

Nancy was accused of being unscientific, an enemy of free speech, and a "hysteric," in the words of the syndicated columnist George Will.

The *Atlantic* dismissed her and other women who objected to Summers as "feminist careerists" seeking "thinly veiled job preferences or quotas for themselves and their friends." In the *New Republic*, Steven Pinker, the Harvard psychology professor and author whose book had helped inform Summers, worried that efforts against sex discrimination might push young women into "lines of work they don't enjoy."

Almost two decades later, there's still work to do. A landmark report in 2018 by the National Academies of Sciences, Engineering, and Medicine found that 50 percent of female faculty members had experienced sexual harassment, and that the biggest complaint was not "sexual coercion" but put-downs about their intelligence, exclusion, and the kind of marginalization that the women of MIT had described twenty years earlier.

But the progress is also undeniable. The MIT report was followed by an explosion in scholarship examining the reasons for the disproportionately low numbers of women in science, technology, engineering, and math, and in strategies to increase their representation and treat them as full participants in their fields. Those efforts have expanded beyond women to include others from traditionally marginalized groups.

Some of those strategies are basic—not unlike stretching a tape measure the length of a lab floor. For instance, boys score higher on tests of spatial reasoning, but tell girls they are performing an "art task" instead of a "math task" and the gender gap fades. Fields presumed to require raw "brilliance"—physics and math—attract more men, while women are more likely to go into fields they believe require "hard work."

Another strategy that works to disrupt old attitudes about women in science is to counter them with examples of women who have achieved at the highest levels. There are many more of them since the MIT report. In July 1999, Shirley Ann Jackson, who had been the first Black woman to earn a PhD at MIT, became president of Rensselaer Polytechnic Institute, a position she would hold for twenty-three years. By 2001, three women were presidents of Ivy League universities: Penn, Brown, and Princeton, led by Shirley Tilghman, a molecular biologist and a member of the National Academy who had also raised two children as a single mother. In 2007, Harvard had its first female president, as did MIT, with Susan Hockfield, a neuroscientist. In 2016, Marcia McNutt, one of the sixteen women who signed the letter to Bob Birgeneau in 1994, became the first woman to lead the National Academy of Sciences. Three others—Penny Chisholm, Ann Graybiel, and JoAnne Stubbe—have joined Millie Dresselhaus as recipients of the National Medal of Science.

In 2020, Nergis Mavalvala, an astrophysicist and MacArthur "Genius Grant" winner, became the first woman to lead the School of Science at MIT. In 2021, the White House named the first two women to lead the President's Council of Advisors on Science and Technology: Maria Zuber, a geophysicist who was one of Birgeneau's hires following the formation of the women's committee—she had previously been the first woman to lead a NASA planetary mission and the first woman to head a science department at MIT—and Frances Arnold, a chemical engineer and Nobel laureate. (Eric Lander, serving as the president's science adviser, was the third cochair; he stepped down in February 2022 after a White House investigation found that he had demeaned and disrespected colleagues.)

As scientists worldwide rushed at warp speed to develop vaccines against COVID-19, they found the key in the lab of a sixty-five-year-old woman, Katalin Karikó, who had persisted in developing technology to harness mRNA despite being ignored and underpaid for years.

Nancy, too, continued in her accomplishments and accolades: by 2004 she had completed the genetic screen in the zebrafish, identifying 25 percent of the genes critical to embryonic development. She had been elected to the American Academy of Arts and Sciences, the National Academy of Medicine, and the National Academy of Sciences.

She had also fallen in love with Dinny Adams, a Harvard man and New York lawyer. They were set up over a rare weekend away at the home of a mutual friend and married near his home on the banks of the Hudson River in July 2007.

Her now famous tape measure is in the MIT Museum. But her women's work is still not done. In 2017, she and Harvey Lodish collaborated on an opinion piece for the *Globe* diagnosing the "woman problem in Boston biotech," a disproportionately low number of women starting or serving on boards of biotechnology companies. That led her to a new initiative with Susan Hockfield and Sangeeta Bhatia, a professor of biochemical engineering and biotechnology entrepreneur who calls Nancy a mentor. Their Boston Biotech Working Group announced its efforts with a report in the *MIT Faculty Newsletter* in early 2021, modeled on Nancy's 1999 report. Through a Future Founders training program and a pledge from venture capital firms, they are working to overcome the decades-long exclusion of women among the entrepreneurs of biotechnology.

In public remarks Nancy likes to say that the women's work allowed her to study human behavior, just as she dreamed of as a nineteen-year-old in that first Bio 2 lecture. Lately, she has turned her interest back to the neuroscience behind it—studying informally with a Harvard neurobiologist and a sixteen-hundred-page textbook. She is newly astonished by the prospect of understanding the complex functioning of the human brain in terms of cells and molecules. Six decades later, what she once called the obsession of science still drives her: the next question, the serendipitous and often improbable thrill of the search.

The Sixteen

The women who signed the letter to Dean Robert Birgeneau in July 1994 are an exceptionally accomplished group by many measures. To cite a few: Of the 857,200 PhD scientists employed in the United States in 2019, less than one-half of 1 percent were elected members of the National Academy of Sciences. Among the MIT 16, the proportion of National Academy members is 68 percent; one is the academy's president, the first woman to serve in the role. An even smaller percentage of scientists in the United States have been awarded the National Medal of Science: 506 people since the medal was first awarded in 1962. Four of the MIT women have been given that high honor. And at any time, just twelve of the roughly one thousand faculty members at MIT are given the rank of Institute Professor, reserved for what the university calls "the great men and women" of the faculty, those who have made "extraordinary contributions to our intellectual and collegial life." Three of these women hold or have held that rank.

SUSAN CAREY is a psychologist who studies the origins of uniquely human concepts over the course of evolution, history, and the development of individuals. Her case studies have included the emergence of mathematical concepts (especially *integer, rational number*), intuitive and scientific theories of the physical and biological worlds, and the capacity for logically structured thought. She is known for her work identifying a rich innate repertoire of core concepts as well as the qualitative changes in the course of conceptual development, and for characterizing the learning mechanisms that underlie conceptual change. She is now a professor at Harvard University.

SYLVIA CEYER is a physical chemist who studies the atomic-level interaction of molecules with surfaces of materials. The goal is to unravel the fundamental nature of catalysis, which is critical to the development of less expensive and cleaner energy sources, and the fundamental nature of semiconductor chemistry, which is essential to the production of electrical and mechanical nanodevices. In so doing, her work uncovered phenomena responsible for differences in surface chemistry under high-pressure industrial conditions and low-pressure laboratory conditions, including splat chemistry, chemistry with a hammer, and collision-induced processes. The discovery of a transiently excited hydrogen atom emerging from beneath a metal surface changed the picture for the mechanism for catalytic hydrogenation. This fundamental research paved the way for the development of a novel, low-temperature catalyst for oxidation reactions. Demonstration by the Ceyer laboratory that fluorine and xenon difluoride etching of silicon occurred via an atom abstraction mechanism and via gas phase dissociation of a surface reaction product, respectively, resulted in the modification of reaction models used to predict rates and yields for semiconductor etching under realistic conditions.

SALLIE (PENNY) CHISHOLM is a biological oceanographer known for her role in the discovery of *Prochlorococcus*, the smallest and most abundant photosynthetic microorganism in the sea. Her research spans scales from individual genes to ocean ecosystems and has been instrumental in catalyzing the field of ecological genomics. Using *Prochlorococcus* as a guide, her lab has given us insights into the evolutionary forces that shaped Earth's early oceans and atmospheres, and the role of genetic diversity in stabilizing today's ocean ecosystems. She has also written extensively on the risks of ocean geoengineering and has coauthored a series of children's books on the role of photosynthesis in shaping life on Earth.

SUZANNE CORKIN, a neuroscientist, was most well-known for her career-long work with the amnesic patient known as H.M., which helped explain the anatomy of memory and its disorders by linking specific cognitive processes to discrete brain circuits. Her additional research helped identify the regions of the brain involved in degenerative diseases such as Alzheimer's and Parkinson's. She also studied the long-term consequences of head injuries among veterans of World War II and the Korean War. She died in 2016, at seventy-nine.

MILDRED DRESSELHAUS, a solid-state physicist, made revolutionary discoveries about the electronic structure of graphite and other carbon structures. Among her developments was the conception of nanotubes, made by rolling sheets of carbon atoms, that are thinner than human hairs but as strong as steel. Used in space programs and smartphones, the tubes helped usher in an age of smaller and more universal technology. In the 1990s she and a colleague published two theoretical papers showing that reducing the dimensionality of thermoelectric materials could dramatically improve their performance; those papers are credited with reviving the field of thermoelectrics. She died in 2017, at eighty-six, days after starring in a General Electric commercial that asked, "What if we treated great female scientists like they were stars?"

ANN GRAYBIEL, a neuroscientist, was the first to identify the importance of a previously overlooked area of the brain known as the basal ganglia, uncovering the role it plays in learning, motivation, and behavior. Her research led to an understanding of how healthy brains work, and the understanding and treatment of the neurodegenerative and neuropsychiatric disorders caused by dysfunction in the region, including Parkinson's disease, Huntington's disease, Tourette's syndrome, obsessive-compulsive disorder, and depression.

NANCY HOPKINS, a molecular biologist, was among the first to use genetics to map and identify genes responsible for biological properties of nondefective mouse leukemia viruses. Her lab identified genes that determine host range and the type and severity of cancers caused by this class of RNA tumor viruses, including capsid protein P30, envelope genes, and transcriptional elements known as enhancers. Later she changed fields to study genes involved in early vertebrate development, using the zebrafish model. Her lab devised an efficient method of insertional mutagenesis for the fish using pseudo-typed mouse retroviral vectors and a large-scale genetic screen. They identified and cloned 315 genes, about 25 percent of the genes that are genetically essential for early development, including some genes that, when mutated, predispose fish to cancer.

RUTH LEHMANN, a developmental and cell biologist, studies the life cycle of the germ line. She is known for her work studying germ cells, which give rise to the egg and sperm and transmit to the next generation the potential to build a completely new organism. Research in her lab contributed to the first genetic framework for the specification of germ-cell fate in any organism and demonstrated the role of RNA localization and localized translation in the establishment of embryonic polarity. Lehmann also helped uncover how oocyte mitochondria avoid transmitting mutations within their small genomes to offspring. She is the director of the Whitehead Institute for Biomedical Research at MIT.

MARCIA McNUTT is a geophysicist who has studied the dynamics of the Earth's upper mantle and lithosphere on geological time scales. She was the first to document the existence of and provide an explanation for a South Pacific "superswell" that has produced an abnormally shallow seafloor for more than 100 million years. She is a veteran of more than a dozen deep-sea explorations, serving as chief scientist on most. She has also described the "exceptions" to the theory of plate tectonics in places such as the Hawaiian Islands, the Colorado Rocky Mountains, and the volcanoes of French Polynesia, where the eruptions happen in the center of the plates rather than at the margins. She served as director of the United States Geological Survey during the Obama administration, overseeing government response to disasters, including major earthquakes in Haiti, Chile, and Japan and the explosion of the *Deepwater Horizon* oil rig. Since 2016 she has served as president of the National Academy of Sciences, the first woman to lead the organization.

TERRY ORR-WEAVER is a cell and developmental biologist who studied the processes that regulate cell division during early development. Working as a graduate student with Jack Szostak, she revealed the events that lead to chromosomal recombination during meiosis, the process by which genes on maternal and paternal chromosomes are reorganized. In her lab at MIT, she determined how organisms prepare and activate the egg cell for fertilization and the subsequent transition from oocyte to embryo. In her studies, she discovered crucial control proteins for chromosome segregation and DNA replication, and revealed that cells can become extra large by increasing their DNA copy number in a process that is achieved by excessive DNA endoreplication and endomitosis.

MARY-LOU PARDUE, a geneticist and cell biologist, studied the structure and function of chromosomes in eukaryotes. As a graduate student in the 1960s, she worked with Joseph Gall to develop the technique of in situ hybridization, now widely used in laboratory and clinical settings. She studied the mechanisms by which cells cope with stresses of various kinds to promote understanding of heat shock genes and proteins. She studied the protective ends of chromosomes, known as telomeres, in *Drosophila*, discovering that they are maintained by special transposable elements called retrotransposons.

MARY (MOLLY) POTTER is a psychologist who is known for her elegant and counterintuitive contributions to the understanding of the structure of the human mind. Among Potter's contributions were the demonstration that concepts are distinct from mental representations of words and pictures and that pictures and words both have independent direct routes to concepts, that the first few milliseconds of perception include conceptual information, and that there is no simple limit to working memory. Her discoveries played an important role in characterizing the computational mechanisms that are deployed in the earliest stages of information processing.

PAOLA MALANOTTE-RIZZOLI, a physical oceanographer, investigates the dynamical processes of ocean circulation in different oceanic basins and their changes under global warming. In particular, she has long studied rising sea levels in the Mediterranean and her home city of Venice, Italy, and helped to design the movable barriers known as the MOSE in the lagoon inlets to protect the city from the increasing extreme floods.

LEIGH ROYDEN is a geologist and geophysicist whose research focuses on the mechanics that control plate tectonics, continental collision, mountain building, subduction, and deformation of the Earth's crust and mantle. She was among the first to describe the process of thermal subsidence in sedimentary basins and the critical role of advancing and retreating plate boundaries during continental collision and the growth of mountain ranges. Her work has contributed to understanding the formation and long-term stability of continental plateaus, including the Tibetan Plateau. Her recent studies of subduction dynamics explain the record-breaking motion of the Indian plate prior to collision with Eurasia and provide the first explanation for the angle at which subducting slabs descend through the upper mantle.

LISA STEINER, an immunologist, studied the structural variation of antibodies. Her work demonstrated that effective antibodies, while sharing certain basic features such as variable and constant regions, can be unexpectedly diverse in structure, varying in number of polypeptide chains or in pattern of disulfide bridging. It also showed that the polypeptide chains in nonmammalian vertebrates do not generally fall into classifications developed for mammalian antibodies.

JOANNE STUBBE has revealed the mechanisms of controlled radical-based reactions in biology, including the enzymes ribonucleotide reductases (RNRs) and the natural product bleomycins (BLMs). RNRs make the deoxynucleotide building blocks essential for DNA biosynthesis and repair. BLMs site-specifically degrade DNA. Stubbe in collaboration has unraveled the mechanisms of Gemcitabine, a mechanism-based inhibitor of RNR, and the BLMs. Both are used therapeutically in the treatment of a wide range of cancers.

Acknowledgments

It has been my great privilege and even better fortune to have encountered the people whose stories shaped this book. My greatest debt is, of course, to Nancy Hopkins. We first met over a phone call placed from my desk at the old *Globe* building on Morrissey Boulevard in 1999. Six years later, I watched from a distance as Nancy's doubters in the Larry Summers episode accused her of being a publicity hound, and knew that nothing could be further from the truth; she has always wanted to be known as a scientist, not an activist. "I don't expect to come out of this looking like a hero," she said to me in the writing of this book, even as so many other people I was talking to described her as just that. I realize it has been painful for her to relive some of the episodes in the book, and I am grateful for the grace and good humor she has shown in tolerating my questions and intrusions. My thanks also to Dinny Adams, who had to live with the consequences of those intrusions yet consistently welcomed me with his trademark good cheer.

Like Nancy, the other women of MIT—in the sixteen and beyond—have taught me about science, and life, and by example reminded me of the wisdom of Ellen Swallow Richards, MIT's first female graduate: "Keep thinking." Lotte Bailyn and Penny Chisholm were involved in conversations about this book from the beginning and entertained my inquiries with candor, curiosity, and patience, as did Sylvia Ceyer, Ruth Lehmann, Marcia McNutt, Mary-Lou Pardue, Molly Potter, Paola Rizzoli, Leigh Royden, Lisa Steiner, and JoAnne Stubbe. My enormous thanks to them, and to Bob Birgeneau, who took my call in 1999 and has been answering my questions about physics and other subjects ever since. I was very glad to finally meet the late Millie Dresselhaus in 2011, and to have shared conversations with Chuck Vest on several occasions before his death in 2013.

My thanks also, extremely belated, to Hazel Sive, who revisited the status of women in the School of Science for a 2011 report and suggested then that I might want to write more about it, and to Ruth Perry for sharing her recollections and her "Palace Coup" files. And to others at MIT who helped me understand the institute and the story: Jerry Friedman, Lorna Gibson, David Housman, David Kaiser, Eric Lander, Harvey Lodish, Tom Magnanti, Ken Manning, June Matthews, Mary Rowe, Bob Sauer, Robin Scheffler, Phil Sharp, Susan Silbey, Gigliola Staffilani, and Kathryn Willmore.

Susan Gerbi, Joan Steitz, and especially Mark Ptashne were generous not only in sharing their memories, but in reading and offering thoughts and expertise on parts of the manuscript. I am grateful to many others for their generosity with interviews: Larry Bacow, David Baltimore, Mahzarin Banaji, Sue Berget, Bob Bosselman, David Botstein, Rafael Bras, Don Brown, Steve Burden, Marianne Dresselhaus Cooper, Shoshi Cooper, Ruth Schwartz Cowan, Titia DeLange, Ann and Jan Dubois (and Alice, and Anna in utero, for keeping me in touch with Nancy), Joe Gall, Nancy Gertner, Andrea Ghez, Terri Grodzicker, Barbara Grosz, Evelynn Hammonds, Deming Holleran, Alice Huang, Chuck and Reida Kimmel, Dan Luria, Shirley Malcom, Sandy Masur, Miriam McKendall, Claire Moore, Boyce Rensberger, Joan Ruderman, Sylvia Sanders, Ed Scolnick, Kimberlee Shauman, Allan Spradling, Larry Summers, Shirley Tilghman, Dan Vapnek, Leslie Vosshall, Jim, Liz, and Duncan Watson, Serita Winthrop, Ting Wu, and Virginia Zakian. A small number of people requested confidentiality, and I thank them for trusting me with their recollections.

The physicist Richard Feynman famously referred to "the pleasure of finding things out," and what's true for physics is also true for reporting. Particular thanks to the archivists Alex McGee at MIT, Mila Pollock and Stephanie Satalino at Cold Spring Harbor, Sarah Hutcheon at the Schlesinger Library at Radcliffe, and those who helped search the Luria and McClintock papers at the American Philosophical Society. And to several people who provided scientific explanations and helped me track down information or fill in gaps: Susan Beachy, Susan Carey, Joss Corkin, Debbie Douglas, Peter Fritschel, Martin Harwit, Linda Hamilton Kreiger, Eric Masur, Janet Mertz, and Virginia Valian. My thanks to Kathy Bleier Eisenberg for permission to quote from her father's letter, and for directing me to the work of her mother, Ruth Bleier, on gender and science.

The Rijksuniversiteit Groningen, my grandfather's scientific home and my father's alma mater, invited me to give three talks at their lustrum in June 2019. Those talks, and my conversations with the recipients of the Rosalind Franklin fellowships, helped my thinking about women in science. I am grateful to Rick Lifton, the president of Rockefeller University, for including me at a dinner the following week celebrating Nancy's honorary degree. Also to Amy Brand, for inviting me to a reunion of the MIT women in 2018. Thanks also to Anita Zieher, chairwoman of Vienna's portraittheater, for sharing the script of the play *Women at Work*, about the remarkable life of Marie Jahoda, which I was lucky enough to see performed in Cambridge before the world's stages went dark in 2020.

This book took me back to Boston, where my career began, and made me think about the importance of mentors—I've had a few. At the *Globe*, my biggest thanks to Walter V. Robinson, who hired me and still works a story harder than anyone else I know. I have reflected often on a night in early spring 1997, up against deadline on a sensitive story, with the big four of the newsroom—Matt Storin, Helen Donovan, Greg Moore, and Al Larkin—standing around my desk like the points of a compass as I tried to coax a source on the phone. I kept thinking that they'd call in the Big Reporter to write the story—until I realized they expected me to be that reporter. I thank them for their confidence on that and so many other stories, and for the tremendous opportunities they gave me. My thanks also (and a sheet cake!) to my Globe Gang of editors, colleagues, friends, especially Geeta Anand, Tina Cassidy, Steve Kurkjian, Scot Lehigh, Frank Phillips, and Adrian Walker. Friends from my earliest days in Boston—Jeremy Crockford,

Mary Freed, Jimmy Golen, and Carolyn Ryan—helped make it the best place to learn how to be a reporter. Thanks to Marcella Bombardieri, a successor on the higher education beat at the *Globe*, for sharing her recollections of the events around the Larry Summers speech. At the *Times*, my professional home for twenty-two years now, my thanks to my editors, mentors, friends, and coconspirators, especially Ethan Bronner, who hired me, Abby Goodnough, who was my first friend, Monica Davey, Paul Fishleder, Ruth Fremson, Marty Gottlieb, Wendell Jamieson, Adam Liptak, Danielle Mattoon, and Jim Rutenberg. I learned so much from the women of the old Week in Review: Katy Roberts, Mary Jo Murphy, and Mary Suh. As a young reporter my dream was to be a national correspondent for the *New York Times*; I thank the succession of editors who brought and kept me to the National Desk— Jim Roberts, Suzanne Daley, Rick Berke, and the indefatigable Alison Mitchell—and Jia Lynn Yang for welcoming me back in 2022.

Dean Baquet has offered great advice and an open door over the years. I thank him for that as well as for his leadership and, not least, granting me leave to write this book. Matt Purdy has been a tremendous mentor and friend ever since the day shortly after 9/11 when he cheerfully asked if I was aware that working for the *New York Times* was sort of like joining the army. I was fortunate to have Arthur Gregg Sulzberger—who appears in the kicker to *The Girls in the Balcony*—as my editor, and we are all fortunate to have him and the Sulzberger family as champions of independent journalism. A special edition of thanks to the (near-) daily miracle of TWAC: Ginia Bellafante, Penelope Green, Alexandra Jacobs, Sarah Lyall, Julia Moskin, and Robin Pogrebin.

My thanks swell around the team of women who have been with this book since it was a glimmer of an idea. First, my agent, Elyse Cheney, with her tough mind and big heart, and Claire Gillespie at the Cheney Agency for her insights. Shortly after I began the book another agent asked me who my editor was, and when I told him it was Kathy Belden, he said, "Ah, one of the greats." I soon discovered he was understating it. Kathy pushed, encouraged, and taught me and became a friend. She had the inspiration to bring in another generation, with Sarah Goldberg, who was with us for half the book and in that time asked two or three questions that altered the direction of my thinking and reporting for the better. Nan Graham's enthusiasm for the project softened the edges around a big birthday two days before our first meeting and carried me over the finish line. Thanks also to others at Scribner: Jaya Miceli for the elegant cover design; Carolyn Levin for her careful read; Rebekah Jett and Katie Rizzo for production and more.

Among the friends who helped: Alison Franklin has been my first call on so many things over the last twenty-five years, and so she was the first call when I first considered doing this book, and as I faced the daunting task of writing the first lines. I am grateful for her early reconnaissance in Cambridge, her edits, and, as ever, her wisdom. Kit Seelye provided housing that was practically on campus and her good company. Susan Durkee gave me the great gift of a quiet place to write in Wellfleet, where Dan and Becky Okrent read, listened to ideas, and kept me fed and entertained. Dale Russakoff nourished my thoughts about the project over many lunches and phone calls and brought her exquisite mind and red pen to the first chapter. Daryl Levinson was once again an early and encouraging reader. Gerry Laybourne, fairy godmentor, appeared with just the right advice at just the right time.

My father long encouraged me to write a book. I wrote my first at lightning speed, aware that he was close to the end of his life. He died ten months after it was published. This is the book I wish he were still around to read, as without him I would never have written it. He raised me to respect science, and to love stories. My wonderful mother died during the writing of this book. Cleaning out her house afterward, among the things I discovered was a scrap of paper on which she had scratched her edits to a particularly tough section of the book I called her to discuss. She helped me craft so many passages—in work and in life—when I was stuck, and I wish she, too, were here to read this story. It is inspired by hers. I thank Andrea Silber, my mother's longtime oncologist, for reminding me of the mission my mother left me.

I owe enormous thanks to the rest of my family. My brother Harry in particular led the Quaranteam, read, helped with photographs, and fixed my grammar; for that alone he would have made our parents proud. And I count myself exceptionally lucky for the love of those to whom I am W slash M: my sons, Frits and Nico, have delighted me with their humor, wisdom, and natural enthusiasm for this project. And nobody, nobody, has read more pages, cooked more meals, filled in more words, or soothed more nerves than Jonathan.

A Note on Sources

This narrative relies on interviews I conducted over four years, as well as archival material including letters, university reports, oral histories, videos, yearbooks, and photographs. I have also used thousands of pages of Nancy Hopkins's papers, including diaries, date-books, personal communications, records of the meetings that began with the women in the summer of 1994, and the color-coded floor plans. Nancy allowed me free access to the papers, which were then in her office at MIT. Subsequently she gave those papers to the MIT archives, but as of this writing they had not been cataloged. In the notes I have described them as NHP.

Sources labeled CSHL are from the James D. Watson Collection in the Cold Spring Harbor Laboratory Archives. Those labeled MIT are from the Department of Distinctive Collections in Cambridge.

Shortly after the Larry Summers episode, a literary agent suggested Nancy try her hand at a memoir; Nancy wrote several chapters before abandoning the idea, and no attempt was made to sell the book. Where I have relied on that, it is indicated in the notes by the title she gave it, *Fifteen Tenured Women*.

Where I describe conversations, I have taken the dialogue as it was recalled in diaries or other contemporaneous accounts, as noted below, or have confirmed it with the participants in the conversation.

The original MIT report that Nancy wrote in 1996 has never been made public and is not available in any archive, and Nancy did not give me access to that. She and Bob Birgeneau insisted that it remain classified, as it was based on confidential interviews and contains data easily traceable to specific individuals. I obtained a copy of that report independently.

Notes

PROLOGUE

xiv *"A climate change"*: Natasha Loder, "US Science Shocked by Revelations of Sexual Discrimination," *Nature* 405 (June 8, 2000): 713–14.

CHAPTER ONE: AN EPIPHANY ON DIVINITY AVENUE

4 *A tutor*: Student Count, CSHL.

5 *he began grandly*: "Outline of Lectures by J. D. Watson—Biology 2," CSHL.

9 *"The world knows"*: Sally Schwager, "Taking Up the Challenge: The Origins of Radcliffe," in *Yards and Gates: Gender in Harvard and Radcliffe History*, ed. Laurel Ulrich (New York: Palgrave Macmillan, 2004), 138.

9 *"with the taste"*: Radcliffe College, Annual Reports of the President and Treasurer, United States, 1894.

9 *For decades*: Marie Hicks, "Integrating Women at Oxford and Harvard Universities, 1964–1977," in Ulrich, *Yards and Gates*, 371.

10 *It was an exclusive set*: Marcia G. Synnott, "The Changing 'Harvard Student': Ethnicity, Race, and Gender," in Ulrich, *Yards and Gates*, 298.

10 *none in Nancy's class*: *Radcliffe College Freshman Register* (Harvard Yearbook Publications, August 1960).

10 *The president of Harvard*: Synnott, "Changing 'Harvard Student,'" 298.

10 *"only skin deep"*: *The Red Book: Student's Handbook of Radcliffe College* (Cambridge, MA: Radcliffe College, 1950–51), 100.

10 *"detached and thin"*: Radcliffe College, *Report of the President*, 1960–61, 5.

10 *publicly decrying*: Mary I. Bunting, "A Huge Waste: Educated Womanpower," *New York Times Magazine*, May 7, 1961, 23–112.

11 *Friedan had asked Bunting*: Maggie Doherty, *The Equivalents: A Story of Art, Female Friendship, and Liberation in the 1960s* (New York: Alfred A. Knopf, 2020), 65.

11 *"A dissatisfied woman"*: Bunting, "Huge Waste," 112.

11   *on the fringes:* Ibid., 109.

11   *"awfully interesting":* "One Woman, Two Lives," *Time* 78, no. 18 (November 3, 1961).

12   *She faulted: Report of the President,* 94.

12   *And she saw Radcliffe:* Ibid., 7–8.

12   *"wisely and largely":* Ibid., 99.

12   *Bunting started:* Synnott, "Changing 'Harvard Student,'" 304–5.

12   *noted approvingly: Report of the President,* 16.

13   *"fearfully bright":* "One Woman, Two Lives."

13   *Bunting thought the rules:* Ann Karnovsky, "Nostalgia and Promise," in Ulrich, *Yards and Gates,* 351; "The Rules Revolution," in *Radcliffe College Yearbook, Class of 1964* (Cambridge, MA: 1964), 56.

13   *"What would happen if": Report of the President,* 24.

14   *There were 47:* Course Lists, 1901–84 (inclusive), Harvard University Archives.

14   *women declaring biology: Report of the President,* 1960–61, 56; *Report of the President,* 1961–64, 45.

15   *He chafed against:* James D. Watson, *Avoid Boring People: Lessons from a Life in Science* (New York: Alfred A. Knopf, 2007), 118–31.

16   *Wilson called him:* E. O. Wilson, "The Molecular Wars: Changing Paradigms, Clashing Personalities, and the Revolution in Modern Biology," *Harvard Magazine,* May–June 1995, 42–49.

16   *At the Nobel ceremony:* Letter from Nathan M. Pusey to James D. Watson, December 28, 1962, CSHL.

17   *Steitz's first choice:* Richard Panek, "Don't Listen to the Naysayers," *Yale Alumni Magazine,* July/August 2019.

19   *They had formed:* Letter from George Gamow to James D. Watson, November 25, 1954, CSHL.

20   *"Both young men":* Watson, *Avoid Boring People,* 108.

CHAPTER TWO: THE CHOICE

23   *she worked:* Elga Wasserman, *The Door in the Dream: Conversations with Eminent Women in Science* (Washington, DC: Joseph Henry Press, 2000), 5.

25   *The dean of students warned:* John T. Bethell, *Harvard Observed: An Illustrated History of the University in the Twentieth Century* (Cambridge, MA: Harvard University Press, 1998), 213.

26   *"The young women of today":* "Women in Education," *Radcliffe College Yearbook, Class of 1964* (Cambridge, MA: 1964), 76.

26   *students had glimpsed:* "The Academic Year," *Radcliffe Yearbook,* 111.

26   *editors dismissed:* "Women in Education," 76–79.

27   *half of all: 1965 Handbook on Women Workers,* Women's Bureau Bulletin no. 290 (Washington, DC: US Department of Labor, 1965), 5.

27   *The Radcliffe graduates: Radcliffe Yearbook,* 79.

28   *Census surveys:* Kristin Smith, Barbara Downs, and Martin O'Connell, "Mater-

nity Leave and Employment Patterns: 1961–1995," *Current Population Reports* (Washington, DC: US Census Bureau, 2001), 20, 11.

28   *"A bright and extremely pleasant"*: Copy of letter from James D. Watson to Detlev Bronk regarding Nancy Hopkins, CSHL.

29   *Jim recommended Nancy*: Letter of recommendation for Nancy Hopkins, CSHL.

31   *the holy grail*: Horace Freeland Judson, *The Eighth Day of Creation: Makers of the Revolution in Biology, Commemorative Edition* (Cold Spring Harbor, NY: Cold Spring Harbor Laboratory Press, 2013), 560.

32   *"You spoiled me"*: Handwritten letter from Nancy Hopkins to James D. Watson, CSHL.

32   *She immediately judged*: Handwritten letter from James D. Watson to Nancy Hopkins, CSHL.

32   *A few months later*: Handwritten letter from Nancy Hopkins to James D. Watson, CSHL.

33   *She had heard*: Nancy Hopkins, notes regarding isolation of the lambda repressor, CSHL.

34   *she wrote Jim*: Handwritten letter from Nancy Hopkins to James D. Watson, CSHL.

CHAPTER THREE: AN IMMODEST PROPOSAL

37   *The American Academy*: Margaret W. Rossiter, *Women Scientists in America: Volume 2, Before Affirmative Action, 1940–1972* (Baltimore: Johns Hopkins University Press, 1995), 324–25.

38   *When the dean*: "Harvard's First Lady," Harvard T. H. Chan School of Public Health, https://www.hsph.harvard.edu/news/centennial-harvards-first-lady/.

38   *the first white woman*: "Biographies of Women Mathematicians," Agnes Scott College, https://www.agnesscott.edu/lriddle/women/women.htm.

38   *During World War II*: "The Outlook for Women in Science," Women's Bureau Bulletin 223-1 (Washington, DC: US Government Printing Office, 1949), 21.

38   *But after the war*: Rossiter, *Women Scientists*, 31.

38   *The number of female professors*: Ibid., 226.

38   *Male scientists worked around this*: Ibid., 149.

38   *Gerty Cori worked*: "Women in Health Sciences," Washington University in St. Louis School of Medicine, Bernard Becker Medical Library Digital Collection, http://beckerexhibits.wustl.edu/mowihsp/bios/cori.htm.

39   *Women who dared complain*: Rossiter, *Women Scientists*, 19.

39   *Others argued that*: Ibid., 47.

39   *The biggest hurdle*: Helen Hill Miller, "Science: Careers for Women," *Atlantic*, October 1957.

39   *Betty Lou Raskin, a chemist*: Betty Lou Raskin, "Women's Place Is in the Lab, Too," *New York Times Magazine*, April 19, 1959, 17–20.

40   *A headline in*: Dorothy Barclay, "For Bright Girls: What Place in Society?," *New York Times Magazine*, September 13, 1959, 126.

40   *When Maria Goeppert Mayer:* Rossiter, *Women Scientists*, 330.

40   *accounted for just 7 percent:* "Women in Scientific Careers," National Science Foundation NSF 60-51 (Washington, DC: Government Printing Office, 1961).

40   *Women earned about:* "Women in Science and Engineering," *Science*, September 25, 1964, 1389.

40   *At the twenty top:* Rossiter, *Women Scientists*, 129.

41   *"As the need for specialized":* "Women in Scientific Careers," 1.

41   *In view of:* Ibid., 10.

41   *The commission had been the result:* Dorothy Sue Cobble, "Labor Feminists and President Kennedy's Commission on Women," in *No Permanent Waves: Recasting Histories of U.S. Feminism*, ed. Nancy A. Hewitt (New Brunswick, NJ: Rutgers University Press, 2010), 156–60.

42   *"each woman must arrive":* American Women: Report of the President's Commission on the Status of Women (Washington, DC, 1963), 2.

43   *Her paternal grandmother:* Lotte Bailyn, "Four Generations: A Memoir of Women's Lives," Matina S. Horner Distinguished Visiting Professor Lecture Series, Radcliffe Public Policy Institute, May 6, 1997.

43   *Lotte as a toddler:* Sandra Schüddekopf and Anita Zieher, *Women at Work: Kathe Leichter and Marie Jahoda* (Vienna: portraittheater, 2019), 9.

43   *the young couple:* Marie Jahoda, Hans Zeisel, and Paul Lazarsfeld, *Marienthal: The Sociography of an Unemployed Community* (New Brunswick, NJ: Transaction Publishers, 2002), 66; Paul Neurath, "Sixty Years Since Marienthal," *Canadian Journal of Sociology* 20, no. 1 (1995): 91–105.

44   *Paul Lazarsfeld was becoming:* David L. Sills, *Paul F. Lazarsfeld, 1901–1976: A Biographical Memoir* (Washington, DC: National Academy of Sciences, 1987); Hynek Jerabek, "Paul Lazarsfeld—the Founder of Modern Empirical Socilogy: A Research Biography," *International Journal of Public Opinion Research* 13, no. 3 (2001): 229–44.

45   *Her paper argued:* Lotte Bailyn, "Notes on the Role of Choice in the Psychology of Professional Women," *Daedalus* 93, no. 2 (Spring 1964): 700–710.

46   *He acknowledged the stigma:* David Riesman, "Two Generations," *Daedalus* 93, no. 2 (Spring 1964): 711–35.

47   *Erikson called for:* Erik H. Erikson, "Inner and Outer Space: Reflections on Womanhood," *Daedalus* 93, no. 2 (Spring 1964): 582–606.

47   *She asked a male:* Rossiter, *Women Scientists*, 365.

48   *The paper she presented:* Alice S. Rossi, "Equality Between the Sexes: An Immodest Proposal," *Daedalus* 93, no. 2 (Spring 1964): 607–52.

49   *She was called:* "Alice Rossi, 87, Noted Sociologist, Leading Feminist," *Boston Globe*, November 10, 2009.

49   *a two-day symposium:* "Should Science Be for Men Only?," *Technology Review* 67 (December 1964): 28–46; Alice S. Rossi, "Women in Science: Why So Few?," *Science* 148 (May 28, 1965): 1196–202; John Lear, "Will Science Change Marriage?," *Saturday Review*, December 5, 1964, 75–77.

51 *But few:* Timothy Leland, "Over the Din of Babes at MIT / Women Told of Science Role," *Boston Globe*, October 24, 1964, 2.

51 *More scathing:* "A Woman's Place . . . ," *Tech*, October 28, 1964, 4.

51 *The status quo:* E. C. Pollard, "How to Remain in the Laboratory Though Head of a Department," *Science* 145, no. 3636 (September 4, 1964): 1018–21.

CHAPTER FOUR: AT THE FEET OF HARVARD'S GREAT MEN

53 *The demonstrators swarmed:* Robert S. McNamara, *In Retrospect: The Tragedy and Lessons of Vietnam* (New York: Vintage Books, 1996), 256; Stephen D. Lerner, "McNamara Protest On Despite Master's Move to Control Picketers," *Crimson*, November 4, 1966; "1966: The Last Time," *Crimson*, May 3, 1985. McNamara's memoir refers to Frank as an undergraduate; in fact, Frank, class of 1962, was working at the Kennedy School, where he had arranged McNamara's visit.

53 *The next October:* W. Bruce Springer, "300 Stage Sit-In at Mallinckrodt Hall to Halt Dow Chemical Recruitment," *Crimson*, October 26, 1967, https://www.thecrimson.com/article/1967/10/26/300-stage-sit-in-at-mallinckrodt-hall/.

53 *In his annual report:* Report of the President of Harvard College, 1966–67, 22, Harvard University Archives.

54 *His parents':* Jane Gitschier, "Irrepressible: An Interview with Mark Ptashne," *PLOS Genetics*, July 16, 2015.

54 *"younger shock troopers":* Wilson, "Molecular Wars."

56 *Jim invited:* "The Scientist," YouTube, https://www.youtube.com/watch?v=kdO goTl9Fog.

58 *Watson submitted:* Walter Gilbert and Benno Müller-Hill, "Isolation of the Lac Repressor," *Proceedings of the National Academy of Sciences* 56, no. 6 (December 1966): 1891–98.

58 *in February:* Mark Ptashne, "Isolation of the Lambda Phage Repressor," *Proceedings of the National Academy of Sciences* 57, no. 2 (February 1967): 306–13.

58 *they mixed:* Mark Ptashne, *A Genetic Switch: Phage Lambda Revisited*, 3rd ed. (Cold Spring Harbor, NY: Cold Spring Harbor Press, 2004), 72–73.

60 *Jim had grown:* Watson, *Avoid Boring People*, 240–48; Nancy Hopkins, Notes Regarding Isolation of the Lambda Repressor, CSHL.

64 Vogue: James D. Watson, *The Double Helix: A Personal Account of the Discovery of the Structure of DNA* (New York: Simon & Schuster, 1996), 65.

64 *gravity:* Ibid., 90.

65 *"The thought":* Ibid., 20.

65 *Watson realized:* Ibid., 212.

65 *Crick accused:* Francis Crick to James D. Watson, April 13, 1967, Francis Harry Compton Crick Papers, Wellcome Library for the History and Understanding of Medicine, https://profiles.nlm.nih.gov/spotlight/sc/catalog/nlm:nlmuid-101 584582X137-doc.

66 *Richard Feynman:* Michelle Feynman, *Perfectly Reasonable Deviations from the*

Beaten Track: The Letters of Richard P. Feynman (New York: Basic Books, 2005), 236.

66   *"One has only"*: Nancy Hopkins to James D. Watson Sr., Watson Collection, CSHL, http://libgallery.cshl.edu/items/show/81296.

68   *When Radcliffe students:* Hicks, "Integrating Women," 384.

68   *only after Wald:* George Wald to W. Bickel, March 8, 1967, photograph of letter at University of Zurich, shared in email communication with the author by Karl Gademann, November 11, 2019.

68   *"By proudly offering"*: Ruth Hubbard, "Memories of Life at Radcliffe," in Ulrich, *Yards and Gates,* 344.

70   *"Dear Mrs. Brooke"*: L. L. Waters to Nancy Hopkins, May 24, 1971, CSHL.

70   *Jim had recommended her:* James D. Watson to L. L. Waters, November 20, 1970, CSHL.

CHAPTER FIVE: BUNGTOWN ROAD

72   *Within a decade:* Daniel Okrent, *The Guarded Gate: Bigotry, Eugenics, and the Law That Kept Two Generations of Jews, Italians, and Other European Immigrants out of America* (New York: Scribner, 2019), 24–28, 118–24.

72   *surviving through:* Elizabeth L. Watson, *Houses for Science: A Pictorial History of Cold Spring Harbor Laboratory* (Cold Spring Harbor, NY: Cold Spring Harbor Laboratory Press, 1991), 115–19.

72   *As the organizers wrote:* "Conference on Quantitative Biology at Cold Spring Harbor," *Science* 78, no. 2023 (October 6, 1933): 304–5.

73   *new lab director:* Watson, *Houses for Science,* 145.

73   *The announcement:* "Symposium on Quantitative Biology at Cold Spring Harbor," *Science* 93, no. 2416 (April 18, 1941): 370.

74   *the new gospel:* Gunther S. Stent, "That Was the Molecular Biology That Was," *Science* 160, no. 3826 (April 26, 1968): 390–95; William Hayes, *Max Ludwig Henning Delbrück: 1906–1981, a Biographical Memoir* (Washington, DC: National Academy of Sciences, 1993).

76   *When Watson took over:* Watson, *Avoid Boring People,* 271–80.

79   *"foremost investigator"*: Evelyn Fox Keller, *A Feeling for the Organism: The Life and Work of Barbara McClintock* (New York: W. H. Freeman, 1983), 44–45.

79   *"She is sore"*: Ibid., 73–74.

80   *iron-on tape:* Nathaniel Comfort, *The Tangled Field: Barbara McClintock's Search for the Patterns of Genetic Control* (Cambridge, MA: Harvard University Press, 2001), 17.

80   *she refused to have a telephone:* Gina Kolata, "Dr. Barbara McClintock, 90, Gene Research Pioneer, Dies," *New York Times,* September 4, 1992, 1.

81   *Evaluations:* Keller, *Feeling for the Organism,* 74–75.

81   *"Katharine Hepburn"*: Comfort, *Tangled Field,* 247.

CHAPTER SIX: "WOMEN, PLEASE APPLY"

85   *"dedicated"*: Miller, "Science: Careers for Women," *Atlantic*, October 1957.

85   *"I figured"*: Comfort, *Tangled Field*, 65.

90   *cows' brains*: Ralph J. Greenspan, *Seymour Benzer, 1921–2007, a Biographical Memoir* (Washington, DC: National Academy of Sciences, 2009).

94   *had some results*: Joseph G. Gall, "The Origin of In Situ Hybridization—a Personal History," *Methods*, April 2016.

94   *new technique*: Mary Lou Pardue and Joseph G. Gall, "Molecular Hybridization of Radioactive DNA to the DNA of Cytological Preparations," *Proceedings of the National Academy of Sciences* 64 (October 1969): 600–604.

96   *"Please Apply"*: Boris Magasanik, Letter to the Editor, *Science* 171 (February 19, 1971).

96   *the article*: Patricia Albjerg Graham, "Women in Academe," *Science* 169 (September 25, 1970): 1284–90.

98   *"dump"*: Roger W. Bolz, Letter to the Editor, *Science* 170 (December 18, 1970): 1260.

98   *psychology professor*: Sandra Scarr, Letter to the Editor, *Science* 170 (December 18, 1970): 1260.

CHAPTER SEVEN: THE VOW

101   *"untraditional generation"*: Gloria Steinem, "Living the Revolution: Commencement Address at Vassar College," May 31, 1970.

101   *Kistiakowsky*: Chris Kenrick, "Physics Alienates Women," *Tech*, March 7, 1972, 1; Anna Nowogrodzki, "A Binder Full of Physicists," *Technology Review*, April 21, 2015.

102   *similar caucus*: Laura A. Williams, "The History of WICB: The Founding and Early Years," *ASCB Newsletter*, August 1, 1996.

103   American Men of Science: Rossiter, *Women Scientists*, 380.

103   *"trivial hindrances"*: Dora B. Goldstein, Letter to the Editor, *Science* 173 (September 17, 1971): 1080.

103   *fictitious résumés*: Arie Y. Lewin and Linda Duchan, "Women in Academia: A Study of the Hiring Decision in Departments of Physical Science," *Science* 173 (September 3, 1971): 892–95.

104   *"You come on too strong"*: Bernice Resnick Sandler, "Title IX: How We Got It and What a Difference It Made," *Cleveland State Law Review* 55, no. 4 (2007): 473–89.

104   *Mayer*: Rossiter, *Women Scientists*, 381.

105   *"Let us not"*: Susan Tolchin, *Women in Congress* (Washington, DC: Government Printing Office, 1976), 32.

106   *never appeared in the bill*: Sandler, "Title IX," 478.

106   *one line*: Eric Wentworth, "New Programs to Make Mark on Education," *Washington Post*, June 24, 1972, 4.

107   *The survey*: Lotte Bailyn, "Career and Family Orientations of Husbands and

Wives in Relation to Marital Happiness," *Human Relations* 23, no. 2 (1970): 97–113.

111 *War on Cancer:* Siddhartha Mukherjee, *The Emperor of All Maladies: A Biography of Cancer* (New York: Scribner, 2010), 183–88.

112 *$4.4 million:* Sari Kalin, "The Free Thinker," *Technology Review*, August 23, 2011.

112 *"ultimate research objective":* Letter from James D. Watson to Jon R. Beckwith, November 5, 1971, CSHL.

112 *"stick with them":* Letter from James D. Watson to Salvador E. Luria, October 9, 1972, CSHL.

114 *From Philadelphia:* Handwritten letter from Nancy Hopkins to Marie, CSHL.

118 *a nun of science:* Devan Sipher, "Vows: Nancy Hopkins and Dinny Adams," *New York Times*, July 29, 2007.

CHAPTER EIGHT: "WE SHOULD DISTANCE ALL COMPETITORS"

121 *"practical men":* Merritt Roe Smith, "'God Speed the Institute': The Foundational Years, 1861–1894," in *Becoming MIT: Moments of Decision*, ed. David Kaiser (Cambridge, MA: MIT Press, 2010), 19.

121 *"knowledge-seeking":* Emma Barton Rogers, *Life and Letters of William Barton Rogers, Edited by His Wife*, 2 vols. (Boston: Houghton Mifflin, 1896), 1:259.

121 *He'd had six enslaved people:* Craig S. Wilder, "William Barton Rogers: Race and the Founding of MIT," MIT Black History, https://www.blackhistory.mit.edu /story/william-b-rogers.

121 *"sad trials":* Ibid., 2:138.

121 *"to guide":* Ibid., 2:139.

122 *fifteen:* Smith, "'God Speed,'" 15.

122 *by 1916:* Bruce Sinclair, "Mergers and Acquisitions," in Kaiser, *Becoming MIT*, 37–57.

122 *The institute grew:* Deborah Douglas, "MIT and War," in Kaiser, *Becoming MIT*, 84–96.

123 *more money:* David Kaiser, "Elephant on the Charles: Postwar Growing Pains," in Kaiser, *Becoming MIT*, 104.

123 *Hood dairy:* Nidhi Subbaraman, "The Evolution of Cambridge," *Technology Review*, December 21, 2010.

123 *"It is difficult":* Kaiser, "Elephant," 103.

124 *"Our motto":* Samuel Jay Keyser, *Mens et Mania: The MIT Nobody Knows* (Cambridge, MA: MIT Press), 8.

124 HU-GE-EGO: "Interesting Hacks to Fascinate People: The MIT Gallery of Hacks," hacks.mit.edu.

124 *"distance all competitors":* Rogers, *Life and Letters*, 2:276.

125 *Ellen Henrietta Swallow:* David Mindell and Susan Hockfield, "Introductory Remarks: MIT150 Symposium, 'The Women of MIT,'" March 28, 2011.

126 *In 1882:* Robert M. Gray, "Coeducation at MIT, 1950s–1970s," https://ee.stan ford.edu/~gray/Coeducation_MIT.pdf.

127 *"helpless female"*: Amy Sue Bix, *Girls Coming to Tech! A History of American Engineering Education for Women* (Cambridge, MA: MIT Press, 2014), 224.

127 *one in twenty:* Gray, "Coeducation," 15.

127 *"immature lulus"*: Ibid., 24.

127 *"somewhat revolting"*: "Coeds, Even," *Crimson*, March 2, 1956.

128 *the committee:* Ibid., 20–27.

128 *"forgotten men"*: Bix, *Girls Coming to Tech!*, 223.

128 *"plenty of professorial attention"*: "Where the Brains Are," *Time*, October 18, 1963.

128 *"for all her brains"*: Bix, *Girls Coming to Tech!*, 231.

129 *"rather spectacular"*: *President's Report Issue*, 1963–64, 16.

129 *"This rate of increase"*: Ibid., 423.

129 *"they are that"*: Ibid., 240.

129 *class of 1964:* Gray, "Coeducation," 40.

CHAPTER NINE: OUR MILLIE

131 *"sordid mess"*: Mildred Dresselhaus, "Mildred Dresselhaus Discusses Her Life in Science," interview by James Dacey, *Physics World* Video Series, August 13, 2014.

132 *"Any equation"*: Natalie Angier, "A Conversation With: Carbon Catalyst for Half a Century," *New York Times*, July 3, 2012, D1.

133 *Millie repurposed:* Mark Anderson, "Mildred Dresselhaus, the Queen of Carbon," *IEEE Spectrum*, April 28, 2015.

133 *thesis adviser:* Alice Dragoon, "The 'What If?' Whiz," *Technology Review*, April 23, 2013.

133 *They could find:* Mildred Dresselhaus, "Vannevar Bush Award Talk," National Science Foundation, May 13, 2009.

133 *papers together:* Mildred Dresselhaus, "MIT History: The Women of the Institute 1997, Panel Discussion," Cambridge, MA, June 27, 1997.

133 *sick child:* Natalie Angier, "Mildred Dresselhaus, the Queen of Carbon, Dies at 86," *New York Times*, February 24, 2017, B15.

134 *snow day:* Dragoon, " 'What If?' Whiz."

134 *A colleague:* Anderson, "Mildred Dresselhaus."

134 *she figured:* Dresselhaus, "Vannevar Bush."

135 *Nixon White House:* Marianne Dresselhaus Cooper, "My Extended Family: Growing Up as the Daughter of Millie Dresselhaus," Celebrating Our Millie: The Legacy and Impact of Mildred Dresselhaus, MIT, Cambridge, MA, November 26, 2017.

135 *suckling pig:* Paul Dresselhaus, "Growing Up with Millie," Celebrating Our Millie.

135 obey: Albin Krebs, "Abby Rockefeller Mauze, Philanthropist, 72, Is Dead," *New York Times*, May 29, 1976, 26.

136 *weed killer:* Emily Wick, "MIT History: The Women of the Institute 1997," panel discussion.

136 *Quantum Theory:* Aviva Brecher, "Remembering My Mentor, Millie," Celebrating Our Millie.

137 *"hippies":* MIT, *Report of the President,* 1969, 505.

137 *"year of":* MIT, *Report of the President,* 1969–70, 299.

138 *"weatherman":* Ibid., 300.

138 *heard the news:* Shirley Ann Jackson, "Remarks at MIT Black Students' Union Fiftieth Anniversary Celebration," Cambridge, MA, November 2, 2018.

138 *freshman class:* MIT, *Report of the President,* 1969–70, 55.

138 *By 1970:* Ibid., 299.

138 *fundraising:* Dresselhaus, "MIT History."

139 *increased:* Dorothy Bowe, "MIT History: The Women of the Institute 1997."

139 *had forgotten:* Dresselhaus, "MIT History."

140 *"a good man":* Archibald MacLeish, "A Celebration of Jerry Wiesner, 13th President of MIT," MIT Video Productions, YouTube, December 21, 2017.

140 *data and rigor:* Millie Dresselhaus, "Women and MIT: Some History," *MIT Faculty Newsletter* 14, no. 4 (April/May 2002).

140 *released that spring: Role of Women Students at MIT,* Report of the Ad Hoc Committee on the Role of Women at MIT, 1972.

141 *critical mass:* Dresselhaus, "MIT History."

142 *"Remedying injustices": Role of Women,* 57.

142 *self-confidence:* Karen Arenson, interview with Sheila E. Widnall, November 17, 2010, MIT150 Infinite History Project.

CHAPTER TEN: THE BEST HOME FOR A FEMINIST

145 *centennial:* Association of MIT Alumnae Records, MIT; Photographs of AMITA Centennial Convocation banquet, June 2, 1973, in "Women—general photos," MIT Museum.

145 *"perhaps defend":* MIT, *Report of the President and Chancellor,* 1972–73, 12.

145 *"changing issues":* Ibid., 2.

146 *"socially constructive":* MIT, *Report of the President,* 1969, 508.

146 *Within four years:* Stuart W. Leslie, "'Time of Troubles' for the Special Laboratories," in Kaiser, *Becoming MIT,* 138.

146 *"Now is the moment":* MIT, *Report of the President,* 1972–73, 9–10.

146 *"Social progress":* Ibid., 13.

146 *The number:* Ibid., 14.

147 *The brochure:* "MIT: A Place for Women," Women and Scientists and Engineers Oral History Collection, MIT.

147 *borrowed boats:* Roseanna Means, interviewed by Madeleine Kline, June 2, 2017, Margaret MacVicar Memorial AMITA Oral History Project, MIT Libraries.

147 *Luscomb:* Bowe, "MIT History."

147 *"intention to":* MIT, *Report of the President,* 1971–72, 193.

148 *"hoopla":* James D. Watson, "Dedication of the Seeley G. Mudd Building,

MIT—Running Too Fast?," March 6, 1975, CSHL. Watson's comments upset many supporters of the new "war" on cancer, as Walter Rosenblith, the provost of MIT, noted in his annual report that year: "The dedication symposium represented an intersection of historical perspective, up to date science and a few statements which were a trifle controversial on science policy in the area of cancer research. The latter made the newspapers": MIT, *Report to the President*, 1974–75, 71.

149   *Polly Bunting*: Evelyn M. Witkin, "Chances and Choices: Cold Spring Harbor, 1944–1955," *Annual Review of Microbiology* 56 (April 2, 2002).

149   *washing dishes*: "Men in the News: Salvador E. Luria," *New York Times*, October 17, 1969, 24.

149   *"brilliant"*: S. E. Luria, *A Slot Machine, a Broken Test Tube: An Autobiography* (New York: Harper and Row, 1984), 136.

150   *just six*: MIT, *Report of the President*, 63.

150   *the lights*: "Building the Foundation of Modern Cancer Research: Four Decades of Discovery within the CCR at MIT," Koch Institute, MIT, YouTube, April 3, 2012.

152   *"I'm on my way"*: Nancy Hopkins, draft of chapters for memoir titled *Fifteen Tenured Women*, unpublished, 15, NHP.

154   *floor meeting*: "Building the Foundation."

155   *"Survival"*: Handwritten letter from Nancy Hopkins to James D. Watson, CSHL.

156   *lunches*: Dresselhaus, "MIT History."

157   *"You're lucky"*: Sheila E. Widnall, "Millie's Impact on Women at MIT," Celebrating Our Millie.

160   *the front page*: Gene I. Maeroff, "Minority Hiring Said to Hurt Colleges," *New York Times*, June 28, 1974, 1.

161   *In public*: "Affirmative Action: The Negative Side," *Time*, July 15, 1974.

162   *"minutiae"*: Mary P. Rowe, "The Progress of Women in Educational Institutions: The Saturn's Rings Phenomenon," *Graduate and Professional Education of Women: Proceedings of the American Association of University Women Conference*, 1974.

CHAPTER ELEVEN: LIBERATED LIFESTYLES

165   *"Take her"*: David Baltimore, interview by Sara Lippincott, October 13, 2009, transcript, Oral History Project, California Institute of Technology Archives, Pasadena, CA; Hillary Bhaskaran, "Alice Huang: Keeping Science and Life in Focus," *Caltech News* 33, no. 1 (1999): 3.

168   *The authors concluded*: Eva Ruth Kashket, Mary Louise Robbis, Loretta Leive, and Alice S. Huang, "Status of Women Microbiologists," *Science* 183, no. 4124 (February 8, 1974): 488–94.

168   *At Harvard*: Wasserman, *Door in the Dream*, 20.

170   *Like many white women*: Shirley Mahaley Malcom, Paula Quick Hall, and Janet Welsh Brown, *The Double Bind: The Price of Being a Minority Woman in Sci-*

ence (Washington, DC: American Association for the Advancement of Science, 1976).

171 *drive an hour*: Aseem Z. Ansari, Marsha Rich Rosner, and Julius Adler, "Har Gobind Khorana, 1922–2011," *Cell* 147 (December 23, 2011).

172 *Nancy's was*: Nancy Hopkins, "The High Price of Success in Science: A Woman Scientist Disputes the Notion That a Woman Can Be a Successful Wife and Mother as Well as a Successful Scientist," *Radcliffe Quarterly*, June 1976, 16–18.

174 *"disservice"*: Margaret Horton Weiler, Letter to the Editor, *Radcliffe Quarterly*, September 1976, 32.

174 *Goodenough insisted*: Ursula W. Goodenough, Letter to the Editor, *Radcliffe Quarterly*, September 1976, 31.

175 *"absolutely right"*: Nancy Kleckner, Letter to the Editor, *Radcliffe Quarterly*, September 1976, 32.

175 *"truly brilliant"*: Letter from Barbara McClintock to Nancy Hopkins, NHP, September 21, 1976.

176 *news echoed*: Catherine Brady, *Elizabeth Blackburn and the Story of Telomeres: Deciphering the Ends of DNA* (Cambridge, MA: MIT Press, 2009), 44.

177 *haggling*: Jon Cohen, "The Culture of Credit," *Science* 268, no. 23 (1995): 1706–11.

CHAPTER TWELVE: KENDALL SQUARE

183 *shipped them*: Susan E. Maycock, *East Cambridge: Survey of Architectural History of Cambridge* (Cambridge, MA: MIT Press, 1988), 60–68.

183 *Depending on*: Doug Brown, "A Brief History of Zoning in Cambridge," *Cambridge Historian* 16, no. 2 (Fall 2016): 1; Nidhi Subbaraman, "The Evolution of Cambridge," *Technology Review*, December 21, 2010.

183 *migration*: LaDale C. Winling, *Building the Ivory Tower: Universities and Metropolitan Development in the Twentieth Century* (Philadelphia: University of Pennsylvania Press, 2018), 153–57.

183 *once-teeming*: Cambridge Community Development Department, "East Cambridge Riverfront Plan," 1978.

183 *City officials*: Cambridge Redevelopment Authority, "Background of the Kendall Square Urban Renewal Area," cambridgeredevelopmentauthority.org/development-history-of-kendall.

183 *cratered*: Cambridge Redevelopment Authority, Image Gallery, cambridgeredevelopment.org.

185 *Asilomar*: Donald S. Fredrickson, "Asilomar and Recombinant DNA," in *Biomedical Politics*, ed. Kathi E. Hanna (Washington, DC: National Academies Press, 1991).

185 *Vellucci*: David Arnold, "Vellucci Is Back in the Running," *Boston Globe*, August 5, 1991, 13; David Arnold, "Roots of a Quarrel; Vellucci, *Lampoon* Wage Feud over a Tree," *Boston Globe*, April 6, 1991, 25.

185 *Wald, a Nobel*: John Kifner, "'Creation of Life' Experiment at Harvard Stirs Heated Dispute," *New York Times*, June 17, 1976, 22; Barbara J. Culliton,

"Recombinant DNA: Cambridge City Council Votes Moratorium," *Science* 193, no. 4250 (1976): 300–301.

186   *"Refrain":* Cambridge RDNA Hearings, 1976, Oral History Collection on the Recombinant DNA Controversy, MIT.

187   *Kendall Square would become:* Maryann Feldman and Nichola Lowe, "Consensus from Controversy: Cambridge's Biosafety Ordinance and the Anchoring of the Biotech Industry," *European Planning Studies* 16, no. 3 (April 2008).

188   *"diminished":* MIT, *Report of the President and Chancellor*, 1979–80, 8.

188   *doubled: Reports to the President*, 1983, 11.

188   *"unrelenting":* Ibid., 10.

188   *industry sponsorship: Report of the President*, 1981–82, 7.

188   *a sum greater:* Center for Education Statistics, *Endowment Assets, Yield, and Income in Institutions of Higher Education: Fiscal Years 1982–85* (Washington, DC: US Department of Education, 1987).

189   *"not a stranger": Report of the President*, 1981–82, 10.

189   *In 1979:* National Center for Education Statistics, *Digest of Education Statistics: 2019* (Washington, DC: US Department of Education, February 2021).

189   *sixty-two cents: Women's-to-Men's Earnings Ratio, 1979–2008* (Washington, DC: US Bureau of Labor Statistics, July 31, 2009).

190   *nine out of ten:* Evan Thomas, *First: Sandra Day O'Connor: An Intimate Portrait of the First Woman Supreme Court Justice* (New York: Random House, 2019), 141.

191   *"Perhaps":* Handwritten letter from Nancy Hopkins to Jim Watson, CSHL.

193   *The experiment:* P. A. Chatis et al., "Role for the 3′ End of the Genome in Determining Disease Specificity of Friend and Moloney Murine Leukemia Viruses," *Proceedings of the National Academy of Sciences* 80 (July 1983): 4408–11.

193   *"We are now":* Handwritten letter from Nancy Hopkins to James D. Watson, CSHL.

194   *journal editors:* John Maddox, "Preface to the Expanded Edition," in Judson, *Eighth Day of Creation*.

CHAPTER THIRTEEN: "THIS SLOW AND GENTLE ROBBERY"

199   *"I bet I know":* This account is based on interviews with Nancy's friend, separate from Nancy, and checked against datebook entries from the time.

201   *Evelyn insisted otherwise:* Evelyn Fox Keller, "Pot-Holes Everywhere: How (Not) to Read My Biography of Barbara McClintock," in *Writing about Lives in Science: (Auto)biography, Gender, and Genre*, ed. Paola Govoni and Zelda Alice Franceschi (Göttingen, Germany: V&R Unipress, 2014), 33–42.

201   *"Oh, dear":* John Noble Wilford, "Woman in the News: A Brilliant Loner in Love with Genetics," *New York Times*, October 11, 1983, C7.

202   *"problematic addition":* Walter Goodman, "Women's Studies: The Debate Continues," *New York Times Magazine*, April 24, 1984, 39.

203   *"great experimental scientists":* Anne Sayre, *Rosalind Franklin & DNA* (New York: W. W. Norton, 1975), 178.

203   *"termagant"*: Ibid., 21.

203   *"eyesight"*: Ibid.

204   *"big helix"*: Ibid., 128.

204   *"never to have achieved"*: Ibid., 102.

204   *"something important"*: Ibid., 37.

204   *"alone"*: Ibid., 40.

205   *make a choice*: Ibid., 53–54.

205   *"For herself"*: Ibid., 58.

205   *did not entirely blame*: Ibid., 192–97.

205   *"The intelligent use"*: Ibid., 22.

205   *"slow and gentle"*: Ibid., 189.

206   *"one of a select"*: "Rosalind Franklin, Virus Researcher," *New York Times*, April 20, 1958, 85.

CHAPTER FOURTEEN: "FODDER"

207   *one-half*: "Increasing the Participation of Women in Scientific Research: Summary of a Conference Proceedings, October 1977, and Research Study Project Report, March 1978" (Washington, DC: National Science Foundation, 1978).

208   *nineteen*: Report to the President, 1980–81, 263.

208   *27 percent*: Ibid., 18.

209   *"recent performance"*: Reports to the President, 1985–86, 40.

209   *Every spring*: Keyser, *Mens et Mania*, 78–79.

209   *in 1983*: "Barriers to Equality in Academia: Women in Computer Science at M.I.T." Prepared by female graduate students and research staff in the Laboratory for Computer Science and the Artificial Intelligence Laboratory at MIT, February 1983.

212   *first in his family*: MIT Department of Biology, "A Conversation with Frank Solomon," June 17, 2014, https://biology.mit.edu/video-post/a-conversation -with-frank-solomon/.

212   *literature seminar*: Luria, *Slot Machine*, 150.

219   *rarely mentioned*: The scientist, Françoise Barré-Sinoussi, was awarded the Nobel Prize twenty-five years later, in 2008.

220   *"Red Book"*: "Harvard Class of 1964," enclosure from Nancy Hopkins to James D. Watson, CSHL.

CHAPTER FIFTEEN: FUN IN MIDDLE AGE

227   *founder*: Máté Varga, "The Doctor of Delayed Publications—the Remarkable Life of George Streisinger," Node, July 21, 2016, https://thenode.biologists.com /doctor-delayed-publications-remarkable-life-george-streisinger/careers/.

229   *"living breathing fish"*: Fax from Nancy Hopkins to Jane Reece, CSHL.

232   *Arthur gave*: List of Hopkins lab grants, fellowships, other funding, NHP; letter from Nancy Hopkins to Mark Wrighton, April 13, 1993, NHP.

233 *two women*: Kristin Kain, "Using Zebrafish to Understand the Genome: An Interview with Nancy Hopkins," *Disease Models & Mechanisms*, May–June 2009, 214–17.

234 *she called back*: Nancy Hopkins, "Developing Insertional Mutagenesis to Identify Essential Genes in Zebrafish: A Tale of Serendipity and Luck," 2021 George Streisinger Award Lecture, June 22, 2021.

CHAPTER SIXTEEN: THREE HUNDRED SQUARE FEET

235 *"bright young women"*: Eloise Salholz, "The Marriage Crunch: Too Late for Prince Charming?," *Newsweek*, June 2, 1986.

235 *In 1989*: Felice N. Schwartz, "Management Women and the New Facts of Life," *Harvard Business Review*, January–February 1989.

236 *"mommy track"*: Tamar Lewin, " 'Mommy Career Track' Sets Off a Furor," *New York Times*, March 8, 1989, 18.

236 *"mommy trap"*: Beverly Beyette, "A New Career Flap: What's a Mommy Track and Why Are So Many Women Upset About It?," *Los Angeles Times*, March 17, 1989.

236 *"Women Face the 90s"*: Claudia Wallis, "Onward, Women!," *Time*, December 4, 1989.

237 *author revealed*: Nikki Finke, "*Time* Picks on Feminism, Ticks Off Feminists," *Los Angeles Times*, November 30, 1989.

238 *90 percent*: Megan Brenan, "Gallup Vault: Anita Hill's Charges Against Clarence Thomas," Gallup, September 21, 2018.

238 *new Biology: Reports to the President*, 1989–90, 452.

238 *"I came reluctantly"*: Charles A. Radin, "Sharp Jolts MIT, Rejects Top Post," *Boston Globe*, February 21, 1990, 1.

238 *"special priority"*: *Report of the President*, 1989–90, 16.

239 *"rambunctious"*: Luria, *Slot Machine*, 42.

239 *"best lecturers"*: *Report of the President*, 1992–93, 389.

239 *"I know"*: "A history of 7.012," 125, NHP.

240 *the building manager*: Memo from Nancy Hopkins to Phil Sharp, "Recap of my efforts to obtain equipment resources for my lab that are equal to those of the other faculty in the Cancer Center," NHP.

241 *Susumu had taken*: Ibid. Susumu Tonegawa declined to speak with me, saying in an email on November 19, 2020, that he was "not interested in getting involved in this project of yours."

242 *Dartmouth*: Letter from James D. Watson to Dr. Edward Bresnick, November 1, 1991, CSHL.

242 *two postdocs*: "Fish group—Funding Needs," memo from Nancy Hopkins to Jean Dz, May 11, 1992.

242 *time fixing*: Memo from Bob Bellas to Nancy Hopkins, July 24, 1992, NHP.

242 *One of Nancy's postdocs*: Memo from Nancy Hopkins to Phil Sharp, August 25, 1992, NHP.

243 *"perfect agreement"*: Ibid.; sign-up sheet, October 31, 1991–April 8, 1992, copied April 9, 1992, NHP.

244 *"John Doe / Jane Doe"*: Notes from "Conversation with Mary Rowe 7/92," NHP.

244 *six-page letter*: Memo from Nancy Hopkins to Richard Hynes, July 1992, NHP.

244 *Mary had also*: Notes from meeting with Jean Dz and Mary Rowe, July 9, 1992, NHP.

244 *meeting in his office*: Note from Nancy Hopkins to Jean Dz, July 27, 1992, NHP.

245 *"you are quite right"*: Note from Nancy Hopkins to Mary Rowe, July 24, 1992, NHP.

245 *Within four days*: Memo from Nancy Hopkins to Phil Sharp, August 25, 1992.

245 *a new crisis*: Memo on funding for 1992, NHP.

246 *That afternoon*: Typewritten notes, "Monday, 12/21/92," NHP.

246 *he told her she was misinformed*: Ibid.

246 *"It is difficult"*: Note from Nancy Hopkins to Mary Rowe, December 22, 1992, NHP.

246 *"why do the others"*: Notes written January 11, 1993, NHP.

247 *only if he had*: Note from Nancy Hopkins to Phil Sharp, January 12, 1993, NHP.

247 *"He said this was not true"*: Nancy Hopkins, "Report of most recent problem with Richard Hynes as of 1/11/93," NHP.

247 *"extraordinary feeling"*: Note from Nancy Hopkins to John Fresina, January 28, 1993, NHP.

247 *"habitual use"*: Memo from Susumu Tonegawa to Nancy Hopkins, January 28, 1993, NHP.

248 *"In fact"*: Note from Richard Hynes to Nancy Hopkins, February 5, 1993, NHP.

248 *"crazy?"*: Note from Nancy Hopkins to Mary Rowe, February 5, 1993, NHP.

248 *"Miracles"*: Note card from Nancy Hopkins to James D. Watson, CSHL.

249 *Even a junior faculty member*: "Sq. ft. of space available to faculty in the Center for Cancer Research," NHP; color-coded floor plans, NHP.

249 *they joked*: "Historical Account of the Problem," NHP; memo from graduate students to Nancy Hopkins beginning, "Nancy: Although Shuo and Christina have some interesting suggestions," NHP.

249 *"year of the woman"*: Memo from Nancy Hopkins to Richard Hynes, February 28, 1993, NHP.

250 *met with the lawyer*: Handwritten notes, "Mr. Shapiro, Friday 3/12/93," NHP; invoice from Jonathan Shapiro to Nancy Hopkins, March 15, 1993, NHP.

251 *"The snowy weekend"*: Letter from Nancy Hopkins to Mary Rowe, March 17, 1993.

251 *would soon have*: Typewritten list, "People who will be working in the Hopkins lab in summer of 1993," NHP; note from Nancy Hopkins to Robert Birgeneau, May 3, 1993, NHP.

251 *"I'm too upset"*: Typewritten notes, "4/9–10/93," April 10, 1993, NHP.

251 *Her meeting*: Letter from Nancy Hopkins to Mark Wrighton, April 9, 1993, with attached memo, "What I need," NHP.

252   *Wrighton appeared delighted:* Letter from Mark Wrighton to Nancy Hopkins, April 20, 1993, NHP; memo from Nancy Hopkins to Phil Sharp, April 14, 1993, NHP.

252   *The assistant:* Note from Nancy Hopkins to Robert Birgeneau, May 3, 1993, NHP.

253   *"It is remarkable":* Note from Nancy Hopkins to Mark Wrighton, June 6, 1993, NHP; letter from Nancy Hopkins to Mark Wrighton, February 24, 1994, NHP.

253   *a sign:* Note from Nancy Hopkins to Richard Hynes, November 5, 1993, NHP; note from Nancy Hopkins to Carol Browne, October 27, 1993, NHP.

253   *"Can you jot":* Note from Richard Hynes to Nancy Hopkins, November 2, 1993, NHP.

CHAPTER SEVENTEEN: MIT INC.

255   *started poor:* Robert Birgeneau, interviews by Paul Burnett, January 18, 2019, Oral History Center, Bancroft Library, University of California, Berkeley.

256   *path from university:* Ibid., February 7, 2019.

256   *Black PhDs:* Ibid., June 6, 2019.

257   *The gesture: Reports to the President,* 1993–94, 455.

257   *most diverse:* Center for American Women and Politics, *Women Appointed to Presidential Cabinets* (New Brunswick, NJ: Eagleton Institute of Politics, Rutgers University, 2021).

257   *campuses debated:* Jane Gross, "Love or Harassment? Campuses Bar (and Debate) Faculty-Student Sex," *New York Times,* April 14, 1993, B9.

258   *bonfire:* Keyser, *Mens et Mania,* 116.

258   *"We respect": Reports to the President,* 1992–93, 7–10.

260   *high ratings:* "Undergraduate Evaluations of Biology Courses Taught at MIT in the Last Two Years, 1987–1989," NHP.

260   *67 percent: Reports to the President,* 1991–92, 24.

261   *had to do some convincing:* Michelle Hoffman, "The Whitehead Institute Reaches Toward Adulthood," *Science* 256 (April 3, 1992): 26.

261   *among the crop:* Christopher Anderson, "Genome Project Goes Commercial," *Science* 259 (January 15, 1993): 300–302.

262   *"I can't teach":* Nancy Hopkins, "A history of 7.012 and chronology of recent events," 98, NHP.

262   *"like fun":* Ibid., 125.

262   *took the lead:* Ibid., 126.

262   *how students would respond:* Ibid.

263   *scores rose:* Ibid., 100.

263   *scored:* Student evaluations of the introductory biology courses, NHP; Student evaluations of all faculty who have taught introductory biology, NHP.

263   *"Great job!":* Note from Phil Sharp to Nancy Hopkins, NHP.

264   *"Never":* Hopkins, *Fifteen Tenured Women,* 17.

264   *"That's all changed":* Hopkins, "History of 7.012," 101.

265 *"astounding breach"*: Letter from Nancy Hopkins to Phil Sharp, January 21, 1994, NHP.

266 *"If I received"*: Hopkins, "History of 7.012," 129.

266 *"let me come clean"*: Ibid., 125.

267 *"Why aren't you outraged?"*: Ibid., 130.

267 *"It's enough to make me a feminist"*: Ibid.

267 *"Please don't tell anyone"*: Ibid., 131.

268 *"Do you want to write a book"*: Note from Nancy Hopkins to Robert Birgeneau, March 16, 1994, NHP.

268 *"UNPRECEDENTED"*: Notes from conversation with Dan Steiner, Ropes & Gray, March 28, 1994, NHP; letter from Nancy Hopkins to Daniel Steiner, April 14, 1994, NHP.

269 *Glaxo:* Hopkins, "History of 7.012," 122.

269 *"I created"*: Ibid., 105.

270 *"I don't think you'll be happy"*: Ibid., 134.

270 *"I personally"*: Letter from Nancy Hopkins to Robert Birgeneau, May 5, 1994, NHP.

271 *"a disaster"*: Hopkins, "History of 7.012," 107.

271 *"Oh, come on"*: Ibid., 114; letter from Nancy Hopkins to Alan Grossman, January 15, 2017, NHP.

272 *a partnership:* Letter from Mark Fishman to John Potts Jr., May 20, 1994, NHP.

272 *"All I know"*: Hopkins, "History of 7.012," 117.

272 *"this is not the time"*: Ibid., 108.

273 *"heart broken"*: Ibid., 111.

273 *"I am a professor"*: Draft letter from Nancy Hopkins to Charles Vest, June 1994, NHP.

CHAPTER EIGHTEEN: SIXTEEN TENURED WOMEN

281 *"ultimate joke"*: Hopkins, "History of 7.012," 112.

287 *"You're the one who was brave"*: Interview of Terry L. Orr-Weaver, "Data Driven" Film Interviews Collection, MIT.

287 *Terry suspected:* Diary, June 7, 1994–February 27, 1995, 4, NHP.

289 *"shredding"*: Letter from Nancy Hopkins to Marcia McNutt, July 21, 1994, NHP.

290 *"exhausting"*: Note from Nancy Hopkins to Penny Chisholm, NHP.

290 *"widespread perception"*: Letter from tenured women faculty in Science to Robert Birgeneau, July 21, 1994, NHP.

290 *"We believe"*: A Proposal to the MIT Administration for an Initiative to Improve the Status of Women Faculty in the School of Science, July 21, 1994, NHP.

291 *"more traumatic"*: Confidential memo from the tenured women faculty in Science to Robert Birgeneau, July 21, 1994, NHP.

293 *"as long as the eye could see"*: Nan Robertson, *The Girls in the Balcony: Women, Men and* The New York Times (New York: Random House, 1992), 5.

CHAPTER NINETEEN: X AND Y

296   *"My personal life"*: "Notes on the meeting," Diary, 1.

297   *"Our meeting"*: Memo to tenured women faculty in the School of Science from Women faculty who went to see the Dean, August 12, 1994, NHP.

298   *"There is a snag"*: Diary, 2.

298   *All but three:* Paul Selvin, "Does the Harrison Case Reveal Sexism in Math?," *Science* 252 (June 28, 1991): 1781.

300   *"The president said"*: Diary, 4; interview with Charles M. Vest, MIT150 Infinite History Project.

300   *33 percent: The NEA 1995 Almanac of Higher Education* (Washington, DC: National Education Association, 1995), 14.

300   *fewer than twenty:* Jeffrey Mervis, "Efforts to Boost Diversity Face Persistent Problems," *Science* 284 (June 11, 1999): 1757–59.

301   *patterns for girls:* National Center for Education Statistics, *The Educational Progress of Women* (Washington, DC: US Department of Education, November 1995); National Center for Education Statistics, *Women in Mathematics and Science* (Washington, DC: US Department of Education, July 1997).

301   *"What the hell"*: Diary, 3.

302   *Solomon had turned up:* Ibid.

302   *"I told you"*: Ibid., 4.

303   *"startlingly familiar"*: Note from Nancy Hopkins to Robert Birgeneau, September 30, 1994, NHP.

303   *Phil said:* Diary, 6.

304   *"It was stressful"*: Ibid., 3.

305   *"work-personal life integration"*: Lotte Bailyn, *Breaking the Mold: Women, Men, and Time in the New Corporate World* (New York: Free Press, 1993).

306   *"underwhelmed"*: Diary, 5.

306   *a triumph celebrated:* Daniel J. Boyne, *The Red Rose Crew: A True Story of Women, Winning, and the Water* (Guilford, CT: Lyons Press, 2005).

308   *"very nervous"*: Memo: Answers to questions posed by Birgeneau in Memo of 10/11/94, NHP.

308   *As the dean wrote: Reports to the President,* 1994–95, 398.

309   *Wait twenty-four hours:* Diary, 10.

CHAPTER TWENTY: ALL FOR ONE OR ONE FOR ALL

312   *"proceduralists"*: Diary, 6.

313   *outside professional activities: First Report of the Committee on Women Faculty in the School of Science on the Status and Equitable Treatment of Women Faculty* (MIT), I-22, table 4.

313   *the number:* Ibid., I-16–21.

314   *And by spring:* Typewritten memo titled "Accomplishments to date," NHP.

316   *"that one absolutely"*: Note from Nancy Hopkins to Bob Birgeneau, July 20, 1995, NHP.

316   *"You know we're writing"*: Diary, September 5, 1994–October 22, 1995, 136, NHP.

317   *"fifty-five, white, and male"*: Ibid., 137.

317   *"despicable"*: Note from Nancy Hopkins to Molly Potter, March 2, 1995, NHP.

318   *talked past each other:* Ibid., 138.

318   *"What are you doing to Hopkins?"*: Ibid., 139.

319   *"Does one need"*: Letter from Nancy Hopkins to Bob Silbey, May 24, 1995, NHP.

321   *unusually high:* Selvin, "Does the Harrison Case?," 1781.

321   *"pale-skinned"*: Margy Rocklin, "The Mathematics of Discrimination," *Los Angeles Times Magazine*, May 2, 1993, 29.

321   *Readers judged:* C. E. Grubbs and Abigail Thompson, Letters to the Editor, *Los Angeles Times Magazine*, June 6, 1993, 6.

321   *third major result:* Paul Selvin, "Jenny Harrison Finally Gets Tenure in Math at Berkeley," *Science* 261 (July 16, 1993): 286.

323   *"high point"*: Note from Nancy Hopkins to Bob Birgeneau, January 6, 1995, NHP.

324   *"tossing and turning"*: Note from Nancy Hopkins to Penny Chisholm, NHP.

324   *Six women:* Affidavit, February 16, 1996, NHP.

325   *"cautiously thrilled"*: Note from Nancy Hopkins to Bob Birgeneau, February 21, 1996, NHP.

325   *"call it quits"*: Note from Nancy Hopkins to Miriam McKendall, February 22, 1996, NHP.

325   *a new letter:* Note from Nancy Hopkins to Phil Sharp, February 22, 1996, NHP.

325   *"We will keep working"*: Note from Nancy Hopkins to Sylvia Ceyer, February 22, 1996, NHP.

326   *"ready to move on"*: Note from Nancy Hopkins to Jerry Friedman, March 14, 1996, NHP.

326   *"All such problems pose"*: "Personal Statement from the Committee Chair," *First Report of the Committee*, D-1.

327   *"We've come a long way baby"*: Note from Nancy Hopkins to Leigh Royden, August 9, 1996, NHP.

CHAPTER TWENTY-ONE: "THE GREATER PART OF THE BALANCE"

329   *"a little bland"*: Memorandum from Nancy to Committee on Women Faculty, June 10, 1996, NHP.

330   *"a new person"*: Diary, May 9, 1994–October 17, 1996, 5, NHP.

330   *Molly called:* Ibid., 5.

331   *By February 1997:* Letter from Phil Sharp to Bob Birgeneau, November 4, 1996, obtained by author; letter from Frank Solomon to Molly Potter, March 11, 1997, obtained by author.

331   *"definitely educable"*: Nancy Hopkins, "Meetings in Fall, 1995–Winter, 1996," 6, NHP.

332   *Birgeneau credited: Reports to the President*, 1996–97, 443.

332   *"Today all fears"*: Email from Nancy Hopkins to Bob Birgeneau, March 13, 1997, NHP.

334   *"slightly unsuited"*: Virginia Valian, *Why So Slow? The Advancement of Women* (Cambridge, MA: MIT Press, 1998), 2.

334   *"Each time it happens"*: Ibid., 5.

334   *"These exceptions"*: Ibid., 6.

335   *wrote it in three hours:* "A Study on the Status of Women Faculty in Science at MIT," *MIT Faculty Newsletter*, Special Edition, March 1999, https://web.mit.edu /fnl/women/women.html.

338   *solicited checks:* Vera Kistiakowsky, "Does MIT Need a Faculty Newsletter?," *MIT Faculty Newsletter*, March 10, 1988.

339   *"I have always believed"*: "Study on the Status of Women."

340   *number of women:* Center for American Women and Politics, *Women Candidates, 1992–2020* (New Brunswick, NJ: Eagleton Institute of Politics, Rutgers University, 2021).

340   *"Is Feminism Dead?"*: Ginia Bellafante, "Feminism: It's All About Me!," *Time*, June 29, 1998.

341   *"traces of past experience"*: Anthony G. Greenwald and Mahzarin R. Banaji, "Implicit Social Cognition: Attitudes, Self-Esteem, and Stereotypes," *Psychological Review* 102, no. 1 (1995): 4–27.

341   *"The key conclusion"*: "Study on the Status of Women Faculty."

343   *"extraordinary admission"*: Carey Goldberg, "M.I.T. Admits Discrimination Against Female Professors," *New York Times*, March 23, 1999, 1.

343   *"Women on other campuses"*: "Gender Bias on the Campus," *New York Times*, March 28, 1999, sec. 4, 16.

344   *"Anita Hill"*: Robin Wilson, "An MIT Professor's Suspicion of Bias Leads to a New Movement for Academic Women," *Chronicle of Higher Education*, December 3, 1999, 16.

345   *"It is one thing"*: Email from Leon Eisenberg to Charles M. Vest, March 24, 1999.

345   *Ford Foundation:* Lotte Bailyn, "Academic Careers and Gender Equity: Lessons Learned from MIT," *Gender, Work and Organization* 10, no. 2 (March 2003).

346   *"serious difficulties"*: Andrew Lawler, "Tenured Women Battle to Make It Less Lonely at the Top," *Science* 286 (November 12, 1999): 1272–78.

346   *"ridiculously low"*: Committee for the Equality of Women at Harvard, *1996 Report on the Status of Women at Harvard*.

347   *"The difference"*: "Subtle Discrimination Spurs MIT to Change," *San Francisco Chronicle*, March 24, 1999.

EPILOGUE

349   *"different availability of aptitude"*: Lawrence H. Summers, Remarks at NBER Conference on Diversifying the Science and Engineering Workforce, Cambridge, MA, January 14, 2005.

351   *landmark report:* National Academies of Sciences, Engineering, and Medicine,

Sexual Harassment of Women: Climate, Culture, and Consequences in Academic Sciences, Engineering, and Medicine (Washington, DC: National Academies Press, 2018).

351   *Some of those strategies:* See, for example, Tessa Charlesworth and Mahzarin Banaji, "Gender in Science, Technology, Engineering and Mathematics: Issues, Causes, Solutions," *Journal of Neuroscience* 39, no. 37 (September 11, 2019): 7228–43; Meredith Meyer, Andrei Cimpian, and Sarah-Jane Leslie, "Women Are Underrepresented in Fields Where Success Is Believed to Require Brilliance," *Frontiers in Psychology* 6 (March 11, 2015): 235.

352   *she and Harvey:* Harvey Lodish and Nancy Hopkins, "Boston Biotech Has a Woman Problem," *Boston Globe*, November 15, 2017.

Index

About the Author

Kate Zernike has been a reporter for the *New York Times* since 2000. She was a member of the team that won the 2002 Pulitzer Prize for Explanatory Reporting for stories about al-Qaeda before and after the 9/11 terror attacks. She was previously a reporter for the *Boston Globe*, where she broke the story of MIT's admission that it had discriminated against women on its faculty, on which *The Exceptions* is based. The daughter and granddaughter of scientists, she is a graduate of Trinity College at the University of Toronto and the Graduate School of Journalism at Columbia University. She lives in New Jersey with her husband and sons.